国外优秀数学著作
原 版 系 列

代数结构和几何结构的形变理论

Deformation Theory of Algebraic and Geometric Structures

● ［法］珍—弗朗索瓦·蓬马雷特（Jean-Francois Pommaret）著

（英文）

哈尔滨工业大学出版社
HARBIN INSTITUTE OF TECHNOLOGY PRESS

黑版贸审字 08－2019－180 号

图书在版编目(CIP)数据

代数结构和几何结构的形变理论＝Deformation
Theory of Algebraic and Geometric Structures：英
文/(法)珍－弗朗索瓦·蓬马雷特著. —哈尔滨：哈
尔滨工业大学出版社,2022.8
ISBN 978-7-5767-0253-8

Ⅰ.①代… Ⅱ.①珍… Ⅲ.①变形(数学)－英文
Ⅳ.①O186.1

中国版本图书馆 CIP 数据核字(2022)第 123254 号

DAISHU JIEGOU HE JIHE JIEGOU DE XINGBIAN LILUN

策划编辑　刘培杰　杜莹雪
责任编辑　刘春雷　李兰静
封面设计　孙茵艾
出版发行　哈尔滨工业大学出版社
社　　址　哈尔滨市南岗区复华四道街 10 号　邮编 150006
传　　真　0451－86414749
网　　址　http：//hitpress.hit.edu.cn
印　　刷　黑龙江艺德印刷有限责任公司
开　　本　886 mm×1 230 mm　1/32　印张 8.25　字数 196 千字
版　　次　2022 年 8 月第 1 版　2022 年 8 月第 1 次印刷
书　　号　ISBN 978-7-5767-0253-8
定　　价　48.00 元

CONTENTS

1 INTRODUCTION:

In 1953 the physicists E. Inonü and E.P. Wigner (1963 Nobel prize) introduced the concepts of *deformation and contraction of a Lie algebra* by considering the composition law $(u, v) \rightarrow (u + v)/(1 + (uv/c^2))$ for speeds in special relativity (Poincaré group) when c is the speed of light, claiming that the limit $c \rightarrow \infty$ or $1/c \rightarrow 0$ should produce the composition law $(u, v) \rightarrow u + v$ used in classical mechanics (Galilée group) ([22]). However, this result is not correct indeed as $1/c \rightarrow 0$ has no meaning independently of the choice of length and time units. Hence one has to consider the dimensionless numbers $\bar{u} = u/c, \bar{v} = v/c$ in order to get $(\bar{u}, \bar{v}) \rightarrow (\bar{u} + \bar{v})/(1 + \bar{u}\bar{v})$ with no longer any perturbation parameter involved ([32]). Nevertheless, this idea brought the birth of the theory of *deformation of algebraic structures* ([13],[19],[20],[21],[34],[35],[44],[45]), culminating in the use of the Chevalley-Eilenberg cohomology of Lie algebras ([9]) and one of the first applications of computer algebra in the seventies because a few counterexamples can only be found for Lie algebras of dimension ≥ 11 ([2]). Finally, it must also be noticed that the main idea of general relativity is to deform the Minkowski metric $dx^2 + dy^2 + dz^2 - c^2 dt^2$ of space-time by means of the small dimensionless parameter ϕ/c^2 where $\phi = GM/r$ is the gravitational potential at a distance r of a central attractive mass M with gravitational constant G ([11],[33]).

Let G be a Lie group with identity e and Lie algebra $\mathcal{G} = T_e(G)$, the tangent space of G at e. If $a = (a^\tau)$ with $\tau = 1, ..., p$ are local coordinates on G, the *bracket* $[\mathcal{G}, \mathcal{G}] \subset \mathcal{G}$ is defined by $p^2(p-1)/2$ *structure constants* $c = (c^\tau_{\rho\sigma} = -c^\tau_{\sigma\rho})$ satisfying the *Jacobi identities* $J(c) = 0 \Leftrightarrow c^\lambda_{\rho\sigma}c^\mu_{\lambda\tau} + c^\lambda_{\sigma\tau}c^\mu_{\alpha\rho} + c^\lambda_{\tau\rho}c^\mu_{\lambda\sigma} = 0$. If now $c_t = c + tC + ...$ is a deformation of c satisfying $J(c_t) = 0$, the idea is to study the vector space $\frac{\partial J(c)}{\partial c}C = 0$ modulo the fact that c behaves like a 3-tensor under a change of basis of $\mathcal{G}$. The first condition implies $C \in Z_2(\mathcal{G})$ as a cocycle while the second implies $C \in B_2(\mathcal{G})$ as a coboundary, a result leading to introduce the Chevalley-Eilenberg cohomology groups $H_r(\mathcal{G}) = Z_r(\mathcal{G})/B_r(\mathcal{G})$ and to study $H_2(\mathcal{G})$ in particular.

A few years later, a deformation theory was introduced for *structures on manifolds*, generally represented by *fields of geometric objects* like tensors and we may quote riemannian, symplectic, complex analytic or contact structures with works by

5

H. Goldschmidt ([12],[14],[15]), R.E. Greene ([15]), V. Guillemin ([16],[17],[18]), K. Kodaira ([24],[25]), M. Kuranishi ([27]), L. Nirenberg ([24]), D.C. Spencer ([15]), ([25],[47],[48]) or S. Sternberg ([16],[17],[18]). The idea is to make the underlying geometric objects depending on a parameter while satisfying prescribed integrability conditions. The link betwen the two approaches, though often conjectured, has never been exhibited by the above authors and our aim is to sketch the solution of this problem that we have presented in many books, in particular to study the possibility to use computer algebra for such a purpose. It must be noticed that the general concept of natural bundle and geometric object is absent from the work of Spencer and coworkers ([27],[47],[48]]) though it has been discovered by E. Vessiot as early as in 1903 ([51]). It must also be noticed that the introduction of the book "Lie equations" ([26]) dealing with the "classical" structures on manifolds has nothing to do with the remaining of the book dealing with the nonlinear Spencer sequence and where all the vector bundles and nonlinear operators are different (See ([43]) for more details). This work is a natural continuation of symbolic computations done at RWTH-Aachen university by M. Barakat and A. Lorenz in 2008 ([3],[30]).

The starting point is to refer to the three fundamental theorems of Lie, in particular the third which provides a way to realize an analytic Lie group G from the knowledge of its Lie algebra $\mathcal{G}$ and the construction of the left or right invariant analytic 1-forms on G called *Maurer-Cartan forms* $\omega^\tau = (\omega^\tau_\sigma(a)da^\sigma)$ satisfying the well known *Maurer-Cartan equations* $d\omega^\tau + c^\tau_{\rho\sigma}\omega^\rho \wedge \omega^\sigma = 0$ where d is the exterior derivative, provided that $J(c) = 0$.

The general solution involves tools from differential geometry (Spencer operator, δ-cohomology), differential algebra, algebraic geometry, algebraic analysis and homological algebra. However, the key argument is to acknowledge the fact that the *Vessiot structue equations* ([51]) must be used in place of the *Cartan structure equations* ([4],[5]), along the following crucial diagram describing the references ([36],[38],[43]) which are the best ones existing on this subject while providing applications to engineering (continuum mechanics, electromagnetism) and mathematical (Gauge Theory, General Relativity) physics.

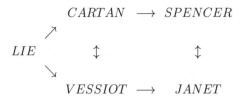

It is important to notice that the *Vessiot structure equations*, first discovered as early as in 1903, are still not known today by the mathematical community for reasons which are not scientific at all, in particular because of an *"affair "* involving the best french mathematicians at the beginning of the last century (H. Poincaré, E. Picard, G. Darboux, J. Drach, E. Cartan, E. Vessiot, P. Painlevé, E. Borel, ...). Such a story, known by everybody at that time but then deliberately forgotten, explains why Cartan never spoke about the work of Vessiot on Lie pseudogroupds and conversely. The corresponding private documents have been given to me by M. Janet, a friend of E. Vessiot, who had also never been quoted by Cartan in his 1930 letters to A. Einstein, as a gift before he died in 1983 , because I made his name and work known ([36]). I have now brought them to the library of Ecole Normale Supérieure in Paris where they can be consulted under request (See my Gordon and Breach books "Differential Galois Theory " ([37], 1983) and "Lie Pseudogroups and Mechanics " ([38], 1988) for more details and many letters).

In a word, one has to replace Lie algebras by Lie algebroids with a bracket now depending on the Spencer operator and use the corresponding canonical *linear Janet sequence* in order to induce a new sequence locally described by finite dimensional vector spaces and linear maps, *even for structures providing infinite dimensional Lie algebroids* (contact and unimodular contact structures are good examples as we shall see). The cohomology of this sequence only depends on the *structure constants* appearing in the Vessiot structure equations (constant riemannian curvature is an example with only one constant having of course nothing to do with any Lie algebra). Finally, the simplest case of a principal homogeneous space for G (for example G itself as before) gives back the Chevalley-Eilenberg cohomology.

In order to motivate the reader and convince him about the novelty of the underlying concepts, we provide in a rather self-contained way and parallel manners the following six striking examples which are the best nontrivial ones we know showing how and why computer algebra could be used and that will be revisited later on. We invite the reader at this stage to try to imagine any possible link that could exist between these examples. We also ask him to look backwards to these explicit examples whenever a new concept will be introduced later on.

We may start with a basic definition, saying that a *Lie pseudogroup of transformations* is a group of transformations solutions of a (linear or even non-linear) system of ordinary or partial differential equations called system of *finite Lie equations*. Of course, any Lie group of transformations $y = f(x, a)$ may be considered as a Lie pseudogroup by differentiating these relations sufficiently enough in order to eliminate the *parameters* a in the language of the last century. Therefore, any classical procedure dealing with the parameters cannot be used any longer.

The starting point of our research has been to understand why, among all these examples, a few corresponding integrability conditions were containing "*constants*" while others were not containing any "*constant*".

EXAMPLE 1.1: (*Principal homogeneous structure*) When Γ is the Lie group of transformations made by the constant translations $y^i = x^i + a^i$ for $i = 1, ..., n$ of a manifold X with $dim(X) = n$, the characteristic object invariant by Γ is a family $\omega = (\omega^\tau = \omega_i^\tau dx^i)$ of n 1-forms with $det(\omega) \neq 0$ in such a way that $\Gamma = \{f \in aut(X) | j_1(f)^{-1}(\omega) = \omega\}$ when $aut(X)$ denotes the pseudogroup of local diffeomorphisms of X, $j_q(f)$ denotes the derivatives of f up to order q and $j_1(f)$ acts in the usual way on covariant tensors. For any vector field $\xi \in T = T(X)$ the tangent bundle to X, introducing the standard Lie derivative $\mathcal{L}(\xi)$ of forms with respect to ξ, we may consider the n^2 *first order Medolaghi equations*:

$$\Omega_i^\tau \equiv (\mathcal{L}(\xi)\omega)_i^\tau \equiv \omega_r^\tau(x)\partial_i\xi^r + \xi^r\partial_r\omega_i^\tau(x) = 0$$

The particular situation is found with the special choice $\omega = (dx^i)$ that leads to the involutive system $\partial_i\xi^k = 0$. Introducing the inverse matrix $\alpha = (\alpha_\tau^i) = \omega^{-1}$, the above equations amount to the bracket relations $[\xi, \alpha_\tau] = 0$ and, using crossed derivatives on the *solved form* $\partial_i\xi^k + \xi^r\alpha_\tau^k(x)\partial_r\omega_i^\tau(x) = 0$, we obtain the $n^2(n-1)/2$

8

zero order equations:

$$\xi^r \partial_r(\alpha^i_\rho(x)\alpha^j_\sigma(x)(\partial_i\omega^\tau_j(x) - \partial_j\omega^\tau_i(x))) = 0$$

The *integrability conditions* (IC) are thus the $n^2(n-1)/2$ *Vessiot structure equations*:

$$\partial_i\omega^\tau_j(x) - \partial_j\omega^\tau_i(x) = c^\tau_{\rho\sigma}\omega^\rho_i(x)\omega^\sigma_j(x)$$

with $n^2(n-1)/2$ *structure constants* $c = (c^\tau_{\rho\sigma} = -c^\tau_{\sigma\rho})$. When $X = G$, these equations can be identified with the *Maurer-Cartan equations* (MC) existing in the theory of Lie groups, on the condition to change the sign of the structure constants involved because we have $[\alpha_\rho, \alpha_\sigma] = -c^\tau_{\rho\sigma}\alpha_\tau$. Writing these equations in the form of the exterior system $d\omega^\tau = c^\tau_{\rho\sigma}\omega^\rho \wedge \omega^\sigma$ and closing this system by applying once more the exterior derivative d, we obtain the quadratic IC:

$$c^\lambda_{\mu\rho}c^\mu_{\sigma\tau} + c^\lambda_{\mu\sigma}c^\mu_{\tau\rho} + c^\lambda_{\mu\tau}c^\mu_{\rho\sigma} = 0$$

also called *Jacobi relations* $J(c) = 0$. Finally, if another family $\bar\omega$ of forms is given, the *equivalence problem* $j_1(f)^{-1}(\omega) = \bar\omega$ cannot be solved even locally if $\bar{c} \neq c$.

EXAMPLE 1.2: (*Riemann structure*) If $\omega = (\omega_{ij} = \omega_{ji})$ is a metric on a manifold X with $dim(X) = n$ such that $det(\omega) \neq 0$, the Lie pseudogroup of transformations preserving ω is $\Gamma = \{f \in aut(X)|j_1(f)^{-1}(\omega) = \omega\}$ and is a Lie group with a maximum number of $n(n+1)/2$ parameters. A special metric could be the Euclidean metric when $n = 1, 2, 3$ as in elasticity theory or the Minkowski metric when $n = 4$ as in special relativity. The *first order Medolaghi equations*:

$$\Omega_{ij} \equiv (\mathcal{L}(\xi)\omega)_{ij} \equiv \omega_{rj}(x)\partial_i\xi^r + \omega_{ir}(x)\partial_j\xi^r + \xi^r\partial_r\omega_{ij}(x) = 0$$

are also called *Killing equations* for historical reasons. The main problem is that *this system is not involutive* unless we prolong it to order two by differentiating once the equations. For such a purpose, introducing $\omega^{-1} = (\omega^{ij})$ as usual, we may define the *Christoffel symbols*:

$$\gamma^k_{ij}(x) = \frac{1}{2}\omega^{kr}(x)(\partial_i\omega_{rj}(x) + \partial_j\omega_{ri}(x) - \partial_r\omega_{ij}(x)) = \gamma^k_{ji}(x)$$

This is a new geometric object of order 2 providing an isomorphism $j_1(\omega) \simeq (\omega, \gamma)$

9

and allowing to obtain the *second order Medolaghi equations*:

$$\Gamma_{ij}^k \equiv (\mathcal{L}(\xi)\gamma)_{ij}^k \equiv \partial_{ij}\xi^k + \gamma_{rj}^k(x)\partial_i\xi^r + \gamma_{ir}^k(x)\partial_j\xi^r - \gamma_{ij}^r(x)\partial_r\xi^k + \xi^r\partial_r\gamma_{ij}^k(x) = 0$$

Surprisingly, the following expression:

$$\rho_{lij}^k(x) \equiv \partial_i\gamma_{lj}^k(x) - \partial_j\gamma_{li}^k(x) + \gamma_{lj}^r(x)\gamma_{ri}^k(x) - \gamma_{li}^r(x)\gamma_{rj}^k(x)$$

is still a first order geometric object and even a 4-tensor with $n^2(n^2-1)/12$ independent components satisfying the purely algebraic relations :

$$\rho_{lij}^k + \rho_{lji}^k = 0, \quad \rho_{lij}^k + \rho_{ijl}^k + \rho_{jli}^k = 0, \quad \omega_{rl}\rho_{kij}^r + \omega_{kr}\rho_{lij}^r = 0$$

Therefore, setting $\rho_{ij} = \rho_{irj}^r$ and $\rho = \omega^{ij}\rho_{ij}$ as usual, we get $\rho_{rij}^r = 0$ and $\rho_{ij} = \rho_{ji}$. Accordingly, the IC must express that the new first order equations:

$$R_{lij}^k \equiv (\mathcal{L}(\xi)\rho)_{lij}^k = 0$$

are only linear combinations of the previous ones, that is to say we must have:

$$\rho_{rij}^k\xi_l^r + \rho_{lrj}^k\xi_i^r + \rho_{lir}^k\xi_j^r - \rho_{lij}^k\xi_r^k + \xi^r\partial_r\rho_{lij}^k = 0$$

whenever $\omega_{rj}\xi_i^r + \omega_{ir}\xi_j^r + \xi^r\partial_r\omega_{ij} = 0$.
In particular, we must therefore have (care):

$$\rho_{rij}^k\xi_l^r + \rho_{lrj}^k\xi_i^r + \rho_{lir}^r\xi_j^r - \rho_{lij}^k\xi_r^k = (\rho_{rij}^k\delta_l^s + \rho_{lrj}^k\delta_i^s + \rho_{lir}^k\delta_j^s - \rho_{lij}^s\delta_r^k)\xi_s^r = 0$$

whenever $\omega_{rj}\xi_i^r + \omega_{ir}\xi_j^r = 0$ or $\xi_{i,j} + \xi_{j,i} = 0$, raising or lowering the indices by means of ω, that is to say, using the Kronecker symbol $\delta_i^k = 0$ if $k \neq i$ or 1 if $k = i$:

$$(\rho_{rij}^k\delta_l^s + \rho_{lrj}^k\delta_i^s + \rho_{lir}^k\delta_j^s - \rho_{lij}^s\delta_r^k)\omega^{rt} - (\rho_{rij}^k\delta_l^t + \rho_{lrj}^k\delta_i^t + \rho_{lir}^k\delta_j^t - \rho_{lij}^t\delta_r^k)\omega^{rs} = 0$$

Contracting in s and l, we get:

$$(n\rho_{rij}^k + \rho_{irj}^k + \rho_{jir}^k - \rho_{sij}^s)\omega^{rt} - \rho_{rij}^k\omega^{rt} - \rho_{srj}^k\delta_i^t\omega^{rs} - \rho_{sir}^k\delta_j^t\omega^{rs} + \rho_{sij}^t\omega^{ks} = 0$$

and thus, using the previous algebraic relations:

$$(n-1)\rho^k_{rij}\omega^{rt} = (\rho^k_{rsj}\delta^t_i - \rho^k_{rsi}\delta^t_j)\omega^{rs} \Rightarrow (n-1)\rho^k_{lij} = (\rho^k_{rsj}\omega_{li} - \rho^k_{rsi}\omega_{lj})\omega^{rs}$$

with $\rho^k_{rsj}\omega^{rs} = \omega^{kt}\rho_{trsj}\omega^{rs} = -\omega^{kt}\rho_{rtsj}\omega^{rs} = -\omega^{kt}\rho_{tj}$. Finally, like above, we must also have $R_{ij} \equiv (\mathcal{L}(\xi)\rho)_{ij} = 0$ whenever $\Omega_{ij} = 0$ and similarly:

$$(\rho_{rj}\delta^s_i + \rho_{ir}\delta^s_j)\omega^{rt} - (\rho_{rj}\delta^t_i + \rho_{ir}\delta^t_j)\omega^{rs} = 0$$

Contracting in s and i, we get $n\rho_{ij} = \rho\omega_{ij}$ and thus $n\rho^k_{rsj}\omega^{rs} = \rho\delta^k_j$. As ρ is a scalar and we must not have any zero order equation, we may set $\rho = n(n-1)c$ where c is a constant and obtain the *Vessiot structure equations* with the only *structure constant c*:

$$\rho^k_{lij}(x) = c(\delta^k_i\omega_{lj}(x) - \delta^k_j\omega_{li}(x))$$

describing the constant Riemannian curvature condition of Eisenhart (compare to [10]). If $\bar{\omega}$ is another nondegenerate metric with structure constant $\bar{c}$, the *equivalence problem* $j_1(f)^{-1}(\omega) = \bar{\omega}$ cannot be solved even locally if $\bar{c} \neq c$.

Treating the case of the *conformal Killing system*:

$$\Omega_{ij} = (\mathcal{L}(\xi)\omega)_{ij} = A(x)\omega_{ij}(x)$$

where $A(x)$ is an arbitrary function, is *much more difficult*. We let the reader prove that the *Vessiot structure equations* only lead to the vanishing of the *Weyl tensor*:

$$\tau^k_{l,ij} \equiv \rho^k_{l,ij} - \tfrac{1}{(n-2)}(\delta^k_i\rho_{lj} - \delta^k_j\rho_{li} + \omega_{lj}\omega^{ks}\rho_{si} - \omega_{li}\omega^{ks}\rho_{sj})$$
$$+ \tfrac{1}{(n-1)(n-2)}(\delta^k_i\omega_{lj} - \delta^k_j\omega_{li})\rho$$

without any structure constant involved.

EXAMPLE 1.3: (*Contact structure*) We only treat the case $dim(X) = 3$ as the case $dim(X) = 2p + 1$ needs much more work ([37]). Let us consider the so-called *contact* 1-form $\alpha = dx^1 - x^3dx^2$ and consider the Lie pseudogroup $\Gamma \subset aut(X)$ of (local) transformations preserving α up to a function factor, that is $\Gamma = \{f \in aut(X) | j_1(f)^{-1}(\alpha) = \rho\alpha\}$ where again $j_q(f)$ is a symbolic way for writing out the derivatives of f up to order q and α transforms like a 1-covariant tensor. It may be

11

tempting to look for a kind of *"object"* the invariance of which should characterize Γ. Introducing the exterior derivative $d\alpha = dx^2 \wedge dx^3$ as a 2-form, we obtain the volume 3-form $\alpha \wedge d\alpha = dx^1 \wedge dx^2 \wedge dx^3$. As it is well known that the exterior derivative commutes with any diffeomorphism, we obtain sucessively:

$$j_1(f)^{-1}(d\alpha) = d(j_1(f)^{-1}(\alpha)) = d(\rho\alpha) = \rho d\alpha + d\rho \wedge \alpha$$

and thus $j_1(f)^{-1}(\alpha \wedge d\alpha) = \rho^2(\alpha \wedge d\alpha)$.

As the volume 3-form $\alpha \wedge d\alpha$ transforms through a division by the Jacobian determinant $\Delta = \partial(f^1, f^2, f^3)/\partial(x^1, x^2, x^3) \neq 0$ of the transformation $y = f(x)$ with inverse $x = f^{-1}(y) = g(y)$, *the desired object is thus no longer a 1-form but a 1-form density* $\omega = (\omega_1, \omega_2, \omega_3)$ transforming like a 1-form but up to a division by the square root of the Jacobian determinant. It follows that the infinitesimal contact transformations are vector fields $\xi \in T = T(X)$ the tangent bundle of X, satisfying the 3 so-called *first order Medolaghi equations*:

$$\Omega_i \equiv (\mathcal{L}(\xi)\omega)_i \equiv \omega_r(x)\partial_i\xi^r - (1/2)\omega_i(x)\partial_r\xi^r + \xi^r\partial_r\omega_i(x) = 0$$

When $\omega = (1, -x^3, 0)$, we obtain the *special* involutive system:

$$\partial_3\xi^3 + \partial_2\xi^2 + 2x^3\partial_1\xi^2 - \partial_1\xi^1 = 0, \partial_3\xi^1 - x^3\partial_3\xi^2 = 0,$$
$$\partial_2\xi^1 - x^3\partial_2\xi^2 + x^3\partial_1\xi^1 - (x^3)^2\partial_1\xi^2 - \xi^3 = 0$$

with 2 equations of class 3 and 1 equation of class 2 obtained by exchanging x^1 and x^3 (see section 4 for the definitions) and thus only 1 *compatibility conditions* (CC) for the second members.

For an arbitrary ω, we may ask about the differential conditions on ω such that all the equations of order $r + 1$ are only obtained by differentiating r times the first order equations, exactly like in the special situation just considered where the system is involutive. We notice that, in a symbolic way, $\omega \wedge d\omega$ is now a scalar $c(x)$ providing the zero order equation $\xi^r\partial_r c(x) = 0$ and the condition is $c(x) = c = cst$. The *integrability condition* (IC) is the *Vessiot structure equation*:

$$\omega_1(\partial_2\omega_3 - \partial_3\omega_2) + \omega_2(\partial_3\omega_1 - \partial_1\omega_3) + \omega_3(\partial_1\omega_2 - \partial_2\omega_1) = c$$

involving the only *structure constant* c.

For $\omega = (1, -x^3, 0)$, we get $c = 1$. If we choose $\bar{\omega} = (1, 0, 0)$ leading to $\bar{c} = 0$, we

12

may define $\bar{\Gamma} = \{f \in aut(X)|j_1(f)^{-1}(\bar{\omega}) = \bar{\omega}\}$ with infinitesimal transformations satisfying the involutive system:

$$\partial_3\xi^3 + \partial_2\xi^2 - \partial_1\xi^1 = 0, \partial_3\xi^1 = 0, \partial_2\xi^1 = 0$$

with again 2 equations of class 3 and 1 equation of class 2. The *equivalence problem* $j_1(f)^{-1}(\omega) = \bar{\omega}$ cannot be solved even locally because this system cannot have any invertible solution. Indeed, studying the system $j_1(g)^{-1}(\bar{\omega}) = \omega$, we have to solve:

$$\frac{\partial g^1}{\partial y^2} + y^3\frac{\partial g^1}{\partial y^1} = 0, \frac{\partial g^1}{\partial y^3} = 0 \Rightarrow \frac{\partial g^1}{\partial y^1} = 0, \frac{\partial g^1}{\partial y^2} = 0, \frac{\partial g^1}{\partial y^3} = 0$$

by using crossed derivatives.

EXAMPLE 1.4: (*Unimodular contact structure*) With similar notations, let us again set $\alpha = dx^1 - x^3dx^2 \Rightarrow d\alpha = dx^2 \wedge dx^3$ but let us now consider the new Lie pseudogroup of transformations preserving α and thus $d\alpha$ too, that is preserving the *mixed* object $\omega = (\alpha, \beta)$ made up by a 1-form α and a 2-form β with $\gamma = \alpha \wedge \beta \neq 0$ and $d\alpha = \beta \Rightarrow d\beta = 0$. Then Γ is a Lie subpseudogroup of the one just considered in the previous example and the corresponding infinitesimal transformations now satisfy the involutive system:

$$\partial_1\xi^1 = 0, \partial_1\xi^2 = 0, \partial_1\xi^3 = 0, \partial_2\xi^1 + x^3\partial_3\xi^3 - \xi^3 = 0,$$
$$\partial_2\xi^2 + \partial_3\xi^3 = 0, \partial_3\xi^1 - x^3\partial_3\xi^2 = 0$$

with 3 equations of class 3, 2 equations of class 2 and 1 equation of class 1 if we exchange x^1 with x^3, a result leading now to 4 CC.

More generally, when $\omega = (\alpha, \beta)$ where α is a 1-form and β is a 2-form satifying $\alpha \wedge \beta \neq 0$, *we may study the same problem as before* for the *general* system $\mathcal{L}(\xi)\alpha = 0, \mathcal{L}(\xi)\beta = 0$ where $\mathcal{L}(\xi)$ is the standard Lie derivative of forms with respect to the vector field ξ, that is $\mathcal{L}(\xi) = i(\xi)d + di(\xi)$ if $i(\xi)$ is the interior multiplication of a form by the vector field ξ. We now provide details on the tedious computation involved as it is at this point that computer algebra may be used. With $\alpha = (\alpha_i), \beta = (\beta_{ij} = -\beta_{ji})$ we may suppose, with no loss of generality, that $\alpha_1 \neq 0, \beta_{23} \neq 0$ in such a way that $\alpha \wedge \beta \neq 0 \Leftrightarrow \alpha_1\beta_{23} + \alpha_2\beta_{31} + \alpha_3\beta_{12} \neq 0$. We can then solve the general equations as in the special situation already considered with respect to the 6 leading principal derivatives $pri = \{\partial_1\xi^1, \partial_1\xi^2, \partial_1\xi^3, \partial_2\xi^1, \partial_2\xi^2, \partial_3\xi^1\}$

as involution is an intrinsic local property and obtain:

$$\partial_1\xi^1 + \frac{\alpha_2}{\alpha_1}\partial_1\xi^2 + \frac{\alpha_3}{\alpha_1}\partial_1\xi^3 + \ldots = 0$$

$$\partial_1\xi^2 + \frac{\beta_{31}}{\beta_{32}}(\partial_1\xi^1 + \partial_3\xi^3) + \frac{\beta_{21}}{\beta_{32}}\partial_3\xi^2 + \ldots = 0$$

$$\partial_1\xi^3 + \frac{\beta_{13}}{\beta_{32}}\partial_2\xi^3 + \frac{\beta_{12}}{\beta_{32}}(\partial_1\xi^1 + \partial_2\xi^2) + \ldots = 0$$

$$\partial_2\xi^1 + \frac{\alpha_2}{\alpha_1}\partial_2\xi^2 + \frac{\alpha_3}{\alpha_1}\partial_2\xi^3 + \ldots = 0$$

$$\partial_2\xi^2 + \partial_3\xi^3 + \frac{\beta_{13}}{\beta_{23}}\partial_2\xi^1 + \frac{\beta_{21}}{\beta_{23}}\partial_3\xi^1 + \ldots = 0$$

$$\partial_3\xi^1 + \frac{\alpha_2}{\alpha_1}\partial_3\xi^2 + \frac{\alpha_3}{\alpha_1}\partial_3\xi^3 + \ldots = 0$$

Solving wth respect to the 6 principal derivatives in a triangular way while introducing the 3 parametric derivatives $par = \{\partial_2\xi^3, \partial_3\xi^2, \partial_3\xi^3\}$, we obtain for example:

$$(1 - \frac{\beta_{13}}{\beta_{23}}\frac{\alpha_2}{\alpha_1})\partial_2\xi^2 - \frac{\beta_{21}}{\beta_{23}}\frac{\alpha_2}{\alpha_1}\partial_3\xi^2 + (1 - \frac{\beta_{21}}{\beta_{23}}\frac{\alpha_3}{\alpha_1})\partial_3\xi^3 - \frac{\beta_{13}}{\beta_{23}}\frac{\alpha_3}{\alpha_1}\partial_2\xi^3 + \ldots = 0$$

Setting now $\bar{\beta} = d\alpha$ and identifying the corresponding coefficients, then $\mathcal{L}(\xi)\bar{\beta}$ is a linear combination of $\mathcal{L}(\xi)\alpha$ and $\mathcal{L}(\xi)\beta$ if and only if we have for example:

$$\frac{\frac{\beta_{13}}{\beta_{23}}\frac{\alpha_3}{\alpha_1}}{1 - \frac{\beta_{13}}{\beta_{23}}\frac{\alpha_2}{\alpha_1}} = \frac{\frac{\bar{\beta}_{13}}{\beta_{23}}\frac{\alpha_3}{\alpha_1}}{1 - \frac{\bar{\beta}_{13}}{\beta_{23}}\frac{\alpha_2}{\alpha_1}}$$

and thus:

$$\frac{\beta_{13}}{\beta_{23}} = \frac{\bar{\beta}_{13}}{\beta_{23}} \Rightarrow \frac{\bar{\beta}_{23}}{\beta_{23}} = \frac{\bar{\beta}_{13}}{\beta_{23}} = \frac{\bar{\beta}_{12}}{\beta_{12}}$$

a result, not evident at first sight, showing that the 2-form $d\alpha$ *must* be proportional to the 2-form β, that is $d\alpha = c'(x)\beta$ and thus $\alpha \wedge d\alpha = c'(x)\alpha \wedge \beta$. As $\alpha \wedge \beta \neq 0$, we *must* have $c'(x) = c' = cst$ and thus $d\alpha = c'\beta$. Similarly, we get $d\beta = c''\alpha \wedge \beta$ and obtain finally the 4 *Vessiot structure equations* $d\alpha = c'\beta, d\beta = c''\alpha \wedge \beta$ involving 2 *structure constants* $c = (c', c'')$. Contrary to the previous situation (but like in the Riemann case !) we notice that we have now 2 structure equations not containing any constant (called *first kind* by Vessiot) and 2 structure equations with the same number of different constants (called *second kind* by Vessiot), namely $\alpha \wedge d\alpha = c'\alpha \wedge \beta, d\beta = c''\alpha \wedge \beta$.

14

Finally, closing this system by taking once more the exterior derivative, we get $0 = d^2\alpha = c'd\beta = c'c''\alpha \wedge \beta$ and thus the unexpected purely algebraic *Jacobi condition* $c'c'' = 0$. For the special choice $\omega = (dx^1 - x^3 dx^2, dx^2 \wedge dx^3)$ we get $c = (1, 0)$, for the second special choice $\bar{\omega} = (dx^1, dx^2 \wedge dx^3)$ we get $\bar{c} = (0, 0)$ and for the third special choice $\bar{\bar{\omega}} = ((1/x^1)dx^1, x^1 dx^2 \wedge dx^3)$ we get $\bar{\bar{c}} = (0, 1)$ with similar comments as before for the possibility to solve the corresponding equivalence problems.

EXAMPLE 1.5: When $n = 2$ and we consider transformations of the plane $(x^1, x^2) \to (y^1, y^2)$ satisfying $y^2 y_1^1 - y^1 y_1^2 = x^2, y^2 y_2^1 - y^1 y_2^2 = -x^1 \Rightarrow y_1^1 y_2^2 - y_2^1 y_1^2 = 1$, no explicit integration can be obtained in order to provide general solutions but another way is to say that the corresponding Lie pseudogroup preserves the 1-form $\alpha = x^2 dx^1 - x^1 dx^2$ and thus the 2-form $d\alpha = -2dx^1 \wedge dx^2$ as we have indeed $y^2 dy^1 - y^1 dy^2 = x^2 dx^1 - x^1 dx^2$ and thus also $dy^1 \wedge dy^2 = dx^1 \wedge dx^2$. The Lie pseudogroup is thus preserving the *geometric object* $\omega = (\alpha, d\alpha)$ made by a 1-form and a 2-form. More generally, we may consider the Lie pseudogroup preserving the geometric object $\omega = (\alpha, \beta)$ where α is a 1-form and β is a 2-form. As $d\alpha$ is also preserved, if we want that the system behaves at least like the preceding one, that is *cannot have any zero order equation*, we *must* have the only *Vessiot structure equation* $d\alpha = c\beta$ for some arbitrary constant c. The two pseudogroups defined by $\omega \to c$ and $\bar{\omega} \to \bar{c}$ can be exchanged by a change of variables bringing ω to $\bar{\omega}$ *if and only if* $\bar{c} = c$. This situation is the simplest example of the celebrated *formal equivalence problem* ([20],[21]).

EXAMPLE 1.6: (*Affine and projective structures*) When $n = 1$ and we consider transformations $y = f(x)$ of the real line, the *affine group* of transformations is defined by the linear system $y_{xx} = 0$ with jet notations saying that any transformation is such that $\partial_{xx} f(x) = 0$ while the *projective group* of transformations is defined by the nonlinear system $\frac{y_{xxx}}{y_x} - \frac{3}{2}\left(\frac{y_{xx}}{y_x}\right)^2 = 0$ with a similar comment. In both cases we have indeed a *Lie group of transformations* depending on a finite number of constant parameters, namely $y = ax + b$ in the first case and $y = \frac{ax+b}{cx+d}$ in the second case. Accordingly, the respective geometric object the invariance of which is characterizing the corresponding Lie pseudogroup are surely not made by tensors because the defining finite Lie equations are not first order. In order to compare the methods used for Lie groups with the ones used for Lie pseudogroups, let us notice that, in the case of the affine transformations, we have successively:

$$y = ax + b \Rightarrow y_x = a \Rightarrow y_{xx} = 0$$

while, in the case of the projective transformations, we have successively:

$$y = \frac{ax+b}{cx+d} \Rightarrow y_x = \frac{ad-bc}{(cx+d)^2} \Rightarrow y_{xx} = \frac{-2c(ad-bc)}{(cx+d)^3} \Rightarrow y_{xxx} = \frac{6c^2(ad-bc)}{(cx+d)^4}$$

Turning now to the Lie group approach and changing slightly our notations for this purpose while considering the affine group of transformations of the real line $y = a^1 x + a^2$, the orbits are defined by $x = a^1 x_0 + a^2$, a definition leading to $dx = da^1 x_0 + da^2$ and thus $dx = ((1/a^1)da^1)x + (da^2 - (a^2/a^1)da^1)$. We obtain therefore the two infinitesimal generators *on the line* (care) $\theta_1 = x\partial_x, \theta_2 = \partial_x \Rightarrow [\theta_1, \theta_2] = -\theta_2$ and the Maurer-Cartan invariant 1-forms *on the group* (care) $\omega^1 = (1/a^1)da^1, \omega^2 = da^2 - (a^2/a^1)da^1 \Rightarrow d\omega^1 = 0, d\omega^2 - \omega^1 \wedge \omega^2 = 0$ providing at once, by inverting the corresponding 2×2 matrix, the invariant vector fields *on the group* (care) $\alpha_1 = a^1\partial_1 + a^2\partial_2, \alpha_2 = \partial_2$ with commutation relation $[\alpha_1, \alpha_2] = -\alpha_2$.

Finally, coming back to the affine and projective examples already presented, we show that the Vessiot structure equations may even exist when $n = 1$. For this, we notice that the only generating differential invariant $\Phi \equiv y_{xx}/y_x$ of the affine case transforms like $u = \bar{u}\partial_x f + (\partial_{xx}f/\partial_x f)$ while the only generating differential invariant $\Psi \equiv (y_{xxx}/y_x) - \frac{3}{2}(y_{xx}/y_x)^2$ of the projective case transforms like $v = \bar{v}(\partial_x f)^2 + (\partial_{xxx}f/\partial_x f) - \frac{3}{2}(\partial_{xx}f/\partial_x f)^2$ when $\bar{x} = f(x)$. If now γ is the geometric object of the affine group $y = ax + b$ and $0 \neq \alpha = \alpha(x)dx \in T^*$ is a 1-form, we consider the object $\omega = (\alpha, \gamma)$ and get at once *one first order and one second order general Medolaghi equations*:

$$\mathcal{L}(\xi)\alpha \equiv \alpha\partial_x\xi + \xi\partial_x\alpha = 0, \qquad \mathcal{L}(\xi)\gamma \equiv \partial_{xx}\xi + \gamma\partial_x\xi + \xi\partial_x\gamma = 0$$

Differentiating the first equation and substituting the second, we get the zero order equation:

$$\xi(\alpha\partial_{xx}\alpha - 2(\partial_x\alpha)^2 + \alpha\gamma\partial_x\alpha - \alpha^2\partial_x\gamma) = 0 \quad \Leftrightarrow \quad \xi\partial_x\left(\frac{\partial_x\alpha}{\alpha^2} - \frac{\gamma}{\alpha}\right) = 0$$

16

and the only *Vessiot structure equation of second kind* $\partial_x \alpha - \gamma \alpha = c\alpha^2$ where c is an arbitrary constant. Equivalently, but without any link with the fact that γ can be considered as a connection, we may also use the two first order equations:

$$\mathcal{L}(\xi)\alpha^2 \equiv 2\alpha^2 \partial_x \xi + \xi \partial_x \alpha^2 = 0,$$
$$\mathcal{L}(\xi)(\partial_x \alpha - \gamma \alpha) \equiv 2(\partial_x \alpha - \gamma \alpha)\partial_x \xi + \xi \partial_x(\partial_x \alpha - \gamma \alpha) = 0$$

With $\alpha = 1, \gamma = 0 \Rightarrow c = 0$ we get the translation subgroup $y = x + b$ while, with $\alpha = 1/x, \gamma = 0 \Rightarrow c = -1$ we get the dilatation subgroup $y = ax$.

Similarly, if ν is the geometric object of the projective group and we consider the new geometric object $\omega = (\gamma, \nu)$, we get at once *one second order and one third order general Medolaghi equations*:

$$\mathcal{L}(\xi)\gamma \equiv \partial_{xx}\xi + \gamma \partial_x \xi + \xi \partial_x \gamma = 0, \qquad \mathcal{L}(\xi)\nu \equiv \partial_{xxx}\xi + 2\nu \partial_x \xi + \xi \partial_x \nu = 0$$

Differentiating the first equation and substracting the second one while taking into account the first one, we get:

$$\mathcal{L}(\xi)(\partial_x \gamma - \frac{1}{2}\gamma^2 - \nu) \equiv 2(\partial_x \gamma - \frac{1}{2}\gamma^2 - \nu)\partial_x \xi + \xi \partial_x(\partial_x \gamma - \frac{1}{2}\gamma^2 - \nu) = 0$$

and obtain therefore the only *Vessiot structure equation of first kind* $\partial_x \gamma - \frac{1}{2}\gamma^2 - \nu = 0$ as there is no longer any first order equation involved.

However, considering now the third order Medolaghi system for $\omega = (\alpha, \gamma, \nu)$:

$$\mathcal{L}(\xi)\alpha \equiv \alpha \partial_x \xi + \xi \partial_x \alpha = 0, \; \mathcal{L}(\xi)\gamma \equiv \partial_{xx}\xi + \gamma \partial_x \xi + \xi \partial_x \gamma = 0,$$
$$\mathcal{L}(\xi)\nu \equiv \partial_{xxx}\xi + 2\nu \partial_x \xi + \xi \partial_x \nu = 0$$

Differentiating the second equation and substracting the third one while taking into account the second one, we get:

$$\mathcal{L}(\xi)(\partial_x \gamma - \frac{1}{2}\gamma^2 - \nu) \equiv 2(\partial_x \gamma - \frac{1}{2}\gamma^2 - \nu)\partial_x \xi + \xi \partial_x(\partial_x \gamma - \frac{1}{2}\gamma^2 - \nu) = 0$$

as before but get now the *two Vessiot structure equations of second kind*:

$$\partial_x \alpha - \gamma \alpha = c\alpha^2, \quad \partial_x \gamma - \frac{1}{2}\gamma^2 - \nu = c'\alpha^2$$

because we have to take into account the first order equation:

$$\mathcal{L}(\xi)\alpha^2 \equiv 2\alpha^2 \partial_x \xi + \xi \partial_x \alpha^2 = 0.$$

REMARK 1.7: Comparing the various Vessiot structure equations containing structure constants that we have just presented and that we recall below in a symbolic way, we notice that *these structure constants are absolutely on equal footing though they have nothing to do with any Lie algebra.*

$$\partial\omega - \partial\omega \;=\; c\;\omega\,\omega$$
$$\partial\gamma - \partial\gamma + \gamma\gamma - \gamma\gamma \;=\; c\;(\delta\omega - \delta\omega)$$
$$\omega \wedge (\partial\omega - \partial\omega) \;=\; c$$

$$\begin{cases} d\alpha \;=\; c'\;\beta \\ d\beta \;=\; c''\;\alpha \wedge \beta \end{cases}$$

$$d\alpha \;=\; c\;\beta$$
$$\partial\alpha - \gamma\alpha \;=\; c\;\alpha^2$$
$$\partial\gamma - \tfrac{1}{2}\gamma^2 - \nu \;=\; 0$$
$$\partial\gamma - \tfrac{1}{2}\gamma^2 - \nu \;=\; c'\alpha^2$$

Accordingly, the fact that the ones appearing in the MC equations are related to a Lie algebra is a pure coincidence and we may even say:

Cartan structure equations have nothing to do with Vessiot structure equations.

Also, as their factors are either constant, linear or quadratic, *any identification of the quadratic terms appearing in the Riemann tensor with the quadratic terms appearing in the MC equations is definitively not correct* even though most of mathematical physics today is based on such a confusion ([43]). Meanwhile, we understand why the torsion is *automatically combined* with curvature in the Cartan structure equations but *totally absent* from the Vessiot structure equations, even though the underlying group (translations + rotations) is the same. In addition, despite the prophetic comments of the italian mathematician Ugo Amaldi in 1909 ([1]), we do believe that it has been a pity that Cartan deliberately ignored the work of Vessiot at the beginning of the last century and that the things did not improve afterwards in the eighties with Spencer and coworkers (Compare MR 720863 (85m:12004) and MR 954613 (90e:58166)).

In the second section of this paper, which is an extended version of a lecture given at the international conference SCA (Symbolic Computation and its Applications) 2012 held in Aachen (Aix-la-Chapelle), Germany, May 17-20, 2012, we shall recall the definition of the Chevalley-Eilenberg cohomology existing for Lie algebras and describe its use in the study of the *deformation theory of Lie algebras* ([28],[29],[45],[49],[50]). We insist on the fact that the challenge solved in the remaining of the paper is not to generalize this result to arbitrary Lie equations but rather to work out a general framework that will provide *exactly* the deformation cohomology of Lie algebras in the particular case of Example 1.

In the third section we study the *"Vessiot structure equations"* in the nonlinear framework and explain why they must contain *"structure constants"* satifying algebraic Jacobi-like conditions. It is however important to notice that the deformation cohomology will also be defined later on in a direct way, even though the fundamental concepts only depend on the non-linear framework.

In the fourth section, we present the minimum amount of differential geometry (jet theory, Spencer operator, δ-cohomology, differential sequences) needed in order to achieve the formal constructions done in the next sections. More technical additional useful technical results and corresponding striking examples will be provided independently in Appendix 1.

In the fifth and longest section we present for the first time a rather self-contained but complete general procedure for exhibiting a *deformation cohomology* for any system of transitive Lie equations, even of infinite type as in the case of the (unimodular) contact structure. A link with the difficult concept of a *truncated Lie algebra* will also be provided (Compare to [14], II).

In the last section, we provide a few explicit computations based on the previous examples while showing out the possibility to use computer algebra techniques.

We finally add three appendices:

The Appendix 1 is giving additional useful technical results and examples on the construction of differential sequences, studying in particular the conformal Killing operator and showing how different can be the cases of dimension $n = 3$, dimension $n = 4$ and dimension $n \geq 5$. We point out that such a fundamental result had not been discovered during the last hundred years and is still unknown.

The Appendix 2 is providing the explicit computations made by one of my former PhD students Alban Quadrat, using a computer algebra package that he developped and we explain the main difficulty met during this work.

The short Appendix 3 is sketching the possibility to apply all these new methods to mathematical physics, in particular General Relativity and Gauge Theory.

2 DEFORMATION THEORY OF LIE ALGEBRAS:

Let V be a finite dimensional vector space over a field k containing the field $\mathbb{Q}$ and set $V^* = hom_k(V, k)$. We shall denote the elements of V by $X, Y, Z, ...$ with components $X = (X^\rho)$ for $\rho = 1, ..., dim(V)$.

DEFINITION 2.1: A *Lie algebra* $\mathcal{G} = (V, c)$ is an algebraic structure on V defined over k by a bilinear map $[\ ] : \wedge^2 V \to V$ called *bracket* through the formula $([X, Y])^\tau = c^\tau_{\rho\sigma} X^\rho Y^\sigma$ where the *structure constants* $c \in \wedge^2 V^* \otimes V$ are satisfying $c^\tau_{\rho\sigma} + c^\tau_{\sigma\rho} = 0$ and the *Jacobi relations*:

$$J(c) = 0 \qquad c^\lambda_{\rho\sigma} c^\mu_{\lambda\tau} + c^\lambda_{\sigma\tau} c^\mu_{\alpha\rho} + c^\lambda_{\tau\rho} c^\mu_{\lambda\sigma} = 0$$

a result leading to the *Jacobi identity* for the bracket, namely:

$$[X, [Y, Z]] + [Y, [Z, X]] + [Z, [X, Y]] \equiv 0 \qquad , \forall X, Y, Z \in V$$

Considering a Lie algebra as a point c of the algebraic set defined by the quadratic equations $J(c) = 0$, we may state:

DEFINITION 2.2: A *deformation* c_t of c is a curve passing through c in this algebraic set, that is to say a set of points c_t indexed by a parameter t and such that $J(c_t) = 0$ with $c_t = c + tC + ...$ As a byproduct, the Lie algebra $\mathcal{G}_t = (V, c_t)$ is called a deformation of $\mathcal{G} = (V, c)$.

Of course, a central problem in the theory of Lie algebras is to exhibit properties of $\mathcal{G}$ that do not depend on the basis chosen for V. In particular, if $a \in aut(V) \subset V^* \otimes V$, we may define an equivalence relation among the structure constants as follows:

DEFINITION 2.3: $\bar{c} \sim c \Leftrightarrow a\bar{c}(X, Y) = c(aX, aY), \forall X, Y \in V$ or equivalently $\bar{c} = aaa^{-1}c$ in a symbolic tensorial notation showing that c transforms like a (2-covariant, 1-contravariant)-tensor under the action of a. In particular, we get at once $\bar{c} \sim c \Leftrightarrow J(\bar{c}) = 0$ whenever $J(c) = 0$.

When $a_t \in aut(V)$ is such that $a_0 = id_V$, then the set of points $c_t \sim c$ is a deformation of c called *trivial* and we may state:

DEFINITION 2.4: $a \in aut(V)$ is called an *automorphism* of $\mathcal{G}$ if $\bar{c} = c$ and $A \in end(V) = V^* \otimes V$ is called a *derivation* of $\mathcal{G}$ if $A[X,Y] = [AX,Y] + [X,AY], \forall X,Y \in V$. In particular, if $a_t = a_0 + tA + ... \in aut(\mathcal{G})$ is such that $a_0 = id_V$, then $A = \frac{da_t}{dt}|_{t=0}$ is a derivation of $\mathcal{G}$. Moreover, if we define the *adjoint action* of V on V by $ad(X)Y = [X,Y]$, then it follows from the Jacobi identity that $ad(X)$ is a derivation of $\mathcal{G}$ called *inner derivation*.

DEFINITION 2.5: A Lie algebra $\mathcal{G}$ is said to be *rigid* if it cannot admit a deformation which is not trivial.

As the algebraic set of structure constants may have very bad local properties, it may be interesting to study infinitesimal deformations when t is a small parameter, that is $t \ll 1$. In particular, we should have $\frac{\partial J(c)}{\partial c}C = 0$ for any deformation while a trivial deformation should lead to:

$$(\bar{C})^\tau_{\rho\sigma} = (A)^\mu_\rho (C)^\tau_{\mu\sigma} + (A)^\mu_\sigma (C)^\tau_{\rho\mu} - (A)^\tau_\mu (C)^\mu_{\rho\sigma}$$

with no way to unify all these technical formulas.

By chance, such an infinitesimal study can be studied by means of the *Chevalley-Eilenberg cohomology* that we now describe.. For this, by analogy with the exterior derivative, let us define an application $d : \wedge^r V^* \otimes V \rightarrow \wedge^{r+1} V^* \otimes V$ depending on c by the formula:

$$df(X_1, ..., X_{r+1}) = \sum_{i<j}(-1)^{i+j}f([X_i, X_j], X_1, ..., \hat{X}_i, ..., \hat{X}_j, ..., X_{r+1})$$
$$+ \sum_i(-1)^{i+1}[X_i, f(X_1, ..., \hat{X}_i, ..., X_{r+1})]$$

where a "hat" indicates an omission. Using the Jacobi identity for the bracket and a straightforward but tedious computation left to the reader as an exercise, one can prove that $d \circ d = 0$. Hence we may define in the usual way *coboundaries* $B_r(\mathcal{G})$ as images of d, *cocycles* $Z_r(\mathcal{G})$ as kernels of d and *cohomology groups* $H_r(\mathcal{G}) = Z_r(\mathcal{G})/B_r(\mathcal{G})$ in such a way that $B_r(\mathcal{G}) \subseteq Z_r(\mathcal{G}) \subseteq \wedge^r V^* \otimes V$.

22

LEMMA 2.6: $C = \frac{dc_t}{dt}\big|_{t=0} \in Z_2(\mathcal{G})$ and $c_t \sim c \Leftrightarrow C = \frac{dc_t}{dt}\big|_{t=0} \in B_2(\mathcal{G})$. Accordingly, a sufficient condition of rigidity is $H_2(\mathcal{G}) = 0$.

LEMMA 2.7: A is a derivation $\Rightarrow A \in Z_1(\mathcal{G})$ and A is an inner derivation $\Rightarrow A \in B_1(\mathcal{G})$. Accordingly, the vector space of derivations of $\mathcal{G}$ modulo the vector space of inner derivations is $H_1(\mathcal{G})$.

Up to the moment we have only been looking at infinitesimal deformations. However, a Lie algebra $\mathcal{G}$ may be rigid on the finite level even if $H_2(\mathcal{G}) \neq 0$, that is even if it can be deformed on the infinitesimal level ([28]) and we may state:

DEFINITION 2.8: An element $C \in Z_2(\mathcal{G})$ is said to be *integrable* if there exists a deformation c_t of c such that $\frac{dc_t}{dt}\big|_{t=0} = C$. It is said to be *formally integrable* if there exists a formal power series $c_t = \sum_{\nu=0}^{\infty} \frac{t^\nu}{\nu!} C_\nu$ with $C_0 = c, C_1 = C$.

EXAMPLE 2.9: Even if Example 1.4 has nothing to do at first sight with any Lie algebra as we already said, we may adapt the previous arguments to $c'c'' = 0$. Indeed, we must have:

$$
\begin{aligned}
0 &= (c' + tC_1' + \tfrac{t^2}{2}C_2' + ...)(c'' + tC_1'' + \tfrac{t^2}{2}C_2'' + ...) \\
&= t(c''C_1' + c'C_1'') + \tfrac{t^2}{2}(c''C_2' + 2C_1'C_1'' + c'C_2'') + ...
\end{aligned}
$$

As cocycles are defined by the condition $c''C' + c'C'' = 0$, it follows that the cocycle $(1,1)$ cannot be integrable at $c = (0,0)$ as we should have $C'C'' = 0$.

In order to study the formal integrability of a cocycle $C \in Z_2(\mathcal{G})$, we shall use a trick by introducing two parameters s and t. Then c_{s+t} becomes a deformation of c_t in s and we have:

$$
c_{s+t} = \sum_{\nu=0}^{\infty} \frac{(s+t)^\nu}{\nu!} C_\nu = \sum_{\nu=0}^{\infty} \sum_{\lambda+\mu=\nu} \frac{s^\lambda}{\lambda!} \frac{t^\mu}{\mu!} C_\nu = c_t + s\left(\sum_{\nu=0}^{\infty} \frac{t^\nu}{\nu!} C_{\nu+1}\right) + ...
$$

where one may notice the change $C \to \sum_{\nu=0}^{\infty} \frac{t^\nu}{\nu!} C_{\nu+1}$.

PROPOSITION 2.10: A sufficient condition for the formal integrability of any cocycle is $H_3(\mathcal{G}) = 0$.

23

Proof: By definition we have $d = d(c)$ and we may set $d(t) = d(c_t)$ in order to obtain:

$$dC_1 = 0 \Rightarrow d(t) \sum_{\nu=0}^{\infty} \frac{t^\nu}{\nu!} C_{\nu+1} = 0$$

Differentiating with respect to t and setting $t = 0$ while using the Leibnitz formula:

$$d^\nu(ab) = \sum_{\lambda,\mu \geq 0, \lambda+\mu=\nu} \frac{\nu!}{\lambda!\mu!} (d^\lambda a)(d^\mu b)$$

with $d = d/dt$, we get:

$$\sum_{\lambda,\mu \geq 0, \lambda+\mu=\nu} \frac{\nu!}{\lambda!\mu!} \left(\frac{d^\lambda d(t)}{dt^\lambda}\right)\big|_{t=0} C_{\mu+1} = 0$$

or, in an equivalent way:

$$dC_{\nu+1} + \sum_{\lambda>0, \mu \geq 0, \lambda+\mu=\nu} \frac{\nu!}{\lambda!\mu!} \left(\frac{d^\lambda d(t)}{dt^\lambda}\right)\big|_{t=0} C_{\mu+1} = 0$$

As the first term on the left belongs to $B_3(\mathcal{G})$, it just remains to prove that the sum on the right, which only depends on $C_1, ..., C_\nu$, is in $Z_3(\mathcal{G})$. Otherwise, as shown by the previous example, the study of these inductively related conditions may sometimes bring *obstructions* to the deformation at a certain order.

Now we have $d \circ d \equiv 0 \Rightarrow d(t) \circ d(t) \equiv 0$ and thus again:

$$\sum_{\lambda,\mu \geq 0, \lambda+\mu=\nu} \frac{\nu!}{\lambda!\mu!} \left(\frac{d^\lambda d(t)}{dt^\lambda}\right)\big|_{t=0} \circ \left(\frac{d^\mu d(t)}{dt^\mu}\right)\big|_{t=0} \equiv 0$$

Accordingly, we obtain successively by using this identity (care to the minus sign):

$$-d \circ \sum_{\lambda>0, \mu \geq 0, \lambda+\mu=\nu} \frac{\nu!}{\lambda!\mu!} \left(\frac{d^\lambda d(t)}{dt^\lambda}\right)\big|_{t=0} C_{\mu+1} \equiv - \sum_{\lambda>0, \mu \geq 0, \lambda+\mu=\nu} \frac{\nu!}{\lambda!\mu!} d \circ \left(\frac{d^\lambda d(t)}{dt^\lambda}\right)\big|_{t=0} C_{\mu+1}$$

$$\equiv \sum_{\lambda>0, \mu \geq 0, \lambda+\mu=\nu} \frac{\nu!}{\lambda!\mu!} \left[\sum_{\alpha>0, \beta \geq 0, \alpha+\beta=\lambda} \frac{\lambda!}{\alpha!\beta!} \left(\frac{d^\alpha d(t)}{dt^\alpha}\right)\big|_{t=0} \circ \left(\frac{d^\beta d(t)}{dt^\beta}\right)\big|_{t=0} C_{\mu+1}\right]$$

$$\equiv \sum_{\alpha=1}^{\nu} \frac{\nu!}{\alpha!(\nu-\alpha)!} \left(\frac{d^\alpha d(t)}{dt^\alpha}\right)\big|_{t=0} \circ \left[\sum_{\mu \geq 0, \beta \geq 0, \beta+\mu=\nu-\alpha} \frac{(\nu-\alpha)!}{\beta!\mu!} \left(\frac{d^\beta d(t)}{dt^\beta}\right)\big|_{t=0} C_{\mu+1}\right] \equiv 0$$

24

because we may take into account the integrability conditions up to order $\nu - 1$ as $\alpha > 0$ and $\alpha + \beta + \mu = \lambda + \mu = \nu$.

Q.E.D.

Studying the successive integrability conditions, we get:

$\nu = 0$ $\qquad\qquad\qquad dC_1 = 0$

$\nu = 1$ $\qquad\qquad\qquad dC_2 + \frac{\partial^2 J(c)}{\partial c \partial c} C_1 C_1 = 0$

and we have the following result not evident at first sight:

COROLLARY 2.11: The Hessian of the Jacobi conditions provides a quadratic map:

$$H_2(\mathcal{G}) \to H_3(\mathcal{G}) : C \to \frac{\partial^2 J(c)}{\partial c \partial c} CC$$

Proof: The previous proposition proves that this map takes $C \in Z_2(\mathcal{G})$ to $Z_3(\mathcal{G})$ and we just need to prove that it also takes $C \in B_2(\mathcal{G})$ to $B_3(\mathcal{G})$. For this, let $A \in V^* \otimes V$ be such that $C = dA$. We have:

$$-(\frac{dd(t)}{dt})|_{t=0} C = -(\frac{dd(t)}{dt})|_{t=0} \circ dA = d \circ (\frac{dd(t)}{dt})|_{t=0} A$$

as we wished.

Q.E.D.

3 VESSIOT STRUCTURE EQUATIONS :

If X is a manifold, we denote as usual by $T = T(X)$ the *tangent bundle* of X, by $T^* = T^*(X)$ the *cotangent bundle*, by $\wedge^r T^*$ the *bundle of r-forms* and by $S_q T^*$ the *bundle of q-symmetric tensors*. More generally, let $\mathcal{E}$ be a *fibered manifold*, that is a manifold with local coordinates (x^i, y^k) for $i = 1, ..., n$ and $k = 1, ..., m$ simply denoted by (x, y), *projection* $\pi : \mathcal{E} \to X : (x, y) \to (x)$ and changes of local coordinates $\bar{x} = \varphi(x), \bar{y} = \psi(x, y)$. If $\mathcal{E}$ and $\mathcal{F}$ are two fibered manifolds over X with respective local coordinates (x, y) and (x, z), we denote by $\mathcal{E} \times_X \mathcal{F}$ the *fibered product* of $\mathcal{E}$ and $\mathcal{F}$ over X as the new fibered manifold over X with local coordinates (x, y, z). We denote by $f : X \to \mathcal{E} : (x) \to (x, y = f(x))$ a global *section* of $\mathcal{E}$, that is a map such that $\pi \circ f = id_X$ but local sections over an open set $U \subset X$ may also be considered when needed. Under a change of coordinates, a section transforms like $\bar{f}(\varphi(x)) = \psi(x, f(x))$ and the derivatives transform like:

$$\frac{\partial \bar{f}^l}{\partial \bar{x}^r}(\varphi(x))\partial_i \varphi^r(x) = \frac{\partial \psi^l}{\partial x^i}(x, f(x)) + \frac{\partial \psi^l}{\partial y^k}(x, f(x))\partial_i f^k(x)$$

We may introduce new coordinates (x^i, y^k, y_i^k) transforming like:

$$\bar{y}_r^l \partial_i \varphi^r(x) = \frac{\partial \psi^l}{\partial x^i}(x, y) + \frac{\partial \psi^l}{\partial y^k}(x, y)y_i^k$$

in order to define the 1-*jet bunble* $J_1(\mathcal{E})$ of $\mathcal{E}$.Differentiating q times and proceeding similarly, we shall denote by $J_q(\mathcal{E})$ the *q-jet bundle* of $\mathcal{E}$ with local coordinates $(x^i, y^k, y_i^k, y_{ij}^k, ...) = (x, y_q)$ called *jet coordinates* and sections $f_q : (x) \to (x, f^k(x), f_i^k(x), f_{ij}^k(x), ...) = (x, f_q(x))$ transforming like the sections $j_q(f) : (x) \to (x, f^k(x), \partial_i f^k(x), \partial_{ij} f^k(x), ...) = (x, j_q(f)(x))$ where both f_q and $j_q(f)$ are over the section f of $\mathcal{E}$. Of course $J_q(\mathcal{E})$ is a fibered manifold over X with projection π_q while $J_{q+r}(\mathcal{E})$ is a fibered manifold over $J_q(\mathcal{E})$ with projection $\pi_q^{q+r}, \forall r \geq 0$.

DEFINITION 3.1: A (non-linear) *system* of order q on $\mathcal{E}$ is a fibered submanifold $\mathcal{R}_q \subset J_q(\mathcal{E})$ and a *solution* of $\mathcal{R}_q$ is a section f of $\mathcal{E}$ such that $j_q(f)$ is a section of $\mathcal{R}_q$.

DEFINITION 3.2: When the changes of coordinates have the linear form $\bar{x} = \varphi(x), \bar{y} = A(x)y$, we say that $\mathcal{E}$ is a *vector bundle* over X and denote for simplicity a vector bundle and its set of sections by the same capital letter E. When the changes

26

of coordinates have the form $\bar{x} = \varphi(x), \bar{y} = A(x)y + B(x)$ we say that $\mathcal{E}$ is an *affine bundle* over X and we define the *associated vector bundle* E over X by the local coordinates (x, v) changing like $\bar{x} = \varphi(x), \bar{v} = A(x)v$.

DEFINITION 3.3: If the tangent bundle $T(\mathcal{E})$ has local coordinates (x, y, u, v) changing like $\bar{u}^j = \partial_i \varphi^j(x)u^i, \bar{v}^l = \frac{\partial \psi^l}{\partial x^i}(x,y)u^i + \frac{\partial \psi^l}{\partial y^k}(x,y)v^k$, we may introduce the *vertical bundle* $V(\mathcal{E}) \subset T(\mathcal{E})$ as a vector bundle over $\mathcal{E}$ with local coordinates (x, y, v) obtained by setting $u = 0$ and changes $\bar{v}^l = \frac{\partial \psi^l}{\partial y^k}(x,y)v^k$. Of course, when $\mathcal{E}$ is an affine bundle over X with associated vector bundle E over X, we have $V(\mathcal{E}) = \mathcal{E} \times_X E$.

For a later use, if $\mathcal{E}$ is a fibered manifold over X and f is a section of $\mathcal{E}$, we denote by $f^{-1}(V(\mathcal{E}))$ the *reciprocal image* of $V(\mathcal{E})$ by f as the vector bundle over X obtained when replacing (x, y, v) by $(x, f(x), v)$ in each chart. A similar construction may also be done for any affine bundle over $\mathcal{E}$.

We now recall a few basic geometric concepts that will be constantly used through this paper. First of all, if $\xi, \eta \in T$, we define their *bracket* $[\xi, \eta] \in T$ by the local formula $([\xi, \eta])^i(x) = \xi^r(x)\partial_r\eta^i(x) - \eta^s(x)\partial_s\xi^i(x)$ leading to the *Jacobi identity* $[\xi, [\eta, \zeta]] + [\eta, [\zeta, \xi]] + [\zeta, [\xi, \eta]] = 0, \forall \xi, \eta, \zeta \in T$ allowing to define a *Lie algebra* and to the useful formula $[T(f)(\xi), T(f)(\eta)] = T(f)([\xi, \eta])$ where $T(f) : T(X) \to T(Y)$ is the tangent mapping of a map $f : X \to Y$.

When $I = \{i_1 < ... < i_r\}$ is a multi-index, we may set $dx^I = dx^{i_1} \wedge ... \wedge dx^{i_r}$ for describing $\wedge^r T^*$ and introduce the *exterior derivative* $d : \wedge^r T^* \to \wedge^{r+1} T^* : \omega = \omega_I dx^I \to d\omega = \partial_i \omega_I dx^i \wedge dx^I$ with $d^2 = d \circ d \equiv 0$ in the *Poincaré sequence*:

$$\wedge^0 T^* \xrightarrow{d} \wedge^1 T^* \xrightarrow{d} \wedge^2 T^* \xrightarrow{d} ... \xrightarrow{d} \wedge^n T^* \longrightarrow 0$$

The *Lie derivative* of an r-form with respect to a vector field $\xi \in T$ is the linear first order operator $\mathcal{L}(\xi)$ linearly depending on $j_1(\xi)$ and uniquely defined by the following three properties:

1) $\mathcal{L}(\xi)f = \xi.f = \xi^i \partial_i f, \forall f \in \wedge^0 T^* = C^\infty(X)$.
2) $\mathcal{L}(\xi)d = d\mathcal{L}(\xi)$.
3) $\mathcal{L}(\xi)(\alpha \wedge \beta) = (\mathcal{L}(\xi)\alpha) \wedge \beta + \alpha \wedge (\mathcal{L}(\xi)\beta), \forall \alpha, \beta \in \wedge T^*$.

27

It can be proved that $\mathcal{L}(\xi) = i(\xi)d + di(\xi)$ where $i(\xi)$ is the *interior multiplication* $(i(\xi)\omega)_{i_1...i_r} = \xi^i \omega_{ii_1...i_r}$ and that $[\mathcal{L}(\xi), \mathcal{L}(\eta)] = \mathcal{L}(\xi) \circ \mathcal{L}(\eta) - \mathcal{L}(\eta) \circ \mathcal{L}(\xi) = \mathcal{L}([\xi, \eta]), \forall \xi, \eta \in T$.

Indeed, if $\alpha = \alpha_i(x)dx^i \in T^*$, we have successively:

$$
\begin{aligned}
\mathcal{L}(\xi)\alpha = \xi^r \partial_r \alpha_i dx^i + \alpha_i \mathcal{L}(\xi)dx^i &= \xi^r \partial_r \alpha_i dx^i + \alpha_i \partial_r(\xi)^i dx^r \\
&= \xi^r(\partial_r \alpha_i - \partial_i \alpha_r)dx^i + \partial_r(\alpha_i \xi^i)dx^r.
\end{aligned}
$$

We now turn to group theory and start with two basic definitions:

Let G be a *Lie group*, that is another manifold with local coordinates $a = (a^1, ..., a^p)$ called *parameters*, a *composition* $G \times G \to G : (a, b) \to ab$, an *inverse* $G \to G : a \to a^{-1}$ and an *identity* $e \in G$ satisfying:

$$
(ab)c = a(bc) = abc, \qquad aa^{-1} = a^{-1}a = e, \qquad ae = ea = a, \qquad \forall a, b, c \in G
$$

DEFINITION 3.4: G is said to *act* on X if there is a map $X \times G \to X : (x, a) \to y = ax = f(x, a)$ such that $(ab)x = a(bx) = abx, \forall a, b \in G, \forall x \in X$ and we shall say that we have a *Lie group of transformations* of X. In order to simplify the notations, we shall use global notations even if only local actions are existing. The set $G_x = \{a \in G \mid ax = x\}$ is called the *isotropy subgroup* of G at $x \in X$ and the action is said to be *effective* if $ax = x, \forall x \in X \Rightarrow a = e$.

DEFINITION 3.5: A *Lie pseudogroup of transformations* $\Gamma \subset aut(X)$ is a group of transformations solutions of a system of OD or PD equations such that, if $y = f(x)$ and $z = g(y)$ are two solutions, called *finite transformations*, that can be composed, then $z = g \circ f(x) = h(x)$ and $x = f^{-1}(y) = g(y)$ are also solutions while $y = x$ is a solution and we shall set $id_q = j_q(id)$.

It becomes clear from Examples $1.1, 1.2$ and 1.6 that Lie groups of transformations are particular cases of Lie pseudogroups of transformations as the system defining the finite transformations can be obtained by eliminating the parameters among the equations $y_q = j_q(f)(x, a)$ when q is large enough. The underlying system may be nonlinear and of high order as we have seen. We shall speak of an *algebraic pseudogroup* when the system is defined by *differential polynomials* that is polynomials

28

in the derivatives. Looking for transformations "close" to the identity, that is setting $y = x + t\xi(x) + ...$ when $t \ll 1$ is a small constant parameter, dividing by t and passing to the limit $t \to 0$, we may linearize the above nonlinear *system of finite Lie equations* in order to obtain a linear *system of infinitesimal Lie equations* of the same order for vector fields. Such a system has the property that, if ξ, η are two solutions, then $[\xi, \eta]$ is also a solution. Accordingly, the set $\Theta \subset T$ of solutions of this new system satisfies $[\Theta, \Theta] \subset \Theta$ and can therefore be considered as the Lie algebra of Γ.

We now turn to the theory proposed by Vessiot in 1903 ([51]) and sketch in a few successive steps the main results that we have obtained in many books ([36],[37],[38]) and ([39]). We invite the reader to follow the procedure on each of the examples provided for this purpose in the introduction.

1) If $\mathcal{E} = X \times X$, we shall denote by $\Pi_q = \Pi_q(X, X)$ the open subfibered manifold of $J_q(X \times X)$ defined independently of the coordinate system by $det(y_i^k) \neq 0$ with *source projection* $\alpha_q : \Pi_q \to X : (x, y_q) \to (x)$ and *target projection* $\beta_q : \Pi_q \to X : (x, y_q) \to (y)$. We shall sometimes introduce a copy Y of X with local coordinates (y) in order to avoid any confusion between the source and the target manifolds. Let us start with a Lie pseudogroup $\Gamma \subset aut(X)$ defined by a system $\mathcal{R}_q \subset \Pi_q$ of order q. In all the sequel we shall suppose that the system is *involutive* (see next section) and that Γ is *transitive* that is $\forall x, y \in X, \exists f \in \Gamma, y = f(x)$ or, equivalently, the map $(\alpha_q, \beta_q) : \mathcal{R}_q \to X \times X : (x, y_q) \to (x, y)$ is surjective.

2) The Lie algebra $\Theta \subset T$ of infinitesimal transformations is then obtained by linearization, that is to say setting $y = x + t\xi(x) + ...$, dividing by t and passing to the limit $t \to 0$ as we already said, in order to obtain the linear involutive system $R_q = id_q^{-1}(V(\mathcal{R}_q)) \subset J_q(T)$ by reciprocal image with $\Theta = \{\xi \in T | j_q(\xi) \in R_q\}$ while taking into account the fact that $T = id^{-1}(V(X \times X))$. From now on we shall suppose that R_q is *transitive*, that is to say the canonical projection $\pi_0^q : J_q(T) \to T$ induces an epimorphism $\pi_0^q : R_q \to T$ with kernel $R_q^0 \subset R_q$ and we have the useful short exact sequence $0 \longrightarrow R_q^0 \longrightarrow R_q \overset{\pi_0^q}{\longrightarrow} T \longrightarrow 0$.

3) Passing from source to target, we may *prolong* the vertical infinitesimal transformations $\eta = \eta^k(y)\frac{\partial}{\partial y^k}$ to the jet coordinates up to order q in order to obtain for

29

any $\eta \in T(Y)$:

$$\eta^k(y)\frac{\partial}{\partial y^k} + \frac{\partial \eta^k}{\partial y^u}y_i^u\frac{\partial}{\partial y_i^k} + (\frac{\partial^2 \eta^k}{\partial y^u \partial y^v}y_i^u y_j^v + \frac{\partial \eta^k}{\partial y^u}y_{ij}^u)\frac{\partial}{\partial y_{ij}^k} + ...$$

where we have replaced $j_q(f)(x)$ by y_q, each component beeing the "formal" derivative of the previous one obtained by introducing $d_i = \partial_i + y_{\mu+1_i}^k\frac{\partial}{\partial y_\mu^k}$. Replacing the derivatives of η with respect to the target y by sections of $R_q(Y)$ over Y, we get for example:

$$\sharp(\eta_2) = \eta^k(y)\frac{\partial}{\partial y^k} + \eta_u^k(y)y_i^u\frac{\partial}{\partial y_i^k} + (\eta_{uv}^k(y)y_i^u y_j^v + \eta_u^k(y)y_{ij}^u)\frac{\partial}{\partial y_{ij}^k}$$

It is possible to prove that $[\sharp(R_q(Y), \sharp(R_q(Y)] = \sharp([R_q(Y), R_q(Y)])$ where the *differential bracket* involved over the target will be defined independently over the source in Definition 5.2.

4) As $[\Theta, \Theta] \subset \Theta$ and thus $[R_q, R_q] \subset R_q$, we may use the Frobenius theorem in order to find a generating fundamental set of *differential invariants* $\{\Phi^\tau(y_q)\}$ up to order q which are such that $\Phi^\tau(\bar{y}_q) = \Phi^\tau(y_q)$ by using the chain rule for derivatives whenever $\bar{y} = g(y) \in \Gamma$ acting now on Y. Of course, in actual practice *one must use sections of R_q instead of solutions* but it is only in section 5 through Definition 5.2 that we shall see why the use of the Spencer operator will be crucial for this purpose. Specializing the Φ^τ at $id_q(x)$ we obtain the *Lie form* $\Phi^\tau(y_q) = \omega^\tau(x)$ of $\mathcal{R}_q$. Finally, if Φ^τ is any differential invariant at the order q, then $d_i\Phi^\tau$ is a differential invariant at order $q+1, \forall i = 1, ..., n$.

5) It has been the clever discovery of Vessiot in 1903 ([51]) to notice that a *natural bundle $\mathcal{F}$* over X could be associated with any Lie pseudogroup $\Gamma \subset aut(X)$, both with a section ω of $\mathcal{F}$ called *geometric object* or *structure* on X, transforming the same way. Of course, as he also pointed out, different sections can provide the same Lie pseudogroup and this will be the starting point of *deformation theory*. Indeed, the prolongation $\flat(j_q(\xi))$ at order q of any horizontal vector field $\xi = \xi^i(x)\frac{\partial}{\partial x^i}$ commutes with the prolongation at order q of any vertical vector field $\eta = \eta^k(y)\frac{\partial}{\partial y^k}$, exchanging therefore the differential invariants. Keeping in mind the well known property of the Jacobian determinant while passing to the finite point of view, any (local) transformation $\bar{x} = \varphi(x)$ can be lifted to a (local) transformation of the differential invariants

30

between themselves of the form $u \to \lambda(u, j_q(\varphi)(x))$ allowing to introduce a *natural bundle* $\mathcal{F}$ over X by patching changes of coordinates $\bar{x} = \varphi(x)$, $\bar{u} = \lambda(u, j_q(\varphi)(x))$. A section ω of $\mathcal{F}$ is called a *geometric object* or *structure* on X and transforms like $\bar{\omega}(f(x)) = \lambda(\omega(x), j_q(f)(x))$ or simply $\bar{\omega} = j_q(f)(\omega)$ for any (local) transformation $y = f(x)$. This is a way to generalize vectors and tensors ($q = 1$) or even connections ($q = 2$). As a byproduct we have $\Gamma = \{f \in aut(X) | \Phi_\omega(j_q(f)) \equiv j_q(f)^{-1}(\omega) = \omega\}$ as a new way to write out the Lie form and we may say that Γ *preserves* ω. Replacing $j_q(f)$ by f_q, we also obtain $\mathcal{R}_q = \{f_q \in \Pi_q | f_q^{-1}(\omega) = \omega\}$. Coming back to the infinitesimal point of view and setting $f_t = exp(t\xi) \in aut(X), \forall \xi \in T$, we may define the *ordinary Lie derivative* with value in $\omega^{-1}(V(\mathcal{F}))$ by the formula :

$$\mathcal{D}\xi = \mathcal{D}_\omega \xi = \mathcal{L}(\xi)\omega = \frac{d}{dt} j_q(f_t)^{-1}(\omega)|_{t=0} \Rightarrow \Theta = \{\xi \in T | \mathcal{L}(\xi)\omega = 0\}$$

We have $x \to \bar{x} = x + t\xi(x) + ... \Rightarrow u^\tau \to \bar{u}^\tau = u^\tau + t\partial_\mu \xi^k L_k^{\tau\mu}(u) + ...$ where $\mu = (\mu_1, ..., \mu_n)$ is a multi-index and we may write down the system of infinitesimal Lie equations in the *Medolaghi form* (care to signs):

$$\Omega^\tau \equiv (\mathcal{L}(\xi)\omega)^\tau \equiv -L_k^{\tau\mu}(\omega(x))\partial_\mu \xi^k + \xi^r \partial_r \omega^\tau(x) = 0$$

as a way to state the invariance of the section ω of $\mathcal{F}$, that is $u^\tau - \omega^\tau(x) = 0 \Rightarrow \bar{u}^\tau - \bar{\omega}^\tau(\bar{x}) = 0$.

Finally, replacing $j_q(\xi)$ by a section $\xi_q \in J_q(T)$ over $\xi \in T$, we may define $R_q \subset J_q(T)$ *on sections* by the linear (non-differential) equations:

$$\Omega^\tau \equiv (L(\xi_q)\omega)^\tau \equiv -L_k^{\tau\mu}(\omega(x))\xi_\mu^k + \xi^r \partial_r \omega^\tau(x) = 0$$

and obtain the first prolongation $R_{q+1} \subset J_{q+1}(T)$ by adding:

$$\begin{aligned}
\Omega_i^\tau &\equiv (L(\xi_{q+1})j_1(\omega))_i^\tau \\
&\equiv -L_k^{\tau\mu}(\omega(x))\xi_{\mu+1_i}^k - \frac{\partial L_k^{\tau\mu}(\omega(x))}{\partial u^\sigma}\partial_i \omega^\sigma(x)\xi_\mu^k + \partial_r \omega^\tau(x)\xi_i^r + \xi^r \partial_r(\partial_i \omega^\tau(x)) = 0
\end{aligned}$$

6) By analogy with "special" and "general" relativity, we shall call the given section *special* and any other arbitrary section *general*. The problem is now to study the formal properties of the linear system just obtained with coefficients only depending on $j_1(\omega)$, exactly like we did in the examples of the introduction. In particular, if any expression involving ω and its derivatives is a scalar object, it must reduce to a constant because Γ is assumed to be transitive and thus cannot be defined by any

zero order equation. Now let us prove that the CC for $\bar{\omega}$, thus for ω too, only depend on the Φ and take the quasi-linear symbolic form $v \equiv I(u_1) \equiv A(u)u_x + B(u) = 0$ with $u_1 = (u, u_x)$, allowing to define an affine subfibered manifold $\mathcal{B}_1 \subset J_1(\mathcal{F})$ over $\mathcal{F}$. Indeed, if $\sum A(y_q)d_x\Phi$ is a minimum sum of formal derivatives of differential invariants of order q not containing any jet coordinate of strict order $q + 1$, we may suppose by division that the first A in the sum is equal to 1. Applying the prolonged distribution of vector fields introduced in the step 3 at order $q + 1$, we obtain a new sum with less terms and a contradiction unless all the A are again differential in-variants at order q and thus functions of the Φ because of the Frobenius theorem. A similar comment can be done for the B. Now, if one has two sections ω and $\bar{\omega}$ of $\mathcal{F}$, the *equivalence problem* is to look for $f \in aut(X)$ such that $j_q(f)^{-1}(\omega) = \bar{\omega}$. When the two sections satisfy the same CC, the problem is sometimes locally possible (Lie groups of transformations, Darboux problem in analytical mechanics,...) but some-times not ([36], p 333).

7) Instead of the CC for the equivalence problem, let us look for the *integra-bility conditions* (IC) of the system of infinitesimal Lie equations and suppose that, for the given section, all the equations of order $q + r$ are obtained by differenti-ating r times only the equations of order q, then it was claimed by Vessiot ([51] with no proof, see [36], p 313,[39], p 207-212) that such a property is held if and only if there is an equivariant section $c : \mathcal{F} \to \mathcal{F}_1 : (x, u) \to (x, u, v = c(u))$ where $\mathcal{F}_1 = J_1(\mathcal{F})/\mathcal{B}_1$ is a natural vector bundle over $\mathcal{F}$ with local coordinates (x, u, v). Moreover, any such equivariant section depends linearly on a finite number of constants c called *structure constants* and the IC for the *Vessiot structure equa-tions* $I(u_1) = c(u)$ are of a polynomial form $J(c) = 0$. It is important to notice that, according to its construction, *the form of the Vessiot structure equations is in-variant under any change of coordinate system.* In actual practice, this study can be divided into two parts according to the following commutative and exact diagram where R_{q+1} is the first prolongation of R_q and the *symbol* g_{q+1} of R_{q+1} is the kernel of the (not necessarily surjective) map $\pi_q^{q+1} : R_{q+1} \to R_q$ induced by the short exact sequence $0 \to S_{q+1}T^* \otimes T \to J_{q+1}(T) \xrightarrow{\pi_q^{q+1}} J_q(T) \to 0$:

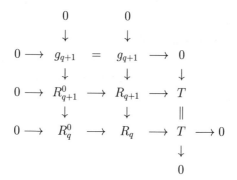

Indeed, chasing in this diagram, we discover that $\pi_q^{q+1} : R_{q+1} \to R_q$ is an epimorphism if and only if $\pi_q^{q+1} : R_{q+1}^0 \to R_q^0$ is an epimorphism *and* $\pi_0^{q+1} : R_{q+1} \to T$ is also an epimorphism. Looking at the form of the corresponding Medolaghi equations $L(\xi_q)\omega = 0$ and $L(\xi_{q+1})j_1(\omega) = 0$, these two conditions respectively bring the Vessiot structure equations of *first kind* $I_*(u_1) = 0$ not invoving any structure constant and the Vessiot stucture equations of *second kind* $I_{**}(u_1) = c$ involving the same number of different structure constants. Such a study, only depending now on linear algebraic techniques, can be achieved by means of computer algebra, in particular when dealing with algebraic pseudogroups ([3],[30]).

Finally, looking at the formal integrability of the system $\mathcal{B}_1 \subset J_1(\mathcal{F})$ defined by the equations $A(u)u_x + B(u) = 0$ and their first prolongation $A(u)u_{xx} + \frac{\partial A(u)}{\partial u}u_x u_x + \frac{\partial B(u)}{\partial u}u_x = 0$, the symbols only depend on $A(u)$ and we may obtain equations of the form $a(u)u_x u_x + b(u)u_x = 0$ by eliminating the jets of order 2. Using local coordinates $(x, u, v = A(u)u_x + B(u))$ for $\mathcal{F}_1$, and substituting (u, v) in place of (u, u_x), we obtain equations of the form $\alpha(u)vv + \beta(u)v + \gamma(u) = 0$. As we may suppose that $c = 0$ for the special section, we finally get equations of the form $\alpha(u)II + \beta(u)I = 0$ and it only remains to set $I(u_1) = c(u)$ in order to get polynomial Jacobi conditions of degree ≤ 2 which may not depend on u anymore because these equations are invariant in form under any change of coordinates.

REMARK 3.6: When $q = 1$, a close examination of the Medolaghi equations and their first prolongation shows at once that we can choose $v = A(u)u_x$ and we get homogeneous Jacobi conditions of degree 2. Such a result explains the difference existing between Examples 1.1, 1.3, 1.4 or 1.5 ($q = 1$) and Examples 1.2 or 1.6 ($q = 2$). Another example will be provided in Section 5. The reader interested may find more details and explicit examples in ([39]).

4 LINEAR AND NONLINEAR JANET SEQUENCES:

Let $\mu = (\mu_1, ..., \mu_n)$ be a multi-index with *length* $|\mu| = \mu_1 + ... + \mu_n$. We shall say that μ is of *class* i if $\mu_1 = ... = \mu_{i-1} = 0, \mu_i \neq 0$ and set $\mu + 1_i = (\mu_1, ..., \mu_{i-1}, \mu_i + 1, \mu_{i+1}, ..., \mu_n)$. We set $y_q = \{y_\mu^k | 1 \leq k \leq m, 0 \leq |\mu| \leq q\}$ with $y_\mu^k = y^k$ when $|\mu| = 0$. If E is a vector bundle over X with local coordinates (x^i, y^k) for $i = 1, ..., n$ and $k = 1, ..., m$, we denote by $J_q(E)$ the q-*jet bundle* of E with local coordinates simply denoted by (x, y_q) and *sections* $f_q : (x) \to (x, f^k(x), f_i^k(x), f_{ij}^k(x), ...)$ transforming like the section $j_q(f) : (x) \to (x, f^k(x), \partial_i f^k(x), \partial_{ij} f^k(x), ...)$ when f is an arbitrary section of E. Then both $f_q \in J_q(E)$ and $j_q(f) \in J_q(E)$ are over $f \in E$ and the *Spencer operator* just allows to distinguish them by introducing a kind of "*difference*" through the operator $D : J_{q+1}(E) \to T^* \otimes J_q(E) : f_{q+1} \to j_1(f_q) - f_{q+1}$ with local components $(\partial_i f^k(x) - f_i^k(x), \partial_i f_j^k(x) - f_{ij}^k(x), ...)$ and more generally $(D f_{q+1})_{\mu,i}^k(x) = \partial_i f_\mu^k(x) - f_{\mu+1_i}^k(x)$. Indeed, the composite map $J_{q+1}(E) \to J_q(E) \xrightarrow{j_1} J_1(J_q(E))$ and the canonical map $J_{q+1}(E) \to J_1(J_q(E))$ identifying $y_{\mu,i}$ with $y_{\mu+1_i}$ when $|\mu| = q$ are both over the identity map $J_q(E) = J_q(E)$ and their difference has therefore an image in $T^* \otimes J_q(E)$ because of the short exact sequence:

$$0 \to T^* \otimes J_q(E) \to J_1(J_q(E)) \xrightarrow{\pi_0^1} J_q(E) \to 0.$$

In a symbolic way, *when changes of coordinates are not involved*, it is sometimes useful to write down the components of D in the form $d_i = \partial_i - \delta_i$ and the restriction of D to the kernel $S_{q+1}T^* \otimes E$ of the canonical projection $\pi_q^{q+1} : J_{q+1}(E) \to J_q(E)$ is *minus* the *Spencer map* $\delta = dx^i \wedge \delta_i : S_{q+1}T^* \otimes E \to T^* \otimes S_q T^* \otimes E$ which can be extended to:

$$\delta : \wedge^r T^* \otimes S_{q+1}T^* \otimes E \to \wedge^{r+1}T^* \otimes S_q T^* \otimes E$$

with $\delta \circ \delta \equiv 0$ if we set $\omega = (\omega_{\mu,I}^k dx^I)$ and define $(\delta\omega)_\mu^k = dx^i \wedge \omega_{\mu+1_i}^k$ when $|\mu| = q$.

The kernel of D is made by sections such that $f_{q+1} = j_1(f_q) = j_2(f_{q-1}) = ... = j_{q+1}(f)$. Finally, if $R_q \subset J_q(E)$ is a *system* of order q on E locally defined by linear equations $\Phi^\tau(x, y_q) \equiv a_k^{\tau\mu}(x) y_\mu^k = 0$ and local coordinates (x, z) for the parametric jets up to order q, the r-*prolongation* $R_{q+r} = \rho_r(R_q) = J_r(R_q) \cap J_{q+r}(E) \subset J_r(J_q(E))$ is locally defined when $r = 1$ by the linear equations $\Phi^\tau(x, y_q) = 0 \Rightarrow$

$d_i \Phi^\tau(x, y_{q+1}) \equiv a_k^{\tau\mu}(x) y_{\mu+1_i}^k + \partial_i a_k^{\tau\mu}(x) y_\mu^k = 0$ and has *symbol* $g_{q+r} = R_{q+r} \cap S_{q+r} T^* \otimes E \subset J_{q+r}(E)$ if one looks at the *top order terms*. If $f_{q+1} \in R_{q+1}$ is over $f_q \in R_q$, differentiating the identity $a_k^{\tau\mu}(x) f_\mu^k(x) \equiv 0$ with respect to x^i and substracting the identity $a_k^{\tau\mu}(x) f_{\mu+1_i}^k(x) + \partial_i a_k^{\tau\mu}(x) f_\mu^k(x) \equiv 0$, we obtain the identity $a_k^{\tau\mu}(x)(\partial_i f_\mu^k(x) - f_{\mu+1_i}^k(x)) \equiv 0$ and thus the restriction $D : R_{q+1} \to T^* \otimes R_q$. This first order operator induces, up to sign, the purely algebraic monomorphism $0 \to g_{q+1} \overset{\delta}{\to} T^* \otimes g_q$ on the symbol level ([36],[40],[49]).

Though surprising it may look like when revisiting the mathematical foundations of physics, let us show how the Spencer operator may interfere with *gauge theory* (GT). For this, having in mind Example 1.6, we notice that the movement of a rigid body can be described by a pair $(a^1(t), a^2(t))$ where $a^1(t)$ is a time depending orthogonal matrix and $a^2(t)$ is a time depending vector. Hence, we have transformed the constant parameters of the group into time depending parameters. More generally, if X is an arbitrary manifold and G is a Lie group *not necessarily acting on X*, we may look at maps $X \to G : (x) \to (a(x))$ and this is just the gauging principle on which GT is based. Now, if G *is acting on X* with an action law $x \to y = f(x, a)$, gauging again G in this new framework provides successively the following section of $\Pi_q(X, X)$:

$$f(x) = f(x, a)|_{a=a(x)}, \quad f_x(x) = \frac{\partial f}{\partial x}(x, a)|_{a=a(x)}, \quad f_{xx}(x) = \frac{\partial^2 f}{\partial x \partial x}(x, a)|_{a=a(x)}, \quad \cdots$$

We obtain therefore, using the chain rule for derivatives:

$$\partial_x f(x) - f_x(x) = \frac{df(x, a(x))}{dx} - \frac{\partial f}{\partial x}(x, a)|_{a=a(x)} = \frac{\partial f}{\partial a}(x, a)|_{a=a(x)} \frac{\partial a(x)}{\partial x}, \cdots$$

and more generally:

$$f_{q+1}(x) = j_{q+1}(f(x, a))|_{a=a(x)} \implies D f_{q+1}(x) = \frac{\partial j_q(f(x, a))}{\partial a}|_{a=a(x)} \frac{\partial a(x)}{\partial x}$$

As q is large enough in such a way that $rk \frac{\partial j_q(f(x,a))}{\partial a}|_{a=a(x)} = dim(G)$, we obtain therefore $D f_{q+1}(x) = 0 \Leftrightarrow \frac{\partial a(x)}{\partial x} = 0 \Leftrightarrow a(x) = a = cst$. The future will decide whether the group must act on the manifold or not, in particular if electromagnetism has to do with $U(1)$ which is not acting on space-time as in classical gauge theory or with the conformal group of space-time as we have claimed in many recent books or papers, along with the idea of H. Weyl ([39], [43], [52]) (See also Appendix 3).

36

DEFINITION 4.1: R_q is said to be *formally integrable* when the restriction π_{q+r}^{q+r+1} : $R_{q+r+1} \to R_{q+r}$ is an epimorphism $\forall r \geq 0$ or, equivalently, when all the equations of order $q+r$ are obtained by r prolongations only, $\forall r \geq 0$. In that case, $R_{q+1} \subset J_1(R_q)$ is a canonical equivalent formally integrable first order system on R_q with no zero order equations, called the *Spencer form*.

DEFINITION 4.2: R_q is said to be *involutive* when it is formally integrable and the symbol g_q is *involutive*, that is all the sequences $\ldots \overset{\delta}{\to} \wedge^s T^* \otimes g_{q+r} \overset{\delta}{\to} \ldots$ are exact $\forall 0 \leq s \leq n, \forall r \geq 0$ or, equivalently, if the image $B_{q+r}^s(g_q)$ of the left δ is equal to the kernel $Z_{q+r}^s(g_q)$ of the right δ, that is to say if all the *Spencer δ-cohomology bundles* $H_{q+r}^s(g_q) = Z_{q+r}^s(g_q)/B_{q+r}^s(g_q) =$ vanish, $\forall r \geq 0$ and $\forall 0 \leq s \leq n$. As we shall see with more details in the Appendix for practical checking, using a linear change of local coordinates if necessary, we may *successively* solve the maximum number $\beta_q^n, \beta_q^{n-1}, \ldots, \beta_q^1$ of equations with respect to the principal jet coordinates of strict order q and class $n, n-1, \ldots, 1$ in order to introduce the *characters* $\alpha_q^i = m\frac{(q+n-i-1)!}{(q-1)!((n-i)!} - \beta_q^i$ for $i = 1, \ldots, n$ with $\alpha_q^n = \alpha$. Then R_q is involutive if R_{q+1} is obtained by only prolonging the β_q^i equations of class i with respect to $d_1, \ldots, d_i$ for $i = 1, \ldots, n$. In that case $dim(g_{q+1}) = \alpha_q^1 + \ldots + \alpha_q^n$ and one can exhibit the *Hilbert polynomial* $dim(R_{q+r})$ in r with leading term $(\alpha/n!)r^n$ when $\alpha \neq 0$. Such a prolongation procedure allows to compute *in a unique way* the principal (pri) jets from the parametric (par) other ones ([23],[46]).

REMARK 4.3 : This definition may also be applied to nonlinear systems as well while using a generic linearization by means of vertical bundles. For example, with $m = 1, n = 2$ and $q = 2$, the nonlinear system $y_{22} - \frac{1}{3}(y_{11})^3 = 0, y_{12} - \frac{1}{2}(y_{11})^2 = 0$ is involutive but the nonlinear system $y_{22} - \frac{1}{2}(y_{11})^2 = 0, y_{12} - y_{11} = 0$ is not involutive.

When R_q is involutive, the linear differential operator $\mathcal{D} : E \overset{j_q}{\to} J_q(E) \overset{\Phi}{\to} J_q(E)/R_q = F_0$ of order q with space of solutions $\Theta \subset E$ is said to be *involutive* and one has the canonical *linear Janet sequence* ([39], p 144):

$$0 \longrightarrow \Theta \longrightarrow T \overset{\mathcal{D}}{\to} F_0 \overset{\mathcal{D}_1}{\to} F_1 \overset{\mathcal{D}_2}{\to} \ldots \overset{\mathcal{D}_n}{\to} F_n \longrightarrow 0$$

where each other operator is first order involutive and generates the *compatibility conditions* (CC) of the preceding one. As the Janet sequence can be cut at any place, *the numbering of the Janet bundles has nothing to do with that of the Poincaré sequence*, contrary to what many people believe.

Equivalently, we have the involutive *first Spencer operator* $D_1 : C_0 = R_q \xrightarrow{j_1} J_1(R_q) \to J_1(R_q)/R_{q+1} \simeq T^* \otimes R_q/\delta(g_{q+1}) = C_1$ of order one induced by $D : R_{q+1} \to T^* \otimes R_q$. Introducing the *Spencer bundles* $C_r = \wedge^r T^* \otimes R_q/\delta(\wedge^{r-1} T^* \otimes g_{q+1})$, the first order involutive $(r+1)$-*Spencer operator* $D_{r+1} : C_r \to C_{r+1}$ is induced by $D : \wedge^r T^* \otimes R_{q+1} \to \wedge^{r+1} T^* \otimes R_q : \alpha \otimes \xi_{q+1} \to d\alpha \otimes \xi_q + (-1)^r \alpha \wedge D\xi_{q+1}$ and we obtain the canonical *linear Spencer sequence* ([39], p 150):

$$0 \longrightarrow \Theta \xrightarrow{j_q} C_0 \xrightarrow{D_1} C_1 \xrightarrow{D_2} C_2 \xrightarrow{D_3} ... \xrightarrow{D_n} C_n \longrightarrow 0$$

as the Janet sequence for the first order involutive system $R_{q+1} \subset J_1(R_q)$. Introducing the other Spencer bundles $C_r(E) = \wedge^r T^* \otimes J_q(E)/\delta(\wedge^{r-1} T^* \otimes S_{q+1} T^* \otimes E)$ with $C_r \subset C_r(E)$, the linear Spencer sequence is induced by the *linear hybrid sequence*:

$$0 \longrightarrow E \xrightarrow{j_q} C_0(E) \xrightarrow{D_1} C_1(E) \xrightarrow{D_2} C_2 \xrightarrow{D_3} ... \xrightarrow{D_n} C_n \longrightarrow 0$$

which is at the same time the Janet sequence for j_q and the Spencer sequence for $J_{q+1}(E) \subset J_1(J_q(E))$ ([39], p 153):

We have the following commutative and exact diagram allowing to relate the Spencer bundles C_r and $C_r(E)$ to the *Janet bundles* $F_r = \wedge^r T^* \otimes F_0/\delta(\wedge^{r-1} T^* \otimes h_1)$ if we start with the short exact sequence $0 \to g_{q+1} \to S_{q+1} T^* \otimes E \to h_1 \to 0$ where $h_1 \subset T^* \otimes F_0$:

38

$$
\begin{array}{ccccccc}
0 & & 0 & & 0 & & \\
\downarrow & & \downarrow & & \downarrow & & \\
\wedge^{r-1}T^* \otimes g_{q+1} & \xrightarrow{\;\delta\;} & \wedge^r T^* \otimes R_q & \longrightarrow & C_r & \longrightarrow 0 & \\
\downarrow & & \downarrow & & \downarrow & & \\
\wedge^{r-1}T^* \otimes S_{q+1}T^* \otimes E & \xrightarrow{\;\delta\;} & \wedge^r T^* \otimes J_q(E) & \longrightarrow & C_r(E) & \longrightarrow 0 & \\
\downarrow & & \downarrow \Phi & \searrow & \downarrow \Phi_r & & \\
\wedge^{r-1}T^* \otimes h_1 & \xrightarrow{\;\delta\;} & \wedge^r T^* \otimes F_0 & \longrightarrow & F_r & \longrightarrow 0 & \\
\downarrow & & \downarrow & & \downarrow & & \\
0 & & 0 & & 0 & &
\end{array}
$$

and obtain the following *crucial commutative diagram* with exact columns:

$$
\begin{array}{ccccccccccc}
 & & 0 & & 0 & & 0 & & & & 0 \\
 & & \downarrow & & \downarrow & & \downarrow & & & & \downarrow \\
0 \to & \Theta & \xrightarrow{j_q} & C_0 & \xrightarrow{D_1} & C_1 & \xrightarrow{D_2} & C_2 & \xrightarrow{D_3} \ldots \xrightarrow{D_n} & C_n & \to 0 \\
 & & & \downarrow & & \downarrow & & \downarrow & & \downarrow & \\
0 \to & E & \xrightarrow{j_q} & C_0(E) & \xrightarrow{D_1} & C_1(E) & \xrightarrow{D_2} & C_2(E) & \xrightarrow{D_3} \ldots \xrightarrow{D_n} & C_n(E) & \to 0 \\
 & & \| & \downarrow \Phi_0 & & \downarrow \Phi_1 & & \downarrow \Phi_2 & & \downarrow \Phi_n & \\
0 \to \Theta \to & E & \xrightarrow{\mathcal{D}} & F_0 & \xrightarrow{D_1} & F_1 & \xrightarrow{D_2} & F_2 & \xrightarrow{D_3} \ldots \xrightarrow{D_n} & F_n & \to 0 \\
 & & & \downarrow & & \downarrow & & \downarrow & & \downarrow & \\
 & & & 0 & & 0 & & 0 & & 0 &
\end{array}
$$

In this diagram, only depending on the linear differential operator $\mathcal{D} = \Phi \circ j_q$, the epimorhisms $\Phi_r : C_r(E) \to F_r$ for $0 \le r \le n$ are induced *step by step* by the canonical projection $\Phi = \Phi_0 : C_0(E) = J_q(E) \to J_q(E)/R_q = F_0$ if we start with the knowledge of $R_q \subset J_q(E)$ or from the knowledge of an epimorphism $\Phi : J_q(E) \to F_0$ if we set $R_q = ker(\Phi)$. It follows that the hybrid sequence projects onto the Janet sequence and that the kernel of this projection is the Spencer sequence. Also, chasing in the previous diagram, we may finally define the Janet bundles, up to an isomorphism, by the formula:

$$F_r = \wedge^r T^* \otimes J_q(E) / (\wedge^r T^* \otimes R_q + \delta(\wedge^{r-1} T^* \otimes S_{q+1} T^* \otimes E))$$

and $\mathcal{D}_r : F_{r-1} \to F_r$ is thus induced by $D : \wedge^{r-1} T^* \otimes J_{q+1} \to \wedge^r T^* \otimes J_q(E)$ ([36], p 391). This result will be crucially used in the next section dealing with the deformation theory of Lie equations when $E = T$, $R_q \subset J_q(T)$ is a transitive involutive system of infinitesimal Lie equations of order q and the operator $\mathcal{D}$ is a *Lie operator*, that is if $\mathcal{D}\xi = 0, \mathcal{D}\eta = 0 \Rightarrow \mathcal{D}[\xi, \eta] = 0, \forall \xi, \eta \in \Theta \in T$.

DEFINITION 4.4: The Janet sequence is said to be *locally exact at F_r* if any local section of F_r killed by $\mathcal{D}_{r+1}$ is the image by $\mathcal{D}_r$ of a local section of F_{r-1} over a convenient open subset. It is called *locally exact* if it is locally exact at each F_r for $0 \le r \le n$. The Poincaré sequence is locally exact but counterexamples may exist ([36], p 202). As another useful definition for applications, we shall also say that a differential sequence is *formally exact* if each operator involved generates *all* the CC of the preceding one. The Poincaré, Janet and Spencer sequences are formally exact by construction but we shall see other examples in the Appendix 1.

EXAMPLE 4.5: When $n = 1$ and we consider the affine transformaions of Example 1.7, we have the following Janet/Spencer diagram:

$$
\begin{array}{ccccccccc}
 & & & & 0 & & 0 & & \\
 & & & & \downarrow & & \downarrow & & \\
0 & \to & \Theta & \xrightarrow{j_2} & 2 & \xrightarrow{D_1} & 2 & \to 0 & \quad Spencer \\
 & & & & \downarrow & & \| & & \\
0 & \to & 1 & \xrightarrow{j_2} & 3 & \xrightarrow{D_1} & 2 & \to 0 & \\
 & & \| & & \downarrow \Phi & & \downarrow & & \\
0 \to & \Theta & \to & 1 & \xrightarrow{\mathcal{D}} & 1 & \longrightarrow 0 & & \quad Janet \\
 & & & & \downarrow & & & & \\
 & & & & 0 & & & &
\end{array}
$$

EXAMPLE 4.6: When $n = 2$ and we consider the situation of Example 1.2, we have $dim(g_1) = 1, g_2 = 0$ and the sequence $0 \to \wedge^2 T^* \otimes g_1 \to 0$ cannot be exact and therefore g_1 cannot be 2-acyclic, that is $H_1^2(g_1) \ne 0$. Accordingly, in order to construct the Janet sequence, we *must* start with R_2 such that $dim(R_2) =$

$3(2\,translations + 1\,rotation)$ and the Spencer bundles becomes $C_r = \wedge r T^* \otimes R_2$. The preceding diagram becomes:

$$
\begin{array}{ccccccccc}
& & & \textit{Spencer} & & & & & \\
& & & 0 & & 0 & & 0 & \\
& & & \downarrow & & \downarrow & & \downarrow & \\
0 & \to & \Theta & \xrightarrow{j_2} 3 & \xrightarrow{D_1} & 6 & \xrightarrow{D_2} & 3 & \to 0 \\
& & & \downarrow & & \downarrow & & \downarrow & \\
0 & \to & 2 & \xrightarrow{j_2} 12 & \xrightarrow{D_1} & 16 & \xrightarrow{D_2} & 6 & \to 0 \\
& & \| & \downarrow \Phi_0 & & \downarrow \Phi_1 & & \downarrow \Phi_2 & \\
0 \to \Theta & \to & 2 & \xrightarrow{D} 9 & \xrightarrow{D_1} & 10 & \xrightarrow{D_2} & 3 & \to 0 \\
& & & \downarrow & & \downarrow & & \downarrow & \\
& & & 0 & & 0 & & 0 & \\
& & & \textit{Janet} & & & & & \\
\end{array}
$$

Hence, increasing n, the bigger is the group involved, the bigger are the dimensions of the Spencer bundles, contrary to what happens in the Janet sequence where the first Janet bundle has only to do with differential invariants. This rather philosophical comment, namely to replace the Janet sequence by the Spencer sequence, must be considered as the key for understanding the work of the brothers E. and F. Cosserat in 1909 for elasticity or of H. Weyl in 1916 for electromagnetism, the best picture being that of two children playing at see-saw ([41]). These results will be presented in Appendix 3.

5 DEFORMATION THEORY OF LIE EQUATIONS:

Having in mind the application of computer algebra to the local theory of Lie pseudogroups, we want first of all to insist on two points which have never been emphasized up to our knowledge.

In order to motivate the *first point*, we sketch it on an example.

EXAMPLE 5.1: Let $\alpha = \alpha_i(x)dx^i$ be a 1-form as in Example 1.1 or 1.4. If we look for infinitesimal transformations preserving α, we have to cancel the Lie derivative as follows:

$$(\mathcal{L}(\xi)\alpha)_i \equiv \alpha_r(x)\partial_i\xi^r + \xi^r\partial_r\alpha_i(x) = 0$$

As a byproduct, we have a well defined *Lie operator* $\mathcal{D} : T \to T^* : \xi \to \mathcal{D}\xi = \mathcal{L}(\xi)\alpha$ such that $\mathcal{D}\xi = 0, \mathcal{D}\eta = 0 \Rightarrow \mathcal{D}[\xi, \eta] = 0$ because of the well known property of the Lie derivative $[\mathcal{L}(\xi), \mathcal{L}(\eta)] \equiv \mathcal{L}(\xi) \circ \mathcal{L}(\eta) - \mathcal{L}(\eta) \circ \mathcal{L}(\xi) = \mathcal{L}([\xi, \eta])$ and such a property can be extended to tensors or even any geometric object. Accordingly, it is usual to say that, if we have two solutions of the system, then their bracket is again a solution. However such a result is coming from mathematics and *cannot* be recognized by means of computer algebra, contrary to what is sometimes claimed. Surprisingly, the underlying reason has to do with formal integrability. Indeed, if we study the first derivative of the bracket $[\xi, \eta]$, it involves in fact the *second derivatives* of ξ and η and sometimes things may change a lot. For example, if $d\alpha \neq 0$, as $\mathcal{L}(\xi)d\alpha = di(\xi)d\alpha = d\mathcal{L}(\xi)\alpha = 0$, then the first order equations brought by $\mathcal{L}(\xi)d\alpha = 0$ may not be linear combinations of the first order equations brought by $\mathcal{L}(\xi)\alpha = 0$. In particular, if $n = 2$ and $\alpha = x^2dx^1$, then $d\alpha = -dx^1 \wedge dx^2$ and the new first order equation $\partial_1\xi^1 + \partial_2\xi^2 = 0$, which is *automatically* satisfied by any solution of the system $R_1 \subset J_1(T)$ defined by $\mathcal{L}(\xi)\alpha = 0$, is not a linear combination of the equations $x^2\partial_1\xi^1 + \xi^2 = 0, \partial_2\xi^1 = 0$ defining R_1. Accordingly, R_1 is not involutive as it is not even formally integrable because $R_1^{(1)} = \pi_1^2(R_2) \subset R_1$ with a strict inclusion. *Hence it becomes a challenge to define a kind of bracket for sections of $R_1 \subset J_1(T)$ and not for solutions as usual, that is to say independently of formal integrability.* This idea, which is a crucial one indeed as it will lead to the concept of Lie algebroid, is to replace the *classical Lie derivative* $\mathcal{L}(\xi)$ for any $\xi \in T$ by a *formal Lie deriva-*

tive $L(\xi_1)$ for any $\xi_1 \in J_1(T)$ over $\xi \in T$ in such a way that $\mathcal{L}(\xi) = L(j_1(\xi))$ and to compute the bracket $[L(\xi_1), L(\eta_1)] \equiv L(\xi_1) \circ L(\eta_1) - L(\eta_1) \circ L(\xi_1)$ in the operator sense in order to be sure that the new bracket on $J_1(T)$ will satisfy the desired Jacobi identity. We obtain successively:

$$(L(\xi_1)\alpha)_i \equiv \alpha_r(x)\xi_i^r + \xi^r \partial_r \alpha_i(x) = 0, \qquad (L(\eta_1)\alpha)_j \equiv \alpha_s(x)\eta_j^s + \eta^s \partial_s \alpha_j(x) = 0$$

$$L([\xi_1, \eta_1])\alpha \equiv \alpha_k(x)(\xi^r \partial_r \eta_i^k + \xi_i^r \eta_r^k - \eta_i^s \xi_s^k - \eta^s \partial_s \xi_i^k) + ([\xi, \eta])^k \partial_k \alpha_i(x) = 0$$

as a way to *define*:

$$([\xi_1, \eta_1])^k = ([\xi, \eta])^k = \xi^r \partial_r \eta^k - \eta^s \partial_s \xi^k$$

$$([\xi_1, \eta_1])_i^k = \xi^r \partial_r \eta_i^k + \xi_i^r \eta_r^k - \eta_i^s \xi_s^k - \eta^s \partial_s \xi_i^k$$

The induced property $[R_1, R_1] \subset R_1$ *can therefore be checked linearly on sections and no longer on solutions.* In particular, we may exhibit the section $\{\xi^1 = 0, \xi^2 = 0, \xi_1^1 = 0, \xi_2^1 = 0, \xi_1^2 = 0, \xi_2^2 = 1\}$ of R_1, even if $\xi_1^1 + \xi_2^2 \neq 0$.

Now, using the *algebraic bracket* $\{j_{q+1}(\xi), j_{q+1}(\eta)\} = j_q([\xi, \eta]), \forall \xi, \eta \in T$, we may obtain by bilinearity a *differential bracket* on $J_q(T)$ extending the bracket on T:

$$[\xi_q, \eta_q] = \{\xi_{q+1}, \eta_{q+1}\} + i(\xi)D\eta_{q+1} - i(\eta)D\xi_{q+1}, \qquad \forall \xi_q, \eta_q \in J_q(T)$$

which does not depend on the respective lifts ξ_{q+1} and η_{q+1} of ξ_q and η_q in $J_{q+1}(T)$. Applying j_q to the Jacobi identity for the ordinary bracket, we obtain:

$$\{\xi_{q+1}, \{\eta_{q+2}, \zeta_{q+2}\}\} + \{\eta_{q+1}, \{\zeta_{q+2}, \xi_{q+2}\}\} + \{\zeta_{q+1}, \{\xi_{q+2}, \eta_{q+2}\}\} \equiv 0$$

As we shall see later on, this bracket on sections satisfies the Jacobi identity and *the following definition is the only one that can be tested by means of computer algebra*:

DEFINITION 5.2: We say that a vector subbundle $R_q \subset J_q(T)$ is a *system of infinitesimal Lie equations* or a *Lie algebroid* if $[R_q, R_q] \subset R_q$, that is to say $[\xi_q, \eta_q] \in R_q, \forall \xi_q, \eta_q \in R_q$. The kernel R_q^0 of the projection $\pi_0^q : R_q \to T$ is the *isotropy Lie algebra bundle* of $\mathcal{R}_q^0 = id^{-1}(\mathcal{R}_q)$ and $[R_q^0, R_q^0] \subset R_q^0$ does not contain derivatives, being thus defined fiber by fiber.

Of course, another difficulty to overcome in this new setting, is that we have no longer an identity like $d\mathcal{L}(\xi)\alpha - \mathcal{L}(\xi)d\alpha = 0$ but it is easy to check in local coordinates that:

$$(dL(\xi_1)\alpha - L(\xi_1)d\alpha)_{ij} = \alpha_r(x)(\partial_i \xi_j^r - \partial_j \xi_i^r) + (\partial_i \xi^r - \xi_i^r)\partial_r \alpha_j(x) - (\partial_j \xi^r - \xi_j^r)\partial_r \alpha_i(x)$$

and the Spencer operator allows to factorize the formula if we notice that:

$$(\partial_i \xi_j^r - \partial_j \xi_i^r) = (\partial_i \xi_j^r - \xi_{ij}^r) - (\partial_j \xi_i^r - \xi_{ij}^r)$$

We finally obtain:

$$i(\zeta_{(1)})i(\zeta_{(2)})(dL(\xi_1)\alpha - L(\xi_1)d\alpha) = i(\zeta_{(2)})L(i(\zeta_{(1)})D\xi_2)\alpha - i(\zeta_1)L(i(\zeta_{(2)}D\xi_2)\alpha$$

and more generally:

LEMMA 5.3: When $\alpha \in \wedge^{r-1}T^*$ we have the formula:

$$i(\zeta_{(1)})...i(\zeta_{(r)})(dL(\xi_1)\alpha - L(\xi_1)d\alpha) = \sum_{s=1}^{r}(-1)^{s+1}i(\zeta_{(1)})...i(\hat{\zeta}_{(s)})...i(\zeta_{(r)})L(i(\zeta_{(s)}D\xi_2)\alpha$$

which does not depend on the lift $\xi_2 \in J_2(T)$ of $\xi_1 \in J_1(T)$.

In order to understand the *second point*, we have to revisit the work of Vessiot. Indeed, if we have a geometric object, that is a section ω of a natural bundle $\mathcal{F}$ of order q, then we may consider the system $\mathcal{R}_q = \{f_q \in \Pi_q \mid f_q^{-1}(\omega) = \omega\}$ of finite Lie equations and the corresponding linearized system $R_q = \{\xi_q \in J_q(T) \mid L(\xi_q)\omega = 0\}$ of infinitesimal Lie equations, both with the particular way to write them out, namely the Lie form and the Medolaghi form respectively, as it becomes clear from Example 1.1 to 1.7. As a byproduct, when constructing the Janet sequence, we can write $F_0 = J_q(T)/R_q$ but we can also use the isomorphic definition $F_0 = \omega^{-1}(V(\mathcal{F}))$ *depending on whether we want to pay attention to the system or to the object.* The main idea of deformation theory will be to begin with the second point of view and finish with the first. Starting with a system $R_q \subset J_q(T)$, we shall suppose that R_q is transitive with a short exact sequence $0 \rightarrow R_q^0 \rightarrow R_q \rightarrow T \rightarrow 0$ and, whatever is the definition of F_0, introduce an epimorphism $\Phi : J_q(T) \rightarrow F_0$ while considering the following commutative and exact diagram:

44

$$\begin{array}{ccccccccc}
& & 0 & & 0 & & 0 & & \\
& & \downarrow & & \downarrow & & \downarrow & & \\
0 & \longrightarrow & R_q^0 & \longrightarrow & J_q^0(T) & \longrightarrow & F_0 & \longrightarrow & 0 \\
& & \downarrow & & \downarrow & & \| & & \\
0 & \longrightarrow & R_q & \longrightarrow & J_q(T) & \xrightarrow{\Phi} & F_0 & \longrightarrow & 0 \\
& & \downarrow & & \downarrow & & \downarrow & & \\
0 & \longrightarrow & T & = & T & \longrightarrow & 0 & & \\
& & \downarrow & & \downarrow & & & & \\
& & 0 & & 0 & & & &
\end{array}$$

The next definition will also be crucial for our purpose and generalizes the standard definition:

$$\mathcal{L}(\xi)\omega = \frac{d}{dt} j_q(exp\ t\xi)^{-1}(\omega)|_{t=0}.$$

DEFINITION 5.4: We say that a vector bundle F is *associated* with R_q if there exists a first order differential operator $L(\xi_q) : F \to F$ called *formal Lie derivative* and such that:

1) $L(\xi_q + \eta_q) = L(\xi_q) + L(\eta_q) \qquad \forall \xi_q, \eta_q \in R_q.$
2) $L(f\xi_q) = fL(\xi_q) \qquad \forall \xi_q \in R_q, \forall f \in C^\infty(X).$
3) $[L(\xi_q), L(\eta_q)] \equiv L(\xi_q) \circ L(\eta_q) - L(\eta_q) \circ L(\xi_q) = L([\xi_q, \eta_q]) \qquad \forall \xi_q, \eta_q \in R_q.$
4) $L(\xi_q)(f\eta) = fL(\xi_q)\eta + (\xi.f)\eta \qquad \forall \xi_q \in R_q, \forall f \in C^\infty(X), \forall \eta \in F$ where $\xi.f = i(\xi)df.$

As a byproduct, if E and F are associated with R_q, we may set on $E \otimes F$:

$$L(\xi_q)(\eta \otimes \zeta) = L(\xi_q)\eta \otimes \zeta + \eta \otimes L(\xi_q)\zeta \qquad \forall \xi_q \in R_q, \forall \eta \in E, \forall \zeta \in F$$

REMARK 5.5: If $\Theta \subset T$ denotes the solutions of R_q, then $\mathcal{L}(\xi) = L(j_q(\xi))$ is simply called the *classical Lie derivative* but cannot be used in actual practice as we already said because Θ may be infinite dimensional as in Examples 1.3, 1.4, and 1.5. We obtain at once:

1) $\mathcal{L}(\xi + \eta) = \mathcal{L}(\xi) + \mathcal{L}(\eta) \qquad \forall \xi, \eta \in \Theta.$
3) $[\mathcal{L}(\xi), \mathcal{L}(\eta)] = \mathcal{L}([\xi, \eta]) \qquad \forall \xi, \eta \in \Theta.$
4) $\mathcal{L}(\xi)(f\eta) = f\mathcal{L}(\xi)\eta + (\xi.f)\eta \qquad \forall \xi \in \Theta, \forall f \in C^\infty(X), \forall \eta \in F.$

45

The extension to tensor products is well known.

The following technical proposition and its corollary will be of constant use later on:

PROPOSITION 5.6: We have:

$$i(\zeta)D\{\xi_{q+1}, \eta_{q+1}\} = \{i(\zeta)D\xi_{q+1}, \eta_q\} + \{\xi_q, i(\zeta)D\eta_{q+1}\}$$

Proof: We have:

$$(\{\xi_{q+1}, \eta_{q+1}\})_\nu^k = \sum_{\lambda+\mu=\nu} (\xi_\lambda^r \eta_{\mu+1_r}^k - \eta_\lambda^s \xi_{\mu+1_s}^k)$$

Now, caring only about ξ_{q+1}, we get:

$$\partial_i(\{\xi_{q+1}, \eta_{q+1}\})_\nu^k - (\{\xi_{q+1}, \eta_{q+1}\})_{\nu+1_i}^k = \sum_{\lambda+\mu=\nu} (\partial_i \xi_\lambda^r - \xi_{\lambda+1_i}^r)\eta_{\mu+1_r}^k - (\partial_i \xi_{\mu+1_s}^k - \xi_{\mu+1_s+1_i}^k)\eta_\lambda^s + \dots$$

and the Proposition follows by bilinearity.

Q.E.D.

The proof of the following proposition is similar and left to the reader as an exercise:

PROPOSITION 5.7: We have the formula:

$$i(\zeta)D[\xi_{q+1}, \eta_{q+1}] = [i(\zeta)D\xi_{q+1}, \eta_q] + [\xi_q, i(\zeta)D\eta_{q+1}] + i(L(\eta_1)\zeta)D\xi_{q+1} - i(L(\xi_1)\zeta)D\eta_{q+1}$$

COROLLARY 5.8: If $R_q \subset J_q(T)$ is such that $[R_q, R_q] \subset R_q$, then $R_{q+1} \subset J_{q+1}(T)$ satisfies $[R_{q+1}, R_{q+1}] \subset R_{q+1}$ even if R_q is not formally integrable.

EXAMPLE 5.9: T and T^* both with any tensor bundle are associated with $J_1(T)$. The case of T^* has been treated at the beginning of this section while for T we may define $L(\xi_1)\eta = [\xi, \eta] + i(\eta)D\xi_1 = \{\xi_1, j_1(\eta)\}$. We have indeed $\xi^r \partial_r \eta^k - \eta^s \partial_s \xi^k + \eta^s (\partial_s \xi^k - \xi_s^k) = -\eta^s \xi_s^k + \xi^r \partial_r \eta^k$ and the four properties of the formal Lie derivative can be checked directly as we did for T^*. Of course, we find back

46

$$\mathcal{L}(\xi)\eta = [\xi, \eta], \forall \xi, \eta \in T.$$

More generally, we have in a coherent way:

PROPOSITION 5.10: $J_q(T)$ is associated with $J_{q+1}(T)$ if we define:

$$L(\xi_{q+1})\eta_q = \{\xi_{q+1}, \eta_{q+1}\} + i(\xi)D\eta_{q+1} = [\xi_q, \eta_q] + i(\eta)D\xi_{q+1}$$

and thus R_q is associated with R_{q+1}.

Proof: It is easy to check the properties 1, 2, 4 and it only remains to prove property 3 as follows.

$$\begin{aligned}
[L(\xi_{q+1}), L(\eta_{q+1})]\zeta_q &= L(\xi_{q+1})(\{\eta_{q+1}, \zeta_{q+1}\} + i(\eta)D\zeta_{q+1}) \\
&\quad -L(\eta_{q+1})(\{\xi_{q+1}, \zeta_{q+1}\} + i(\xi)D\zeta_{q+1}) \\
&= \{\xi_{q+1}, \{\eta_{q+2}, \zeta_{q+2}\}\} - \{\eta_{q+1}, \{\xi_{q+2}, \zeta_{q+2}\}\} \\
&\quad +\{\xi_{q+1}, i(\eta)D\zeta_{q+2}\} - \{\eta_{q+1}, i(\xi)D\zeta_{q+2}\} \\
&\quad +i(\xi)D\{\eta_{q+2}, \zeta_{q+2}\} - i(\eta)D\{\xi_{q+2}, \zeta_{q+2}\} \\
&\quad +i(\xi)D(i(\eta)D\zeta_{q+2}) - i(\eta)D(i(\xi)D\zeta_{q+2}) \\
&= \{\{\xi_{q+2}, \eta_{q+2}\}, \zeta_{q+1}\} + \{i(\xi)D\eta_{q+2}, \zeta_{q+1}\} \\
&\quad -\{i(\eta)D\xi_{q+2}, \zeta_{q+1}\} \\
&\quad +i([\xi, \eta])D\zeta_{q+1} \\
&= \{[\xi_{q+1}, \eta_{q+1}], \zeta_{q+1}\} + i([\xi, \eta])D\zeta_{q+1}
\end{aligned}$$

by using successively the Jacobi identity for the algebraic bracket and the last proposition.

Q.E.D.

COROLLARY 5.11: The differential bracket satisfies the Jacobi identity :

$$[\xi_q, [\eta_q, \zeta_q]] + [\eta_q, [\zeta_q, \xi_q]] + [\zeta_q, [\xi_q, \eta_q]] \equiv 0 \qquad \forall \xi_q, \eta_q, \zeta_q \in J_q(T)$$

PROPOSITION 5.12: We have the formula:

$$i(\zeta)(DL(\xi_{q+2})\eta_{q+1} - L(\xi_{q+1})D\eta_{q+1}) = L(i(\zeta)D\xi_{q+2})\eta_q$$

Proof: Using Proposition 5.6, we have:

$$i(\zeta)DL(\xi_{q+2})\eta_{q+1} = i(\zeta)D\{\xi_{q+2},\eta_{q+2}\} + i(\zeta)Di(\xi)D\eta_{q+2}$$
$$= \{i(\zeta)D\xi_{q+2},\eta_{q+1}\} + \{\xi_{q+1},i(\zeta)D\eta_{q+2}\} + i(\zeta)Di(\xi)D\eta_{q+2}$$

and we must substract:

$$i(\zeta)L(\xi_{q+1})D\eta_{q+1} = L(\xi_{q+1})(i(\zeta)D\eta_{q+1}) - i(L(\xi_1)\zeta)D\eta_{q+1}$$
$$= \{\xi_{q+1},i(\zeta)D\eta_{q+2}\} + i(\xi)Di(\zeta)D\eta_{q+2} - i(L(\xi_1)\zeta)D\eta_{q+1}$$

in order to obtain for the difference:

$$L(i(\zeta)D\xi_{q+2})\eta_q - i(i(\zeta)D\xi_1)D\eta_{q+1} + i(L(\xi_1)\zeta)D\eta_{q+1} + i(\zeta)Di(\xi)D\eta_{q+2} - i(\xi)Di(\zeta)D\eta_{q+2}$$

Finally, the last four terms vanish because $L(\xi_1)\zeta - i(\zeta)D\xi_1 = [\xi,\zeta]$ and:

$$i(\zeta)Di(\xi)D\eta_{q+2} - i(\xi)Di(\zeta)D\eta_{q+2} = -i([\xi,\zeta])D\eta_{q+1}.$$

Q.E.D.

Combining this proposition and Lemma 5.3, we obtain:

PROPOSITION 5.13: When $A_{q+1}^{r-1} \in \wedge^{r-1}T^* \otimes J_{q+1}(T)$, we have the formula:

$$i(\zeta_{(1)})...i(\zeta_{(r)})(DL(\xi_{q+2})-L(\xi_{q+1})D)A_{q+1}^{r-1} = \sum_{s=1}^{r}(-1)^{s+1}i(\zeta_{(1)})...i(\hat{\zeta}_{(s)})...i(\zeta_{(r)})L(i(\zeta_{(s)})D\xi_{q+2})A_q^{r-1}$$

where $A_q^{r-1} \in \wedge^{r-1}T^* \otimes J_q(T)$ is the projection of A_{q+1}^{r-1}.

Proof: With $\alpha \in \wedge^{r-1}T^*$ and $\eta_{q+1} \in J_{q+1}(T)$, we obtain successively:

$$DL(\xi_{q+2})(\alpha \otimes \eta_{q+1}) = D(L(\xi_1)\alpha \otimes \eta_{q+1} + \alpha \otimes L(\xi_{q+2})\eta_{q+1})$$
$$= dL(\xi_1)\alpha \otimes \eta_q + (-1)^{r-1}(L(\xi_1)\alpha) \wedge D\eta_{q+1}$$
$$+ d\alpha \otimes L(\xi_{q+1})\eta_q + (-1)^{r-1}\alpha \wedge DL(\xi_{q+2})\eta_{q+1}$$

$$L(\xi_{q+1})D(\alpha \otimes \eta_{q+1}) = L(\xi_{q+1})(d\alpha \otimes \eta_q + (-1)^{r-1}\alpha \wedge D\eta_{q+1})$$
$$= L(\xi_1)d\alpha \otimes \eta_q + d\alpha \otimes L(\xi_{q+1})\eta_q$$
$$+ (-1)^{r-1}L(\xi_1)\alpha \wedge D\eta_{q+1} + (-1)^{r-1}\alpha \wedge L(\xi_{q+1})D\eta_{q+1}$$

and obtain y substraction:

$$(DL(\xi_{q+2}) - L(\xi_{q+1})D)(\alpha \otimes \eta_{q+1}) = (dL(\xi_1) - L(\xi_1)d)\alpha \otimes \eta_q$$
$$+ (-1)^{r-1}\alpha \wedge (DL(\xi_{q+2}) - L(\xi_{q+1})D)\eta_{q+1}$$

and the proposition follows by skewlinearity.

Q.E.D.

PROPOSITION 5.14: We have the formula:

$$L(\xi_q)\{\eta_q, \zeta_q\} = \{L(\xi_{q+1})\eta_q, \zeta_q\} + \{\eta_q, L(\xi_{q+1})\zeta_q\}$$

which does not depend on the lift $\xi_{q+1} \in J_{q+1}(T)$ of $\xi_q \in J_q(T)$.

Proof: Using the Jacobi identity for the algebraic bracket and proposition 5.6, we obtain:

$$
\begin{aligned}
L(\xi_q)\{\eta_q, \zeta_q\} &= \{\xi_q, \{\eta_{q+1}, \zeta_{q+1}\} + i(\xi)D\{\eta_{q+1}, \zeta_{q+1}\} \\
&= \{\{\xi_{q+1}, \eta_{q+1}\}, \zeta_q\} + \{\eta_q, \{\xi_{q+1}, \zeta_{q+1}\}\} \\
&\quad + \{i(\xi)D\eta_{q+1}, \zeta_q\} + \{\eta_q, i(\xi)D\zeta_{q+1}\} \\
&= \{L(\xi_{q+1})\eta_q, \zeta_q\} + \{\eta_q, L(\xi_{q+1})\zeta_q\}
\end{aligned}
$$

Q.E.D.

Finally, using Proposition 5.12, we obtain at once:

COROLLARY 5.15: We have the formula:

$$
\begin{aligned}
L(\xi_{q+1})[\eta_q, \zeta_q] &= [L(\xi_{q+1})\eta_q, \zeta_q] + [\eta_q, L(\xi_{q+1})\zeta_q] \\
&\quad + L(i(\zeta)D\xi_{q+2})\eta_q - L(i(\eta)D\xi_{q+2})\zeta_q
\end{aligned}
$$

which does not depend on the lift $\xi_{q+2} \in J_{q+2}(T)$ of $\xi_{q+1} \in J_{q+1}(T)$.

Before going ahead, let us stop for a moment and wonder how we could proceed for generalizing the deformation theory of Lie algebras by using the Vessiot structure equations even though we know that the structure constants have nothing to do in general with any Lie algebra. Of course we could start similarly from the Jacobi relations but, if we do want to exhibit a kind of cohomology, we should be able to define a trivial deformation, that is the analogue of a change of basis of the underlying vector space V of the Lie algebra $\mathcal{G}$ in such a natural way that it could induce a change of the structure constants which is surely not of a tensorial nature anymore. The following "trick", already known to Vessiot in 1903 ([51], p 445), is still ig-

49

nored today. For this, assuming that the natural bundle $\mathcal{F}$ is known, let us consider two sections ω and $\bar{\omega}$ giving rise respectively to the systems R_q and $\bar{R}_q$ of infinitesimal Lie equations:

$$R_q \qquad\qquad \Omega^\tau \equiv -L_k^{\tau\mu}(\omega(x))\xi_\mu^k + \xi^r\partial_r\omega^\tau(x) = 0$$

$$\bar{R}_q \qquad\qquad \bar{\Omega}^\tau \equiv -L_k^{\tau\mu}(\bar{\omega}(x))\xi_\mu^k + \xi^r\partial_r\bar{\omega}^\tau(x) = 0$$

and define the following equivalence relation:

DEFINITION 5.16: $\bar{\omega} \sim \omega \Leftrightarrow \bar{R}_q = R_q$

The study of such an equivalence relation is not evident at all and we improve earlier presentations (compare to [36], p 336). First of all, having in mind what we did for Example 1.4, we shall use a solved form of the system obtained by choosing principal jets or, equivalently, choosing a square submatrix $M = (M(u))$ of rank $dim(F_0)$ in the matrix $L = (L(u))$ defining R_q^0 with $dim(J_q^0(T))$ columns which describe the infinitesimal generators of prolongations of changes of coordinates on X acting on the fibers of $\mathcal{F}$ and $dim(F_0) = m$ rows. Of course, a major problem will be to obtain intrinsic results not depending on this choice. The columns of L are thus made by vector fields $L_k^\mu = L_k^{\tau\mu}(u)\frac{\partial}{\partial u^\tau}$ that we can therefore separate into two parts, namely the vectors $Ł_\sigma = M_\sigma^\tau(u)\frac{\partial}{\partial u^\tau}$ for $\sigma = 1, ..., dim(F_0)$ and the vectors $L_{m+r} = \mathcal{E}_{m+r}^\sigma(u)L_\sigma$ for $r = 1, ..., dim(R_q^0)$ obtained by introducing the *stationary functions* $\mathcal{E}(u)$, also called *Grassmann determinants*, while describing the matrix $M^{-1}L = (id_{F_0}, \mathcal{E}(u))$. We are therefore led to look for transformations $\bar{u} = g(u)$ of the fibers of $\mathcal{F}$ such that:

$$(M^{-1})_\tau^\sigma(\bar{u})d\bar{u}^\tau = (M^{-1})_\tau^\sigma(u)du^\tau, \qquad \mathcal{E}_{m+r}^\sigma(\bar{u}) = \mathcal{E}_\tau^\sigma(u)$$

In order to study such a system and to prove that it is defining a Lie pseudogroup of transformations, let us notice that the first conditions are equivalent to saying that the transformations $\bar{u} = g(u)$ preserve the vector fields L_σ and also the vector fields L_{m+r} according to the second conditions. It follows that the transformations $\bar{u} = g(u)$ preserves the vector fields L_k^μ, a property thus not depending on the choice of the principal jets. In addition, we have:

50

PROPOSITION 5.17: The Lie pseudogroup of transformations of the fibers of $\mathcal{F}$ that we have exhibited is in fact a Lie group of transformations, namely the *reciprocal* of the lie group of transformations describing the natural structure of $\mathcal{F}$.

Proof: The defining system is finite type with a zero first order symbol. If $W = W^\tau(u)\frac{\partial}{\partial u^\tau}$ is an infinitesimal transformation, we obtain therefore the Lie operator $[W, L_k^\mu] = 0, \forall 1 \leq |\mu| \leq q, \forall k = 1, ..., m$. Indeed, if W_1 and W_2 are two solutions, then $[W_1, W_2]$ is also a solution because of the Jacobi identity for the bracket and there are at most $dim(F_0)$ linearly independent such vector fields denoted by W_α. It follows that $[W_1, W_2] = \rho_{12}^\alpha(u)W_\alpha$ and we deduce from the Jacobi identity again that $L_k^\mu \cdot \rho_{12}^\alpha(u) = 0 \Rightarrow \rho_{12}^\alpha(u) = \rho_{12}^\alpha = cst$ because $rk(L_k^\mu) = dim(F_0)$. Accordingly, the W_α are the infinitesimal generators of a Lie group of transformations of the fibers of $\mathcal{F}$ and the effective action does not depend on the coordinate system.

Q.E.D.

DEFINITION 5.18: These finite transformations will be called *label transformations* and will be noted $\bar{u} = g(u, a)$ where the number of parameters a is $\leq dim(F_0)$.

If R_q is formally integrable/involutive, then $\bar{R}_q = R_q$ is also formally integrable/involutive and thus $I(j_1(\omega)) = c(\omega) \Leftrightarrow I(j_1(\bar{\omega})) = \bar{c}(\bar{\omega})$ with eventually different structure constants.

COROLLARY 5.19: Any finite label transformation $\bar{u} = g(u, a)$ induces a finite transformation $\bar{c} = h(c, a)$ of the structure constants which is not effective in general and we may set $\bar{\omega} \sim \omega \Rightarrow \bar{c} \sim c$.

It now remains to exhibit a deformation cohomology coherent with the above results.

DEFINITION 5.20: When F is a vector bundle associated with R_q, we may define $\Upsilon = \Upsilon(F) = \{\eta \in F | L(\xi_q)\eta = 0, \forall \xi_q \in R_q\}$ and the sub-vector bundle $E = \{\eta \in F | L(\xi_q^0)\eta = 0, \forall \xi_q^0 \in R_q^0\} \subseteq F$ in such a way that $\Upsilon \subset E \subseteq F$.

In order to look for Υ in general, we shall decompose this study into two parts, exactly as we did in section 3.7, by using a splitting of the short exact sequence $0 \to R_q^0 \to R_q \overset{\pi_0^q}{\to} T \to 0$ called R_q-connection, namely a map $\chi_q : T \to R_q$ such that $\pi_0^q \circ \chi_q = id_T$, in order to have $R_q \simeq R_q^0 \oplus \chi_q(T)$. Such a proce-dure does not depend on the choice of χ_q because, if $\bar{\chi}_q$ is another R_q-connection, then $(\bar{\chi}_q - \chi_q)(T) \in R_q^0$. An R_q-connection may also be considered as a section $\chi_q \in T^* \otimes R_q$ over $id_T \in T^* \otimes T$ and $\bar{\chi}_q - \chi_q \in T^* \otimes R_q^0$ in this case. It follows that we have equivalently $\Upsilon = \{\eta \in E | L(\chi_q(\xi))\eta = 0, \forall \xi \in T\}$ and we may define a first order operator $\nabla : E \to T^* \otimes E$ with zero symbol, called *covariant derivative*, by the formula $(\nabla \cdot \eta)(\xi) = \nabla_\xi \eta = L(\chi_q(\xi))\eta$. We may now extend ∇ to a first order operator $\nabla = \wedge^r T^* \otimes E \to \wedge^{r+1} T^* \otimes E$ by the formula:

$$\nabla(\alpha \otimes \eta) = d\alpha \otimes \eta + (-1)^r \alpha \wedge \nabla\eta, \qquad \forall \alpha \in \wedge^r T^*, \forall \eta \in E$$

LEMMA 5.21: With $\nabla^2 = \nabla \circ \nabla$, we have:

$$(\nabla^2 \eta)(\xi, \bar{\xi}) = L([\chi_q(\xi), \chi_q(\bar{\xi})] - \chi_q([\xi, \bar{\xi}]))\eta = 0 \Rightarrow \nabla^2 = 0$$

Proof: We have:

$$
\begin{aligned}
\nabla^2(\alpha \otimes \eta) &= \nabla(d\alpha \otimes \eta + (-1)^r \alpha \wedge \nabla\eta) \\
&= d^2\alpha + (-1)^{r+1} d\alpha \wedge \nabla\eta + (-1)^r d\alpha \wedge \nabla\eta + (-1)^r \alpha \wedge \nabla^2\eta \\
&= (-1)^r \alpha \wedge \nabla^2\eta
\end{aligned}
$$

Setting $\nabla\eta = dx^i \nabla_i \eta$, we get:

$$\nabla^2\eta = \nabla(dx^i \nabla_i \eta) = dx^i \wedge dx^j \nabla_j \nabla_i \eta = -\frac{1}{2} dx^i \wedge dx^j (\nabla_i \nabla_j - \nabla_j \nabla_i)\eta$$

and we have just to use the fact that $\nabla_i = L(\chi_q(\partial_i))$ with $[\partial_i, \partial_j] = 0$. Indeed, we have:

$$
\begin{aligned}
[\chi_q(\xi), \chi_q(\bar{\xi})] - \chi_q([\xi, \bar{\xi}]) &= \{\chi_{q+1}(\xi), \chi_{q+1}(\bar{\xi}\} + i(\xi)D\chi_{q+1}(\bar{\xi}) \\
&\quad -i(\bar{\xi})D\chi_{q+1}(\xi) - \chi_q([\xi, \bar{\xi}]) \\
&= \{\chi_{q+1}(\xi), \chi_{q+1}(\bar{\xi}\} + D\chi_{q+1}(\xi, \bar{\xi})
\end{aligned}
$$

The first term in the right member is linear in ξ and $\bar{\xi}$ while the sum of the others becomes also linear in ξ and $\bar{\xi}$ because:

$$\xi^i(\partial_i(\chi_{\mu,j}^k \bar{\xi}^j) - \chi_{\mu+1,j}^k \bar{\xi}^j) - \bar{\xi}^j(\partial_j(\chi_{\mu,i}^k \xi^i) - \chi_{\mu+1,i}^k \xi^i) - \chi_{\mu,r}^k(\xi^i \partial_i \bar{\xi}^r - \bar{\xi}^j \partial_j \xi^r)$$

52

$$= (\partial_i \chi_{\mu,j}^k - \partial_j \chi_{\mu,i}^k + \chi_{\mu+1_j,i}^k - \chi_{\mu+1_i,j}^k)\xi^i\bar{\xi}^j$$

The proposition follows from the fact that $\chi_0 = id_T$ and E is R_q^0-invariant.

Q.E.D.

Hence, we obtain by linearity the ∇-*sequence*:

$$0 \longrightarrow \Upsilon \longrightarrow E \xrightarrow{\nabla} T^* \otimes E \xrightarrow{\nabla} \wedge^2 T^* \otimes E \xrightarrow{\nabla} \dots \xrightarrow{\nabla} \wedge^n T^* \otimes E \longrightarrow 0$$

which does not depend on the choice of the connection and is made by first order involutive operators. It follows that Υ can be locally described by a linear combination with constant coefficients of certain sections of $E \subset F$ and we may therefore set $dim(\Upsilon) = dim(E) \leq dim(F)$. The use of computer algebra will essentially be to compute these dimensions by using linear algebra combined with homological algebra techniques.

We now provide a few definitions:

DEFINITION 5.22: When Θ is given, we may define:

Centralizer $C(\Theta) = \{\eta \in T | [\xi, \eta] = 0, \forall \xi \in \Theta\}$.

Center $Z(\Theta) = \{\eta \in \Theta | [\xi, \eta] = 0, \forall \xi \in \Theta\}$.

Normalizer $N(\Theta) = \{\eta \in T | [\xi, \eta] \subset \Theta, \forall \xi \in \Theta\}$.

It is essential to notice that these definitions are not very useful at all in actual practice when Θ is infinite dimensional.

PROPOSITION 5.23: $C(\Theta) = \Upsilon(T)$.

Proof: If $R_1 = \pi_1^q(R_q) \subset J_1(T)$, it follows from Example 5.9 that $\Upsilon(T) = \{\eta \in T | L(\xi_1)\eta = 0, \forall \xi_1 \in R_1\}$, that is to say $\{\Upsilon(T) = \{\eta \in T | \{\xi_1, j_1(\eta)\} = 0, \forall \xi_1 \in R_1\}$ if we choose $j_1(\eta)$ as a lift of η in $J_1(T)$. In particular, if $\xi \in \Theta$ and thus $j_1(\xi) \in R_1$, we have $\{j_1(\xi), j_1(\eta)\} = [\xi, \eta]$ and thus $\Upsilon(T) \subseteq C(\Theta)$.

Now, $j_{q-1}([\xi, \eta]) = \{j_q(\xi), j_q(\eta)\}$ and thus $C(\Theta) = \{\eta \in T | \{\xi_q, j_q(\eta)\} = 0, \forall \xi_q \in R_q\}$, providing by projection $\{\xi_1, j_1(\eta)\} = 0$, that is $C(\Theta) \subseteq \Upsilon(T)$ and thus $C(\Theta) = \Upsilon(T)$.

It follows that $Z(\Theta) = \Theta \cap C(\Theta) \Rightarrow Z(\Theta) = \{\eta \in T | \mathcal{D}\eta = 0, L(\xi_1)\eta = 0, \forall \xi_1 \in R_1\}$ and $Z(\Theta)$ is made by sections of $\Upsilon(T)$ killed by $\mathcal{D}$. The study of $N(\Theta)$ is much more delicate and we first need the next proposition where we notice the importance of involution or at least formal integrability.

PROPOSITION 5.24: The Lie operator $\mathcal{D} : T \longrightarrow F_0$ induces a homomorphism of Lie algebras $\mathcal{D} : \Upsilon(T) \longrightarrow \Upsilon(F_0)$ where the bracket on $\Upsilon(T)$ is induced by the ordinary bracket on T and the bracket on $\Upsilon(F_0)$ is induced by the differential bracket on $J_q(T)$.

Proof: We already know that T inherits a structure of Lie algebra on sections from the ordinary bracket of vector fields and the situation is similar for $J_q(T)$ with the differential bracket. Now, if $L(\xi_1)\eta = 0$ and $L(\xi_1)\zeta = 0$, it follows from Corollary 5.15 that $L(\xi_1)[\eta, \zeta] = 0$ and $[\Upsilon(T), \Upsilon(T)] \subset \Upsilon(T)$.

Similarly, as $F_0 = J_q(T)/R_q = J_q^0(T)/R_q^0$, if $L(\xi_q)\eta_q^0 \equiv [\xi_q, \eta_q^0] \in R_q^0$ and $L(\xi_q)\zeta_q^0 \equiv [\xi_q, \zeta_q^0] \in R_q^0$, then $L(\xi_q)[\eta_q^0, \zeta_q^0] \in R_q^0$ according to the Jacobi identity for the bracket on $J_q^0(T)$ and the fact that $[R_q^0, R_q^0] \subset R_q^0$. Also, if $L(\xi_{q+1})\eta_q \in R_q$ and $L(\xi_{q+1})\zeta_q \in R_q$, then it is less evident to prove that $L(\xi_{q+1})[\eta_q, \zeta_q] \in R_q$. For this, using Corollary 5.15, if we set $L(\xi_{q+1})\eta_q = \theta_q \in R_q$, it is sufficient to notice that $[\theta_q, \zeta_q] = L(\theta_{q+1})\zeta_q - i(\zeta)D\theta_{q+1} \in R_q$ because both terms do belong to R_q. Thus, in any case, we obtain $[\Upsilon(F_0), \Upsilon(F_0)] \subset \Upsilon(F_0)$.

Finally, we have $j_q([\eta, \zeta]) = [j_q(\eta), j_q(\zeta)]$ and we may take $j_q(\eta)$ as a representative of $\mathcal{D}\eta$ in $J_q(T)$. We shall prove that, if $\eta \in \Upsilon(T)$, that is if $L(\xi_1)\eta = 0, \forall \xi_1 \in R_1$, then $j_q(\eta) \in J_q(T)$ is such that $L(\xi_{q+1})j_q(\eta) = 0$. Introducing $R_2 = \pi_2^q(R_q) \subseteq \rho_1(R_1)$ and choosing any $\xi_2 \in R_2$ over $\xi_1 \in R_1$, we have $\pi_0^1(L(\xi_2)j_1(\eta)) = L(\xi_1)\eta = 0 \Rightarrow L(\xi_2)j_1(\eta) \in T^* \otimes T$. However, using Proposition 5.11 and the fact that $Dj_q(\eta) = 0$, we obtain:

$$i(\zeta)D(L(\xi_2)j_1(\eta)) = L(i(\zeta)D\xi_2)\eta = 0$$

because $R_2 \subset \rho_1(R_1) \Rightarrow DR_2 \subset T^* \otimes R_1$ and thus $L(\xi_2)j_1(\eta) = 0$ because there is a monomorphism (even an isomorphism) $0 \to T^* \otimes T \xrightarrow{\delta} T^* \otimes T$. Supposing by induction that $L(\xi_q)j_{q-1}(\eta) = 0$, we should obtain in the same way:

$$i(\zeta)D(L(\xi_{q+1})j_q(\eta)) = L(i(\zeta)D\xi_{q+1})j_{q-1}(\eta) = 0$$

because $R_{q+1} \subseteq \rho_1(R_q) \Rightarrow DR_{q+1} \subset T^* \otimes R_q$ and thus $L(\xi_{q+1})j_q(\eta) = 0$ because there is a monomorphism $0 \to S_{q+1}T^* \otimes T \xrightarrow{\delta} T^* \otimes S_qT^* \otimes T$ and the restriction of D to a symbol is $-\delta$.

$$\text{Q.E.D.}$$

As the reader will discover in the last computational section, *the study of the normalizer is much more delicate.*

DEFINITION 5.25: The *normalizer* $\tilde{\Gamma} = N(\Gamma)$ of Γ in $aut(X)$ is the biggest Lie pseudogroup in which Γ is *normal*, that is (roughly) $N(\Gamma) = \tilde{\Gamma} = \{\tilde{f} \in aut(X)|\tilde{f} \circ f \circ \tilde{f}^{-1} \in \Gamma, \forall f \in \Gamma\}$ and we write $\Gamma \lhd N(\Gamma) \subset aut(X)$.

Of course, $N(\Theta)$ will play the part of a Lie algebra for $N(\Gamma)$ exactly like Θ did for Γ. However, we shall see that $N(\Gamma)$ may have many components different from the connected component of the identity, for example two in the case of the algebraic Lie pseudogroup of contact transformations where $N(\Gamma)/\Gamma$ is isomorphic to the permutation group of two objects, a result not evident at fist sight. Passing to the jets, we get $j_q(\tilde{f} \circ f \circ \tilde{f}^{-1})^{-1}(\omega) = j_q(\tilde{f}) \circ j_q(f)^{-1} \circ j_q(\tilde{f})^{-1}(\omega) = \omega \Leftrightarrow j_q(f)^{-1}(j_q(\tilde{f})^{-1}(\omega)) = j_q(\tilde{f})^{-1}(\omega)$, that is to say $j_q(f)^{-1}(\bar{\omega}) = \bar{\omega}$ if we set $j_q(\tilde{f})^{-1}(\omega) = \bar{\omega}$ and we find back the equivalence relation of Definition 5.16. It follows that $\tilde{\Gamma} = \{\tilde{f} \in aut(X)|j_q(\tilde{f})^{-1}(\omega) = g(\omega,a), h(c,a) = c\}$ is defined by the system $\tilde{\mathcal{R}}_{q+1} = \{\tilde{f}_{q+1} \in \Pi_{q+1}|\tilde{f}_{q+1}(R_q) = R_q\}$ with linearization $\tilde{R}_{q+1} = \{\tilde{\xi}_{q+1}|L(\tilde{\xi}_{q+1})\eta_q \in R_q, \forall \eta_q \in R_q\}$, that is to say $\{\tilde{\xi}_{q+1}, \eta_{q+1}\} + i(\tilde{\xi})D\eta_{q+1} \in R_q \Leftrightarrow \{\tilde{\xi}_{q+1}, \eta_{q+1}\} \in R_q$. Accordingly, the system of infinitesimal Lie equations defining $\tilde{\Theta} = N(\Theta)$ can be obtained by purely algebraic techniques from the system defining Θ. In particular, we notice that $\pi_0^{q+1} : \tilde{R}_{q+1} \to T$ is an epimorphism because $\pi_0^{q+1} : R_{q+1} \to T$ is an epimorphisme by assumption and $R_{q+1} \subseteq \tilde{R}_{q+1}$. We obtain on the symbol level $\{\tilde{g}_{q+1}, \eta_{q+1}\} \subset g_q$ and thus $\delta\tilde{g}_{q+1} \subset T^* \otimes g_q$ leading to $\tilde{g}_{q+1} \subseteq g_{q+1} = \rho_1(g_q)$ and thus $\tilde{g}_{q+1} = g_{q+1}$ because $R_{q+1} \subseteq \tilde{R}_{q+1} \Rightarrow g_{q+1} \subseteq \tilde{g}_{q+1}$. Using arguments from δ-cohomology, it can be proved that $\tilde{R}_{q+1}$ is involutive when R_q is involutive ([36], p 351, 390). Another proof will be given in Corollary 5.52. With more details, using the result of Proposition 5.17, we get the following important local result:

PROPOSITION 5.26: $\Upsilon_0 = \Upsilon(F_0) = \{\Omega^\tau(x) = A^\alpha W_\alpha^\tau(\omega(x)) | A = cst\}$

Proof: Recalling that $F_0 = \omega^{-1}(V(\mathcal{F}))$, we shall first study the natural bundle $\mathcal{F}_0 = V(\mathcal{F})$ of order q. Adopting local coordinates (x, u, v), any infinitesimal change of source $\bar{x} = x + t\xi(x) + ...$ can be lifted to $\mathcal{F}_0$ with $\bar{u}^\tau = u^\tau + t\xi_\mu^k(x)L_k^{\tau\mu}(u) + ...$, $\bar{v}^\tau = v^\tau + t\frac{\partial L_k^{\tau\mu}(u)}{\partial u^\sigma} v^\sigma + ...$, according to the definition of a vertical bundle provided by Definition 3.3. The corresponding infinitesimal generators on $\mathcal{F}_0$ will be:

$$\xi^i(x)\frac{\partial}{\partial x^i} + \xi_\mu^k(x)(L_k^{\tau\mu}\frac{\partial}{\partial u^\tau} + \frac{\partial L_k^{\tau\mu}(u)}{\partial u^\sigma}v^\sigma\frac{\partial}{\partial v^\tau}) \quad 1 \leq |\mu| \leq q, \forall \xi_q \in J_q(T)$$

It follows that a section $\epsilon : \mathcal{F} \to \mathcal{F}_0 : (x, u) \to (x, u, v = \epsilon(x, u))$ will be equivariant, that is $v - \epsilon(x, u) = 0 \Rightarrow \bar{v} - \epsilon(\bar{x}, \bar{u}) = 0$ if and only if $v = \epsilon(u)$ satisfies:

$$L_k^{\tau\mu}(u)\frac{\partial \epsilon^\sigma(u)}{\partial u^\tau} - \frac{\partial L_k^{\tau\mu}(u)}{\partial u^\sigma}\epsilon^\sigma(u) = 0 \quad , \quad 1 \leq |\mu| \leq q$$

Hence, we have $[L_k^\mu, \epsilon] = 0 \Rightarrow \epsilon^\tau(u) = A^\alpha W_\alpha^\tau(u)$ with $A = cst$.

As another approach, working directly with F_0, we may consider the invariance of the section $u - \omega(x) = 0, v - \Omega(x) = 0$ and get (care to the sign):

$$-L_k^{\tau\mu}(\omega(x))\xi_\mu^k + \xi^r\partial_r\omega^\tau(x) = 0 \quad , \quad -\frac{\partial L_k^{\tau\mu}(\omega(x))}{\partial u^\sigma}\Omega^\sigma(x)\xi_\mu^k + \xi^r\partial_r\Omega^\tau(x) = 0$$

The first condition brings at once $\xi_q \in R_q$ and we obtain therefore the following central local result for F_0:

$$\Upsilon_0 = \{\Omega \in F_0 | -\frac{\partial L_k^{\tau\mu}(\omega(x))}{\partial u^\sigma}\Omega^\sigma\xi_\mu^k + \xi^r\partial_r\Omega^\tau = 0 \ , \forall \xi_q \in R_q\}$$

As usual, the study of this system can be cut into two parts. First of all, we have to look for:

$$E_0 = \{\Omega \in F_0 | \frac{\partial L_k^{\tau\mu}(\omega(x))}{\partial u^\sigma}\Omega^\sigma\xi_\mu^k = 0, \forall \xi_q^0 \in R_q^0\} \subseteq F_0$$

In a symbolic way with $pri(R_q^0) = \{\xi^\sigma\}$ and $par(R_q^0) = \{\xi^{m+r}\}$, we get:

$$\frac{\partial L_\sigma}{\partial u}\Omega\xi^\sigma + \frac{\partial(\mathcal{E}_{m+r}^\sigma L_\sigma)}{\partial u}\Omega\xi^{m+r} = 0$$

whenever $\xi^\sigma + \mathcal{E}_{m+r}^\sigma\xi^{m+r} = 0$ and thus $\frac{\partial \mathcal{E}}{\partial u}\Omega = 0$ in agrement with Proposition 5.17.

Then we have to introduce an R_q-connection, that is a section $\chi_q \in T^* \otimes R_q$ over $id_T \in T^* \otimes T$ such that:

$$-L_k^{\tau\mu}(\omega(x))\chi_{\mu,i}^k + \partial_i\omega^\tau(x) = 0 \Rightarrow -\frac{\partial L_k^{\tau\mu}(\omega(x))}{\partial u^\sigma}\chi_{\mu,i}^k\Omega^\sigma + \partial_i\Omega^\tau = 0$$

and it just remains to study this last system for $\Omega^\tau(x) = A^\alpha(x)W_\alpha^\tau(\omega(x))$ as it does not depend on the choice of the connection χ_q. Substituting, we obtain successively:

$$(-\frac{\partial L_k^{\tau\mu}}{\partial u^\sigma}\chi_{\mu,i}^k W_\alpha^\sigma + \frac{\partial W_\alpha^\tau}{\partial u^\sigma}\partial_i\omega^\sigma)A^\alpha + (\partial_i A^\alpha)W_\alpha^\tau = 0$$

$$([L_k^\mu, W_\alpha])^\tau \chi_{\mu,i}^k A^\alpha + (\partial_i A^\alpha)W_\alpha^\tau = 0$$

As $[L_k^\mu, W_\alpha] = 0$ and the action is effective, we finally obtain $\partial_i A^\alpha = 0$ that is $A = cst$.

Q.E.D.

The following corollary is a direct consequence of the above proposition and explains many classical results as we shall see in the last computational section.

COROLLARY 5.27: We have the relation:

$$\Theta = \{\xi \in T | \mathcal{D}\xi \equiv \mathcal{L}(\xi)\omega = 0\} \Rightarrow N(\Theta) = \{\mathcal{D}\xi \equiv \mathcal{L}(\xi)\omega = AW(\omega) \in \Upsilon_0\}$$

REMARK 5.28: If we set $\tilde{R}_q = \pi_q^{q+1}(\tilde{R}_{q+1})$, then $\tilde{R}_{q+1}$ is obtained by a procedure with two steps. First of all, we obtain $\tilde{R}_q = \{\xi_q \in J_q(T)|L(\xi_q)\omega = A(x)W(\omega) \in E_0\}$ by eliminating the infinitesimal parameters A by means of pure linear algebra. Then, we have to take into account that $A = cst$ in the above corollary and another reason for understanding that $g_{q+1} = \tilde{g}_{q+1}$ though we have only $g_q \subseteq \tilde{g}_q$ in general. Such a situation is well known in physics where the Poincaré group is of codimension one in its normalizer which is the Weyl group obtained by adding a dilatation.

According to the definition of the Janet bundles at the end of section 4, using the inclusion $\wedge^r T^* \otimes S_q T^* \otimes T \subset \wedge^r T^* \otimes J_q^0(T) \subset \wedge^r T^* \otimes J_q(T)$, in the case of an involutive system $R_q \subset J_q(T)$ of infinitesimal transitive Lie equations, we have:

$$F_r = \wedge^r T^* \otimes J_q^0(T)/(\wedge^r T^* \otimes R_q^0 + \delta(\wedge^{r-1} T^* \otimes S_{q+1} T^* \otimes T))$$

As R_q^0 is associated with R_q and $J_q^0(T)$ is associated with $R_q \subset J_q(T)$, we obtain:

LEMMA 5.29: The Janet bundles are associated with R_q and we set $\Upsilon_r = \Upsilon(F_r)$.

From this lemma we shall deduce the following important but difficult theorem:

THEOREM 5.30: The first order operators $\mathcal{D}_r : F_{r-1} \to F_r$ induce maps $\mathcal{D}_r : \Upsilon_{r-1} \to \Upsilon_r$ in the *deformation sequence*:

$$0 \longrightarrow Z(\Theta) \longrightarrow C(\Theta) \xrightarrow{\mathcal{D}} \Upsilon_0 \xrightarrow{\mathcal{D}_1} \Upsilon_1 \xrightarrow{\mathcal{D}_2} ... \xrightarrow{\mathcal{D}_n} \Upsilon_n \longrightarrow 0$$

which is locally described by finite dimensional vector spaces and linear maps.

Proof: Any section of F_{r-1} can be lifted to a section $A_q^{r-1} \in \wedge^{r-1}T^* \otimes J_q(T)$ modulo $\wedge^{r-1}T^* \otimes R_q + \delta(\wedge^{r-2}T^* \otimes S_{q+1}T^* \otimes T)$ where the second component is in the image of D. Then we can lift again this section to a section $A_{q+1}^{r-1} \in \wedge^{r-1}T^* \otimes J_{q+1}(T)$, modulo a section of $\wedge^{r-1}T^* \otimes R_{q+1} + D(\wedge^{r-2}T^* \otimes J_{q+2}^q(T)) + \wedge^{r-1}T^* \otimes S_{q+1}T^* \otimes T$ where J_{q+2}^q is the kernel of the projection $\pi_q^{q+2} : J_{q+2}(T) \to J_q(T)$. The image by $\mathcal{D}_r$ is obtained by applying $D : \wedge^{r-1}T^* \otimes J_{q+1}(T) \to \wedge^r T^* \otimes J_q(T)$ and projecting $DA_{q+1}^{r-1} \in \wedge^r T^* \otimes J_q(T)$ thus obtained to F_r while taking into account successively the restriction $D : \wedge^{r-1}T^* \otimes R_{q+1} \to \wedge^r T^* \otimes R_q$, the fact that $D \circ D = D^2 = 0$ and the restriction δ providing an element in $\delta(\wedge^{r-1}T^* \otimes S_{q+1}T^* \otimes T)$. Of course, with such a choice we need to use associations with R_{q+1} *even though finally only R_q is involved* because of the above lemma. The reason is that the restriction of D to $J_{q+1}^0(T)$ has an image in $T^* \otimes J_q(T)$ and *not* in $T^* \otimes J_q^0(T)$. Hence, starting with a section of $\Upsilon_{r-1} \subset F_{r-1}$, we must have:

$$L(\xi_{q+1})A_q^{r-1} \in \wedge^{r-1}T^* \otimes R_q + \delta(\wedge^{q-2}T^* \otimes S_{q+1}T^* \otimes T), \quad \forall \xi_{q+1} \in R_{q+1} = \rho_1(R_q)$$

Accordingly, looking at the right member in the formula of Proposition 5.13, the section in $\wedge^{r-1}T^* \otimes R_q$ is contracted with $r-1$ vectors in order to provide a section of R_q and the skewsymmetrized summation finally produces a section of $\wedge^r T^* \otimes R_q$. Similarly, as $\delta(\wedge^{r-2}T^* \otimes S_{q+1}T^* \otimes T) \subset \wedge^{r-1}T^* \otimes S_q T^* \otimes T$ we should obtain a section of $\wedge^r T^* \otimes S_q T^* \otimes T$ but it is not evident at all that such a section is in $\delta(\wedge^{r-1}T^* \otimes S_{q+1}T \otimes T)$. For this, let us notice that, if the multi-index I has length

$r-2$ and $|\mu| = q$, we have the local formula:

$$\alpha_{\mu+1_i,I,j}dx^i \wedge (dx^j \wedge dx^I) = -\alpha_{\mu+1_i,I,j}dx^j \wedge (dx^i \wedge dx^I)$$

explaining why the following diagram is commutative and exact:

$$
\begin{array}{ccc}
T^* \otimes \wedge^{r-2}T^* \otimes S_{q+1}T^* & \xrightarrow{(-1)^{r-2}\delta} & \wedge^{r-1}T^* \otimes S_{q+1}T^* \longrightarrow 0 \\
\downarrow \delta & & \downarrow \delta \\
T^* \otimes \wedge^{r-1}T^* \otimes S_qT^* & \xrightarrow{(-1)^{r-1}\delta} & \wedge^rT^* \otimes S_qT^* \quad \longrightarrow 0
\end{array}
$$

where the upper map induced by $\delta : \wedge^{r-2}T^* \otimes T^* \to \wedge^{r-1}T^*$ and the lower map induced by $\delta : \wedge^{r-1}T^* \otimes T^* \to \wedge^rT^*$ are both epimorphisms while the left vertical map is induced by $\delta : \wedge^{r-2}T^* \otimes S_{q+1}T^* \to \wedge^{r-1}T^* \otimes S_qT^*$.

We obtain therefore:

$$DL(\xi_{q+2})A_{q+1}^{r-1} - L(\xi_{q+1})DA_{q+1}^{r-1} \in \wedge^rT * \otimes R_q + \delta(\wedge^{r-1}T^* \otimes S_{q+1}T^* \otimes T)$$

Now, from the construction of $\mathcal{D}_r$ already explained, we have:

$$F_{r-1} = \wedge^{r-1}T^* \otimes J_{q+1}(T)/(\wedge^{r-1}T^* \otimes R_{q+1} + D(\wedge^{r-2}T^* \otimes J_{q+2}^q(T)) + \wedge^{r-1}T^* \otimes S_{q+1}T^* \otimes T)$$

It follows that:

$$L(\xi_{q+2})A_{q+1}^{r-1} \in \wedge^{r-1}T^* \otimes R_{q+1} + D(\wedge^{r-2}T^* \otimes J_{q+2}^q(T)) + \wedge^{r-1}T^* \otimes S_{q+1}T^* \otimes T$$

and thus:

$$DL(\xi_{q+2})A_{q+1}^{r-1} \in \wedge^rT^* \otimes R_q + \delta(\wedge^{r-1}T^* \otimes S_{q+1}T^* \otimes T)$$

that is finally:

$$L(\xi_{q+1})DA_{q+1}^{r-1} \in \wedge^rT^* \otimes R_q + \delta(\wedge^{r-1}T^* \otimes S_{q+1}T^* \otimes T)$$

Accordingly, $DA_{q+1}^{r-1} \in \wedge^rT^* \otimes J_q(T)$ is the representative of a section of Υ_r.

$$\text{Q.E.D.}$$

DEFINITION 5.31: The deformation sequence is not necessarily exact and only depends on R_q. We can therefore define as usual *coboundaries* $B_r(R_q)$, *cocycles* $Z_r(R_q)$ and *cohomology groups* $H_r(R_q) = Z_r(R_q)/B_r(R_q)$ with $B_r(R_q) \subseteq Z_r(R_q) \subseteq$

Υ_r for $r = 0, 1, ..., n$.

PROPOSITION 5.32: We have $B_0(R_q) = C(\Theta)/Z(\Theta)$, $Z_0(R_q) = N(\Theta)/\Theta$ and the short exact sequence:

$$0 \longrightarrow C(\Theta)/Z(\Theta) \longrightarrow N(\Theta)/\Theta \longrightarrow H_0(R_q) \longrightarrow 0$$

Proof: The monomorphism on the left is induced by the inclusion $C(\Theta) \subset N(\Theta)$ because $Z(\Theta) = \Theta \cap C(\Theta)$. Then $B_0(R_q)$ is the image of $\mathcal{D}$ in Υ_0 because of Proposition 5.23 and Proposition 5.24. Finally, $Z_0(R_q) = N(\Theta)/\Theta$ is just a way to rewrite Corollary 5.27 while taking into account the fact that $\mathcal{D}_1 \circ \mathcal{D} = 0$. The cocycle condition just tells that the label transformations induced by the normalizer do not change the structure constants because the Vessiot structure equations are invariant under *any* natural transformation. This result should be compared to Lemma 2.7.

Q.E.D.

In order to generalize Proposition 5.26 and to go further on, we need a few more concepts from differential geometry.

DEFINITION 5.33: A chain $\mathcal{E} \overset{\Phi}{\longrightarrow} \mathcal{E}' \overset{\Psi}{\longrightarrow} \mathcal{E}''$ of fibered manifolds is said to be a *sequence with respect to a section f'' of $\mathcal{E}''$* if $im(\Phi) = ker_{f''}(\Psi)$, that is with local coordinates (x, y) on $\mathcal{E}$, (x, y') on $\mathcal{E}'$, (x, y'') on $\mathcal{E}''$ and $y' = \Phi(x, y), y'' = \Psi(x, y')$, we have $\Psi(x, \Phi(x, y)) \equiv f''(x), \forall (x, y) \in \mathcal{E}$.

Differentiating this identity with respect to y, we obtain:

$$\frac{\partial \Psi}{\partial y'}(x, \Phi(x, y)).\frac{\partial \Phi}{\partial y}(x, y) \equiv 0$$

and we have therefore a sequence $V(\mathcal{E}) \overset{V(\Phi)}{\longrightarrow} V(\mathcal{E}') \overset{V(\Psi)}{\longrightarrow} V(\mathcal{E}'')$ of vector bundles pulled back over $\mathcal{E}$ by reciprocal images.

DEFINITION 5.34: A sequence of fibered manifolds is said to be an *exact sequence* if $im(\Phi) = ker_{f''}(\Psi)$ *and* the corresponding vertical sequence of vector bundles is exact.

PROPOSITION 5.35: If $\mathcal{E}, \mathcal{E}', \mathcal{E}''$ are affine bundles over X with corresponding model vector bundles E, E', E'' over X, a sequence of such affine bundles is exact if and only if the corresponding sequence of model vector bundles is exact. In that case, there is an exact sequence $\mathcal{E} \xrightarrow{\Phi} \mathcal{E}' \longrightarrow E''$ which allows to avoid the use of a section of $\mathcal{E}''$ while replacing it by the zero section of E''. Finally, the map Φ is injective if and only if the map $V(\Phi)$ is injective and the map Ψ is surjective if and only if the map $V(\Psi)$ is surjective.

Proof: We have successively $y' = \Phi(x, y) = A(x)y + B(x), y'' = \Psi(x, y') = C(x)y' + D(x)$ and by composition $y'' = C(x)A(x)y + C(x)B(x) + D(x) = f''(x)$. Accordingly, we *must* have $C(x)A(x) \equiv 0, C(x)B(x) + D(x) = f''(x), \forall x \in X$ and we obtain the following commutative diagram:

$$
\begin{array}{ccccc}
E & \xrightarrow{V(\Phi)} & E' & \xrightarrow{V(\Psi)} & E'' \\
\vdots & & \vdots & \nearrow & \vdots \\
\mathcal{E} & \xrightarrow{\Phi} & \mathcal{E}' & \xrightarrow{\Psi} & \mathcal{E}'' \\
\pi \downarrow & & \pi' \downarrow & & \pi'' \downarrow\uparrow f'' \\
X & = & X & = & X
\end{array}
$$

If $(x, y') \in \mathcal{E}'$ is such that $C(x)y' + D(x) = f''(x)$, we obtain by substraction $C(x)(y' - B(x)) = 0 \Leftrightarrow C(x)(y' - \Phi(x, y)) = 0$ with $(x, y' - \Phi(x, y)) \in E'$. Supposing the model sequence exact, we may find $(x, v) \in E$ such that $y' - (A(x)y + B(x)) = A(x)v \Leftrightarrow y' = A(x)(y + v) + B(x)$ and thus $(x, y') \in im(\Phi)$. The converse is similar and left to the reader.

Moreover, , we notice that $(x, y'') \in \mathcal{E}'' \Leftrightarrow (x, y'' - f''(x)) \in E'', \forall x \in X$ with $y'' - f''(x) = C(x)(y' - B(x)) = C(x)(y' - \Phi(x, y))$ where $(x, y) \in \mathcal{E}$ is *any* point over $x \in X$. It follows that the upper diagonal arrow on the right may be defined by $C(x)y' - C(x)B(x) = v''$ and we have indeed $v'' = 0 \Rightarrow \exists(x, y) \in \mathcal{E}, y' = A(x)y + B(x)$ as claimed.

Finally, if $(x, y_1), (x, y_2) \in \mathcal{E}$ and $V(\Phi)$ is injective, then $y' = A(x)y_1 + B(x) = A(x)y_2 + B(x) \Rightarrow A(x)(y_1 - y_2) = 0 \Rightarrow y_1 = y_2$. Also, if $(x, y'') \in \mathcal{E}''$ and $V(\Psi)$ is surjective, then we can find $v' \in E'$ such that $y'' - f''(x) = C(x)v' \in E'' \Rightarrow y'' = f''(x) + C(x)v' = C(x)(B(x) + v') + D(x) = C(x)(A(x)y + B(x) + v') + D(x) = C(x)y' + D(x)$ for *any* $(x, y) \in \mathcal{E}$ and we just need to set $y' = A(x)y + B(x) + v'$.

Q.E.D.

Coming back to the construction of the Vessiot structure equations from the knowledge of the generating differential invariants at order q, in particular for the affine, projective and Rimann structures, we have already exhibited the system $\mathcal{B}_1 \subset J_1(\mathcal{F})$ locally defined by affine equations of the form $I(u_1) \equiv A(u)u_x + B(u) = 0$. The symbol $\mathcal{H}_1 \subset T^* \otimes V(\mathcal{F})$ of this system is defined by linear equations of the form $A(u)v_x = 0$ and the last proposition provides at once the following commutative and exact diagram of affine bundles and model vector bundles *over* $\mathcal{F}$ (care) where we decide, with a slight abuse of notations, that the left central arrow is injective (0 on the left) while right central arrow is surjective (0 on the right):

$$
\begin{array}{ccccccccc}
0 & \longrightarrow & \mathcal{H}_1 & \longrightarrow & T^* \otimes V(\mathcal{F}) & \longrightarrow & \mathcal{F}_1 & \longrightarrow & 0 \\
 & & \vdots & & \vdots & \nearrow & \| & & \\
0 & \longrightarrow & \mathcal{B}_1 & \longrightarrow & J_1(\mathcal{F}) & \longrightarrow & \mathcal{F}_1 & \longrightarrow & 0 \\
 & & \downarrow & & \downarrow & & \downarrow & & \\
 & & \mathcal{F} & = & \mathcal{F} & = & \mathcal{F} & &
\end{array}
$$

More generally, having in mind the diagram at the end of section 4, we may define the *nonlinear Janet bundles* $\mathcal{F}_r = \wedge^r T^* \otimes V(\mathcal{F})/\delta(\wedge^{r-1} T^* \otimes \mathcal{H}_1)$ with $\mathcal{F}_0 = V(\mathcal{F})$ as a family of natural vector bundles *over* $\mathcal{F}$. The following commutative diagram of reciprocal images generalizes Proposition 5.26:

$$
\begin{array}{ccc}
F_r & \longrightarrow & \mathcal{F}_r \\
\textit{section of } \Upsilon_r \;\downarrow\uparrow & & \downarrow\uparrow \;\textit{equivariant section} \\
X & \xrightarrow{\;\omega\;} & \mathcal{F}
\end{array}
$$

Adopting local coordinates (u, v) for $\mathcal{F}_r$, we may generalize Proposition 5.26 by saying that any infinitesimal change of source $\bar{x} = x + t\xi(x) + \ldots$ can be lifted to $\mathcal{F}_r$ with $\bar{u}^\tau = u^\tau + t\partial_\mu \xi^k(x) L_k^{\tau\mu}(u) + \ldots, \bar{v}^\alpha = v^\alpha + t\partial_\mu \xi^k(x) M_{\beta,k}^{\alpha,\mu}(u)v^\beta + \ldots$. Equivalently, the corresponding infinitesimal transformation rules may take the form:

$$
\xi^i(x)\frac{\partial}{\partial x^i} + \partial_\mu \xi^k(x)(L_k^{\tau\mu}(u)\frac{\partial}{\partial u^\tau} + M_{\beta,k}^{\alpha,\mu}(u)v^\beta \frac{\partial}{\partial v^\alpha})
$$

where we may replace $\partial_\mu \xi^k(x)$ by $\xi_\mu^k(x)$ if we want to work formally. However, in the case of $\mathcal{F}_1$ for example, it is essential to notice that setting only $v = A(u)u_x$ is not sufficient for getting a natural vector bundle over $\mathcal{F}$ when $q \geq 2$ as one has to set

$v = A(u)u_x + B(x)$ in general.

Omitting indices for simplicity and using the natural projection $T^* \otimes \mathcal{F}_1 \to \mathcal{F}_2$: $v_x \to \gamma(u)v_x$ over $\mathcal{F}$ in such a way to have:

$$v_x = A(u)u_{xx} + \partial_u A(u)u_x u_x + \partial_u B(u)u_x \Rightarrow \gamma(u)v_x = a(u)u_{xx} + b(u)u_x$$

we may factor the dependence on u_x through a dependence on v as we are dealing with natural bundles and obtain expressions of the form :

$$\gamma(u)v_x = \alpha(u)vv + \beta(u)v$$

because $v = 0$ must be a solution for the special system of Lie equations considered. For the general system, it just remains to use the equivariant section $v = c(u)$ and substitute it in order to obtain polynomial conditions for the structure constants of degree ≥ 2. When $q = 1$, then $B(u), b(u), \beta(u)$ disappear and we get homogeneous Jacobi conditions of degree 2 exactly.

IMPORTANT REMARK 5.36: No classical technique could provide this result because all the known methods of computer algebra do construct the Janet sequence *"step by step"* and never *"as a whole"*, that is from the Spencer operator ([36], p 391). It is also not evident to deal with high order natural bundles in this framework, even when $n = 1$. For example, in the affine case, we have alreday exhibited the transition rules of the second order natural bundle $\mathcal{F}$ and obtain the following transition rules of $J_1(\mathcal{F})$:

$$\bar{x} = \varphi(x), \quad u = \bar{u}\partial_x\varphi + \frac{\partial_{xx}\varphi}{\partial_x\varphi}, \quad u_x = \bar{u}_{\bar{x}}(\partial_x\varphi)^2 + \bar{u}\partial_{xx}\varphi + \left(\frac{\partial_{xxx}\varphi}{\partial_x\varphi} - \frac{(\partial_{xx}\varphi)^2}{(\partial_x\varphi)^2}\right)$$

Hence, the transition rules for $V(\mathcal{F})$ are $v = \bar{v}\partial_x\varphi$ while the transition rules for $T^* \otimes V(\mathcal{F})$ are $v_x = \bar{v}_{\bar{x}}(\partial_x\varphi)^2$. We invite the reader to treat the projective case similarly as an exercise.

PROPOSITION 5.37: The affine bundle:

$$\begin{aligned} A(R_q^0) &= id_T^{-1}(T^* \otimes J_q(T)/(T^* \otimes R_q^0 + \delta(S_{q+1}T^* \otimes T))) \\ &= \{\chi_q \in T^* \otimes J_q(T)/(T^* \otimes R_q^0 + \delta(S_{q+1}T^* \otimes T))|\chi_0 = id_T \in T^* \otimes T\} \end{aligned}$$

is modelled on F_1 and associated with R_q, that is $L(\xi_q)c \in F_1, \forall \xi_q \in R_q, \forall c \in A(R_q^0)$.

Proof: First of all, using the south east arrow in the following commutative and exact diagram:

$$
\begin{array}{ccccccccc}
 & & 0 & & 0 & & 0 & & \\
 & & \downarrow & & \downarrow & & \downarrow & & \\
0 \to & T^* \otimes R_q^0 + \delta(S_{q+1}T^* \otimes T) & \to & T^* \otimes R_q + \delta(S_{q+1}T^* \otimes T) & \to & T^* \otimes T & \to 0 \\
 & \downarrow & \searrow & & \downarrow & & \| & \\
0 \to & T^* \otimes J_q^0(T) & \longrightarrow & T^* \otimes J_q(T) & \to & T^* \otimes T & \to 0 \\
 & \downarrow & & \downarrow & & \downarrow & \\
0 \to & F_1 & = & F_1 & \longrightarrow & 0 & \\
 & \downarrow & & \downarrow & & & \\
 & 0 & & 0 & & &
\end{array}
$$

we see that the above definition is coherent and we shall represent any element $c \in A(R_q^0)$ by a $J_q(T)$-connection $\chi_q \in id_T^{-1}(T^* \otimes J_q(T))$. It follows that $A(R_q^0)$ is modelled on $T^* \otimes J_q^0(T)/(T^* \otimes R_q^0 + \delta(S_{q+1}T^* \otimes T)) = F_1$.

Denoting by $C_1^0(T) = T^* \otimes J_q^0(T)/\delta(S_{q+1}T^* \otimes T)$ the kernel of the canonical projection of $C_1(T) = T^* \otimes J_q(T)/\delta(S_{q+1}T^* \otimes T)$ onto $T^* \otimes T$, it just remains to prove that the affine bundle:

$$A_q = \{\chi_q \in C_1(T)|\chi_0 = id_T \in T^* \otimes T\} = id_T^{-1}(C_1(T))$$

is a natural affine bundle over X modeled on the vector bundle $C_1^0(T) = T^* \otimes J_q^0(T)/\delta(S_{q+1}T^* \otimes T)$ and associated with $J_q(T)$, thus with R_q too, because $T^* \otimes R_q^0$ is associated with R_q.

For this it is sufficient to observe the transition laws of $T^* \otimes J_q(T)$ when $\bar{x} = \varphi(x)$, namely:

$$
\begin{array}{llll}
T^* \otimes T & \bar{\chi}_{,r}^l \partial_i \varphi^r & = & \chi_{,i}^k \partial_k \varphi^l \\
T^* \otimes J_1(T) & \bar{\chi}_{s,r}^l \partial_i \varphi^r \partial_j \varphi^s & = & \chi_{j,i}^k \partial_k \varphi^l + \chi_{,i}^k \partial_{jk} \varphi^l \\
T^* \otimes J_q(T) & \bar{\chi}_{r_1...r_q,r}^l \partial_{i_1} \varphi^{r_1}...\partial_{i_q} \varphi^{r_q} \partial_i \varphi^r + ... & = & \chi_{\mu,i}^k \partial_k \varphi^l + ... + \chi_{,i}^k \partial_{\mu+1_k} \varphi^l
\end{array}
$$

with μ replaced by $(i_1, ..., i_q)$ when $|\mu| = q$. Setting $\chi_{,i}^k = \delta_i^k$, we get $\bar{\chi}_{,r}^l = \delta_r^l$ too and:

$$id_T^{-1}(T^* \otimes J_1(T)) \qquad \bar{\chi}_{s,r}^l \partial_i \varphi^r \partial_j \varphi^s = \chi_{j,i}^k \partial_k \varphi^l + \partial_{ij} \varphi^l$$

$$id_T^{-1}(T^* \otimes J_q(T)) \qquad \bar{\chi}^l_{r_1...r_q,r}\partial_{i_1}\varphi^{r_1}...\partial_{i_q}\varphi^{r_q}\partial_i\varphi^r + ... = \chi^k_{\mu,i}\partial_k\varphi^l + ... + \partial_{\mu+1_i}\varphi^l$$

Finally, we have:

$$i(\eta)L(\xi_{q+1})\chi_q \;=\; L(\xi_{q+1})\chi_q(\eta) - \chi_q(L(\xi_1)\eta)$$
$$=\; [\xi_q, \chi_q(\eta)] + i(\eta)D\xi_{q+1} - \chi_q(L(\xi_1)\eta)$$

and thus $L(\xi_{q+1})\chi_q \in T^* \otimes J_q^0(T)$ because $\chi_0(\eta) = \eta$ (care) and $i(\eta)D\xi_1 - L(\xi_1)\eta = [\xi, \eta]$. If we choose $\xi_{q+1}, \bar{\xi}_{q+1} \in J_{q+1}(T)$ over $\xi_q \in J_q(T)$, the difference will be $i(\eta)D(\bar{\xi}_{q+1} - \xi_{q+1}) = -i(\eta)\delta(\bar{\xi}_{q+1} - \xi_{q+1}) = 0$ by residue because we are in $C_1(T)$. It is important to notice that no formal integrability assumption is needed for R_q.

Q.E.D.

Similarly, we obtain:

PROPOSITION 5.38: $id_q^{-1}(J_1(\Pi_q))$ is an affine natural bundle of order $q + 1$ over X, modelled on $T^* \otimes J_q(T)$.

Proof: We only prove the proposition when $q = 1$ as the remaining of he proof is similar to the previous one. Indeed, if $\bar{x} = \varphi(x), \bar{y} = \psi(y)$ are the changes of coordinates on $X \times Y$, we get:

$$\Pi_1 \qquad\qquad \bar{y}_r^l\partial_i\varphi^r(x) \;=\; \frac{\partial\psi^l}{\partial y^k}(y)y_i^k$$
$$J_1(\Pi_1) \qquad \bar{y}_{r,s}^l\partial_i\varphi^r(x)\partial_j\varphi^s(x) + \bar{y}_r^l\partial_{ij}\varphi^r(x) \;=\; \frac{\partial\psi^l}{\partial y^k}(y)y_{i,j}^k + \frac{\partial^2\psi^l}{\partial y^k\partial y^u}(y)y_i^k y_{,j}^u$$
$$id_1^{-1}(J_1(\Pi_1)) \qquad \bar{y}_{r,s}^l\partial_i\varphi^r\partial_j\varphi^s + \partial_{ij}\varphi^l \;=\; \partial_k\varphi^l y_{i,j}^k + \partial_{ik}\varphi^l y_{,j}^k$$

Then, we just need to set $y = x, y_j^k = \delta_j^k$ and compare to $T^* \otimes J_1(T)$ above.
Finally, as $J_1(\Pi_q)$ is an affine bundle over Π_q modelled on $T^* \otimes V(\Pi_q)$, then $id_q^{-1}(J_1(\Pi_q))$ is an affine bundle over X, modelled on:

$$id_q^{-1}(T^* \otimes V(\Pi_q)) = T^* \otimes id_q^{-1}(V(\Pi_q)) = T^* \otimes J_q(T).$$

Q.E.D.

PROPOSITION 5.39: We have the following exact sequences of affine bundles over X and model vector bundles:

$$0 \longrightarrow \quad T^* \otimes R_q \quad \longrightarrow \quad T^* \otimes J_q(T) \quad \xrightarrow{V(\Phi)} \quad T^* \otimes F_0 \quad \longrightarrow 0$$
$$\vdots \qquad\qquad \vdots \qquad\qquad \vdots$$
$$0 \longrightarrow \ id_q^{-1}(J_1(\mathcal{R}_q)) \ \longrightarrow \ id_q^{-1}(J_1(\Pi_q)) \ \longrightarrow \ \omega^{-1}(J_1(\mathcal{F})) \ \longrightarrow 0$$
$$\downarrow \qquad\qquad \downarrow\uparrow id_{q,1} \qquad\qquad \downarrow\uparrow j_1(\omega)$$
$$X \qquad = \qquad X \qquad = \qquad X$$

Proof: We just need to use $1 \leq\mid \mu \mid\leq q$ and set $y_q = id_q(x)$ in order to obtain:

$$u^\tau = \Phi^\tau(\omega(y), y_\mu) \xrightarrow{id_q^{-1}} \omega^\tau(x)$$

$$u_i^\tau = \frac{\partial \Phi^\tau}{\partial u^\sigma}(y_q)\frac{\partial \omega^\sigma}{\partial y^k}(y)y_{,i}^k + \frac{\partial \Phi^\tau}{\partial y_\mu^k}(y_q)y_{\mu,i}^k \xrightarrow{id_q^{-1}} -L_k^{\tau\mu}(\omega(x))\chi_{\mu,i}^k + \chi_{,i}^k\partial_k\omega^\tau(x)$$

in a coherent way with the upper model sequence and the various distinguished sections.

Q.E.D.

PROPOSITION 5.40: We have the following exact sequences of affine bundles over X and model vector bundles:

$$0 \longrightarrow \quad T^* \otimes R_q^0 \quad \longrightarrow \quad T^* \otimes J_q^0(T) \quad \xrightarrow{V(\Phi)} \quad T^* \otimes F_0 \quad \longrightarrow 0$$
$$\vdots \qquad\qquad \vdots \qquad\qquad \vdots$$
$$0 \longrightarrow \ id_T^{-1}(T^* \otimes R_q) \ \longrightarrow \ id_T^{-1}(T^* \otimes J_q(T)) \ \longrightarrow \ \omega^{-1}(J_1(\mathcal{F})) \ \longrightarrow 0$$
$$\downarrow \qquad\qquad \downarrow \qquad\qquad \downarrow\uparrow j_1(\omega)$$
$$X \qquad = \qquad X \qquad = \qquad X$$

Proof: As id_q is a section of Π_q, then $j_1(id_q)$ is a section of $J_1(\Pi_q)$ over id_q and thus a section of $id_q^{-1}(J_1(\Pi_q))$ denoted by $id_{q,1}$. If $\chi_q = (\delta_i^k, \chi_{\mu,i}^k)$ with $1 \leq\mid \mu \mid\leq q$ is a section of $id_T^{-1}(T^* \otimes J_q(T))$, that is a $J_q(T)$-connection, it is of course a section of $T^* \otimes J_q(T)$ and we may consider the section $y_{q,1} = id_{q,1} + \chi_q$ of $id_q^{-1}(J_1(\Pi_q))$. The image in $\omega^{-1}(J_1(\mathcal{F}))$ is $u_i^\tau = (-L_k^{\tau\mu}(\omega(x))\chi_{\mu,i}^k + \partial_i\omega^\tau(x)) + \partial_i\omega^\tau(x)$ and we obtain: $u_i^\tau = \partial_i\omega^\tau(x) \Leftrightarrow u_1 = j_1(\omega)(x) \Leftrightarrow \chi_q \in id_T^{-1}(T^* \otimes R_q)$.

Q.E.D.

THEOREM 5.41: $\quad A(R_q^0) \simeq \omega^{-1}(J_1(\mathcal{F})/\mathcal{B}_1)$

Proof: The affine sub-bundle $\mathcal{B}_1 \subset J_1(\mathcal{F})$ over $\mathcal{F}$ is the imager of the first prolon-

66

gation $\rho_1(\Phi) : \Pi_{q+1} \longrightarrow J_1(\mathcal{F})$ where Π_{q+1} is an affine bundle over Π_q modelled on $S_{q+1}T^* \otimes V(X \times X)$. We may proceed similarly by introducing the affine bundle $id_q^{-1}(\Pi_{q+1})$ which is modelled on $id_q^{-1}(S_{q+1}T^* \otimes V(X \times X)) = S_{q+1}T^* \otimes id^{-1}(V(X \times X)) = S_{q+1}T^* \otimes T$ and obtain the commutative and exact diagram of affine bundles and model vector bundles where the model sequence is the symbol sequence that has been used in order to introduce F_1:

$$
\begin{array}{ccccccc}
S_{q+1}T^* \otimes T & \longrightarrow & T^* \otimes F_0 & \longrightarrow & F_1 & \longrightarrow & 0 \\
\vdots & & \vdots & & \vdots & & \\
id_q^{-1}(\Pi_{q+1}) & \longrightarrow & \omega^{-1}(J_1(\mathcal{F})) & \longrightarrow & \omega^{-1}(J_1(\mathcal{F})/\mathcal{B}_1) & \longrightarrow & 0 \\
\downarrow & & \downarrow\uparrow j_1(\omega) & & \downarrow & & \\
X & = & X & = & X & &
\end{array}
$$

With $1 \leq |\mu| \leq q$, the first morphism of affine bundles is now described by:

$$u^\tau = \omega^\tau(x),$$
$$u_i^\tau = \frac{\partial \Phi^\tau}{\partial u^\sigma}(y_q)\frac{\partial \omega^\sigma}{\partial y^k}(y)y_i^k + \frac{\partial \Phi^\tau}{\partial y_\mu^k}(y_q)y_{\mu+1_i}^k \xrightarrow{id_q^{-1}} -\sum_{|\mu|=q} L_k^{\tau\mu}(\omega(x))\chi_{\mu+1_i}^k + \partial_i\omega^\tau(x)$$

because x does not appear in $\Phi^\tau(y_q)$, in such a way that:

$$u_i^\tau - \partial_i\omega^\tau(x) = -\sum_{|\mu|=q} L_k^{\tau\mu}(\omega(x))\chi_{\mu+1_i}^k \in T^* \otimes F_0.$$

It just remains to notice that $\delta : S_{q+1}T^* \otimes T \longrightarrow T^* \otimes S_qT^* \otimes T$ is a monomorphism.

Q.E.D.

If $\chi_q \in T^* \otimes J_q(T)$, with projections $\chi_0 = id_T \in T^* \otimes T$ and $\chi_1 \in T^* \otimes J_1(T)$, is a representative of an element $c \in A(R_q^0)$, we may choose a lift $\chi_{q+1} \in T^* \otimes J_{q+1}(T)$ and define $\frac{1}{2}\{\chi_q, \chi_q\} \in \wedge^2 T^* \otimes J_q^0(T)$ by the formula:

$$\frac{1}{2}\{\chi_q, \chi_q\}(\xi, \eta) = \{\chi_{q+1}(\xi), \chi_{q+1}(\eta)\} - \chi_q(\{\chi_1(\xi), \chi_1(\eta)\})$$

PROPOSITION 5.42: The map $\chi_q \longrightarrow \frac{1}{2}\{\chi_q, \chi_q\}$ provides a well defined map $c \longrightarrow \frac{1}{2}\{c, c\}$ from elements of $A(R_q^0)$ invariant by R_q^0 to $F_2 = \wedge^2 T^* \otimes J_q^0(T)/(\wedge^2 T^* \otimes R_q^0 + \delta(T^* \otimes S_{q+1}T^* \otimes T))$.

Proof: First of all, if we change the lift χ_{q+1} to $\bar{\chi}_{q+1}$, then $\bar{\chi}_{q+1} - \chi_{q+1} \in T^* \otimes S_{q+1}T^* \otimes T$ and the difference will therefore be in $\delta(T^* \otimes S_{q+1}T^* \otimes T)$ with a zero projection in F_2.

Then, if we modify χ_q by an element $M_q \in \delta(S_{q+1}T^* \otimes T) \subset T^* \otimes S_qT^* \otimes T \subset T^* \otimes J_q(T)$, the difference in $J_{q-1}(T)$ will be only produced by $\{\chi_q(\xi), \chi_q(\eta)\}$ and will be $\{M_q(\xi), \chi_q(\eta)\} + \{\chi_q(\xi), M_q(\eta)\} = \delta M_q(\xi, \eta) = 0$ because $\delta \circ \delta = 0$. However, when lifting M_q at order $q + 1$, new terms will modify the quadratic application and, in particular, even one more if $q = 1$, namely $\chi_1(\{M_1(\xi), \chi_1(\eta)\} + \{\chi_1(\xi), M_1(\eta)\}) = \chi_1(\delta M_1(\xi, \eta)) = \chi_1(0) = 0$ and such a situation will not differ from the general one we now consider. According to the above results while introducing the lift σ_{q+1}^{q-1} of M_q, the only pertubating term to study is:

$$\{\sigma_{q+1}^{q-1}(\xi), \chi_{q+1}(\eta)\} + \{\chi_{q+1}(\xi), \sigma_{q+1}^{q-1}(\eta)\} - M_q(\{\chi_1(\xi), \chi_1(\eta)\}) \in S_qT^* \otimes T$$

and we just need to prove that such a term comes from an element in $\delta(T^* \otimes S_{q+1}T^* \otimes T)$. With more details and $\mid \mu \mid = q - 1$, we get for the $6 = 3 + 3$ factors of ξ^i, η^j:

$$(\chi_{t,i}^r M_{\mu+1_r+1_j}^k - \chi_{t,j}^r M_{\mu+1_r+1_i}^k) + (M_{\mu+1_t+1_i}^r \chi_{r,j}^k - M_{\mu+1_t+1_j}^r \chi_{r,i}^k) - M_{\mu+1_r+1_t}^k(\chi_{i,j}^r - \chi_{j,i}^r)$$

Setting:

$$N_{q+1} = (N_{\mu+1_t+1_j,i}^k = (\chi_{t,i}^r M_{\mu+1_r+1_j}^k + \chi_{j,i}^r M_{\mu+1_r+1_t}^k) - \chi_{r,i}^k M_{\mu+1_t+1_j}^r) \in T^* \otimes S_{q+1}T^* \otimes T$$

where the first sum in the parenthesis insures the symmetry in t/j, a tedious but straightforward though unexpected calculation proves that all the above terms are just described by δN_{q+1}.

As another proof, we may also consider the 3-cyclic sum obtained by preserving μ but replacing (i, j, t) successively by (j, t, i) and (t, i, j) in order to discover that *all the terms disappear two by two*. It is then sufficient to use the exactness of the δ-sequence:

$$0 \to S_{q+2}T^* \otimes T \xrightarrow{\delta} T^* \otimes S_{q+1}T^* \otimes T \xrightarrow{\delta} \wedge^2 T^* \otimes S_qT \otimes T \xrightarrow{\delta} \wedge^3 T^* \otimes S_{q-1}T^* \otimes T$$

We have thus proved that there is a well defined map from $id_T^{-1}(C_1(T))$ to $C_2^0(T)$ which is sending elements of $T^* \otimes J_q(T)/\delta(S_{q+1}T^* \otimes T)$ projecting onto $id_T \in T^* \otimes T$ to $\wedge^2 T^* \otimes J_q^0(T)/\delta(T^* \otimes S_{q+1}T^* \otimes T)$.

Hence it just remains to modify χ_q by $\sigma_q^0 \in T^* \otimes R_q^0$ with lift $\sigma_{q+1}^0 \in T^* \otimes J_{q+1}^0(T)$ and the difference will be:

$$\{\chi_{q+1}(\xi), \sigma_{q+1}^0(\eta)\} + \{\sigma_{q+1}^0(\xi), \chi_{q+1}(\eta)\} + [\sigma_q^0(\xi), \sigma_q^0(\eta)]$$

$$-\chi_q(\{\sigma_1^0(\xi), \chi_1(\eta)\} + \{\chi_1(\chi), \sigma_1^0(\eta)\}) - \sigma_q^0(\{\chi_1(\xi), \chi_1(\eta)\})$$

As the third and the last terms already belong to R_q^0, we just need consider:

$$\{\chi_{q+1}(\xi), \sigma_{q+1}^0(\eta)\} + \{\sigma_{q+1}^0(\xi), \chi_{q+1}(\eta)\} - \chi_q(\delta\sigma_1^0(\xi, \eta))$$

Now we have:

$$\begin{aligned}
i(\eta)L(\sigma_{q+1}^0(\xi))\chi_q &= L(\sigma_{q+1}^0(\xi))\chi_q(\eta) - \chi_q(L(\sigma_1^0(\xi))\eta) \\
&= \{\sigma_{q+1}^0(\xi), \chi_{q+1}(\eta)\} - \chi_q(L(\sigma_1^0(\xi))\eta)
\end{aligned}$$

and the above remainder becomes:

$$i(\eta)L(\sigma_{q+1}^0(\xi))\chi_q - i(\xi)L(\sigma_{q+1}^0(\eta) \in J_q^0(T)$$

because:

$$L(\sigma_1^0(\xi))\eta - L(\sigma_1^0(\eta))\xi - \delta\sigma_1^0(\xi, \eta) = 0$$

Finally, $L(\xi_q^0)c = 0, \forall \xi_q^0 \in R_q^0$ in $F_1 \Leftrightarrow L(\xi_{q+1}^0)\chi_q \in T^* \otimes R_q^0 + \delta(S_{q+1}T^* \otimes T)$. Hence, modifying σ_{q+1}^0 if necessary by an element in $T^* \otimes S_{q+1}T^* \otimes T$, the above remainder belongs to R_q^0.

It follows that $\frac{1}{2}\{\chi_q, \chi_q\}$ will be modified by an element in $\wedge^2 T^* \otimes R_q^0 + \delta(T^* \otimes S_{q+1}T^* \otimes T)$ that will not change the projection in F_2.

$$\text{Q.E.D.}$$

If we start now with any system $R_q \subset J_q(T)$ of infinitesimal Lie equations satisfying $[R_q, R_q] \subset R_q$, we may choose *any* R_q-connection $\chi_q \in id_T^{-1}(T^* \otimes R_q) \subset id_T^{-1}(T^* \otimes J_q(T))$ and obtain by projection an element $c \in A(R_q^0)$ that we use in the following proposition.

PROPOSITION 5.43: $\pi_q^{q+1} : R_{q+1} \longrightarrow R_q$ is surjective if and only if $L(\xi_q)c = 0, \forall \xi_q \in R_q$. In this case, we have $\frac{1}{2}\{c, c\} = 0$ in F_2.

Proof: We have $\chi_0(\eta) = \eta$ and we may find $\xi_{q+1} \in J_{q+1}(T)$ such that:

$$
\begin{aligned}
i(\eta)L(\xi_{q+1})\chi_q &= L(\xi_{q+1})\chi_q(\eta) - \chi_q(L(\xi_1)\eta) \\
&= [\xi_q, \chi_q(\eta)] + i(\eta)D\xi_{q+1} - \chi_q(L(\xi_1)\eta) \in J_q^0(T)
\end{aligned}
$$

As $[R_q, R_q] \subset R_q$ and $\chi_q \in T^* \otimes R_q$, modifying ξ_{q+1} by an element in $S_{q+1}T^* \otimes T$ if necessary, we get at once:

$$
L(\xi_{q+1})\chi_q \in T^* \otimes R_q^0 \Leftrightarrow L(\xi_q)c = 0
$$
$$
\Leftrightarrow \exists \xi_{q+1} \in J_{q+1}(T), \pi_q^{q+1}(\xi_{q+1}) = \xi_q \in R_q, D\xi_{q+1} \in T^* \otimes R_q
$$

Proceeding backwards in the definition of D given at the beginning of section 4, we obtain therefore:

$$
L(\xi_q)c = 0 \Leftrightarrow \exists \xi_{q+1} \in R_{q+1} = \rho_1(R_q), \pi_q^{q+1}(\xi_{q+1}) = \xi_q \in R_q
$$

It follows that $\pi_q^{q+1} : R_{q+1} \longrightarrow R_q$ is surjective and we can choose a lift $\chi_{q+1} \in T^* \otimes R_{q+1}$ over $\chi_q \in T^* \otimes R_q$ in such a way that $\frac{1}{2}\{\chi_q, \chi_q\} \in \wedge^2 T^* \otimes R_q^0 \Rightarrow \frac{1}{2}\{c, c\} = 0$ in F_2.

Q.E.D.

REMARK 5.44: As already noticed, the IC $L(\xi_q)c = 0$ may be obtained in two steps. First, in order to obtain the surjectivity $\pi_q^{q+1} : R_{q+1}^0 \to R_q^0$, we must have $L(\xi_q^0)c = 0, \forall \xi_q^0 \in R_q^0$, that is c *must be* R_q^0-*invariant*. Second, in order to obtain the surjectivity $\pi_0^{q+1} : R_{q+1} \to T$, we must have $L(\chi_q(\xi))c = 0, \forall \xi \in T$ for any R_q-connection χ_q. Finally, when $q = 1$, we have $\frac{1}{2}\{c, c\}(\xi, \eta) = \{\chi_2(\xi), \chi_2(\eta) - \chi_1(\{\chi_1(\xi), \chi_2(\eta)\})$ with:

$$
\frac{1}{2}(\{c, c\})_{l,ij}^k = \chi_{li,j}^k + \chi_{l,i}^r \chi_{l,j}^k - \chi_{l,j}^r \chi_{l,i}^k - \chi_{lj,i}^k - \chi_{l,r}^k(\chi_{i,j}^r - \chi_{j,i}^r)
$$

Modifying χ_2 by an element of $\delta(T^* \otimes S_2 T^* \otimes T) \subset \wedge^2 T^* \otimes T^* \otimes T$ if necessary, then $\frac{1}{2}\{c, c\}$ only depends on $\chi_{l,i}^r \chi_{l,j}^k - \chi_{l,j}^r \chi_{l,i}^k - \chi_{l,r}^k(\chi_{i,j}^r - \chi_{j,i}^r)$ a result coherent with the fact that the Jacobi relations are described by homogeneous polynomials of degree 2 when $q = 1$. Things may become quite different when $q \geq 2$ as even linear relations may exist.

EXAMPLE 5.45: We revisit the tricky Example 4.14 of ([36], p 329-331) while correcting a few printing mistakes. For this, with $n = 2$, let us consider the Lie pseudogroup $\Gamma = \{y^1 = ax^1 + b, y^2 = cx^2 + d \mid a, b, c, d = cst, \ ac = 1\}$. It is easy to find a generating set of 3 differential invariants of order one but the corresponding non-linear system $\mathcal{R}_1 \subset \Pi_1$ in Lie form:

$$\Phi^1 \equiv \frac{y_2^1}{y_1^1} = 0, \quad \Phi^2 \equiv \frac{y_1^2}{y_2^2} = 0, \quad \Phi^3 \equiv y_1^1 y_2^2 = 1,$$

is formally integrable but not involutive because it is finite type and amounts to the following non-linear system $\mathcal{R}_2 \subset \Pi^2$:

$$y_2^1 = 0, \ y_1^2 = 0, \ y_1^1 y_2^2 = 1 \Rightarrow y_{11}^1 = 0, y_{12}^1 = 0, y_{22}^1 = 0, y_{11}^2 = 0, y_{12}^2 = 0, y_{22}^2 = 0$$

which is nevertheless not in Lie form.

Accordingly, we obtain the following non-linear involutive system $\mathcal{R}_2 \subset \Pi_2$ in Lie form:

$$\Phi^1 \equiv \frac{y_2^1}{y_1^1} = 0, \quad \Phi^2 \equiv \frac{y_1^2}{y_2^2} = 0, \quad \Phi^3 \equiv y_1^1 y_2^2 = 1,$$

$$\Phi^4 \equiv \frac{y_{11}^1}{y_1^1} = 0, \quad \Phi^5 \equiv \frac{y_{12}^1}{y_1^1} = 0, \quad \Phi^6 \equiv \frac{y_{22}^1}{y_1^1} = 0,$$

$$\Phi^7 \equiv \frac{y_{22}^2}{y_2^2} = 0, \quad \Phi^8 \equiv \frac{y_{12}^2}{y_2^2} = 0, \quad \Phi^9 \equiv \frac{y_{11}^2}{y_2^2} = 0$$

We have 6 identities of the form:

$$\begin{aligned}
\Phi^3(1 - \Phi^1\Phi^2)\Phi^4 &= -d_2(\Phi^2\Phi^3) + \Phi^2\Phi^3 d_1\Phi^1 + d_1\Phi^3 \\
\Phi^3(1 - \Phi^1\Phi^2)\Phi^5 &= d_1(\Phi^1\Phi^3) - \Phi^1 d_2(\Phi^2\Phi^3) \\
\Phi^3(1 - \Phi^1\Phi^2)\Phi^6 &= -\Phi^1 d_1(\Phi^1\Phi^3) + (\Phi^1)^2 d_2(\Phi^2\Phi^3) + \Phi^3(1 - \Phi^1\Phi^2)d_2\Phi^1
\end{aligned}$$

and the 3 other ones obtained by the permutation below. These formulas are well defined and can be solved because we have the inequality $(1 - \Phi^1\Phi^2)\Phi^3 \equiv y_1^1 y_2^2 - y_2^1 y_1^2 \neq 0$ and we must therefore have the inequality $(1 - \omega^1\omega^2)\omega^3 \neq 0$ for any general system.

It will be important for the next steps to notice thar this system is invariant under the permutation $(1 \leftrightarrow 2)$ providing the permutations:

$$x^1 \leftrightarrow x^2, y^1 \leftrightarrow y^2 \Leftrightarrow \Phi^1 \leftrightarrow \Phi^2, \phi^3 \leftrightarrow \Phi^3, \Phi^4 \leftrightarrow \Phi^7, \Phi^5 \leftrightarrow \Phi^8, \phi^6 \leftrightarrow \Phi^9$$

with similar changes for the corresponding geometric object ω.

We provide details on the construction of the corresponding natural bundle. For this, we exhibit the following formulas and corresponding Lie forms:

$$u^1 = \frac{y_2^1}{y_1^1} = \frac{\bar{y}_1^1 \partial_2 \varphi^1 + \bar{y}_2^1 \partial_2 \varphi^2}{\bar{y}_1^1 \partial_1 \varphi^1 + \bar{y}_2^1 \partial_1 \varphi^2} = \frac{\partial_2 \varphi^1 + \bar{u}^1 \partial_2 \varphi^2}{\partial_1 \varphi^1 + \bar{u}^1 \partial_1 \varphi^2} \Rightarrow \Phi_\omega^1(y_1) \equiv \frac{y_2^1 + \omega^1(y)y_2^2}{y_1^1 + \omega^1(y)y_1^2} = \omega^1(x)$$

$$\begin{aligned} u^3 = y_1^1 y_2^2 &= (\bar{y}_1^1 \partial_1 \varphi^1 + \bar{y}_2^1 \partial_1 \varphi^2)(\bar{y}_1^2 \partial_2 \varphi^1 + \bar{y}_2^2 \partial_2 \varphi^2) \\ &= \bar{u}^2 \bar{u}^3 \partial_1 \varphi^1 \partial_2 \varphi^1 + \bar{u}^3 \partial_1 \varphi^1 \partial_2 \varphi^2 + \bar{u}^1 \bar{u}^2 \bar{u}^3 \partial_1 \varphi^2 \partial_2 \varphi^1 + \bar{u}^1 \bar{u}^3 \partial_1 \varphi^2 \partial_2 \varphi^2) \end{aligned}$$

$$\Rightarrow \quad \Phi_\omega^3(y_1) \equiv \omega^2(y)\omega^3(y)y_1^1 y_2^1 + \omega^3(y)y_1^1 y_2^2 + \omega^1(y)\omega^2(y)\omega^3(y)y_1^2 y_2^1 + \omega^1(y)\omega^3(y)y_1^2 y_2^2 = \omega^3(x)$$

We let the reader transform the differential invariants at order 2 in order to convince him about the difficulty of the computations involved as he will obtain for example:

$$\Phi_\omega^5(y_2) \equiv \frac{y_{12}^1 + \omega^1(y)y_{12}^2 + \omega^4(y)y_1^1 y_2^1 + \omega^5(y)(y_1^1 y_2^2 + y_2^1 y_1^2) + \omega^6(y)y_1^2 y_2^2}{y_1^1 + \omega^1(y)y_1^2} = \omega^5(x)$$

In a coherent way, the linearized system $R_1 \subset J_1(T)$ defined by $\xi_2^1 = 0, \xi_1^2 = 0, \xi_1^1 + \xi_2^2 = 0$ is finite type with a zero symbol at order 2 and cannot therefore be involutive, even if it is trivially formally integrable as it is made with homogeneous equations. It has 2 equations of order 2 and class 2, 4 equations of order 2 and class 1 and 3 equations of order one. Accordingly the Janet sequence is:

$$0 \to \Theta \to 2 \longrightarrow 9 \longrightarrow 10 \longrightarrow 3 \to 0$$

and we check the Euler-Poincaré formula $2 - 9 + 10 - 3 = 0$.

As a motivation for creating symbolic algebra packages, a straightforward but tedious computation provides the transition laws of the underlying natural bundle with local coordinates:

$$(x^1, x^2, u^1, u^2, u^3, u^4, u^5, u^6, u^7, u^8, u^9)$$

and the corresponding general Medolaghi equations $\Omega \equiv L(\xi_2)\omega = 0$:

$$\Omega^1 \equiv \xi_2^1 + \omega^1\xi_2^2 - \omega^1\xi_1^1 - (\omega^1)^2\xi_1^2 + \xi^r\partial_r\omega^1 = 0$$
$$\Omega^2 \equiv \xi_1^2 + \omega^2\xi_1^1 - \omega^2\xi_2^2 - (\omega^2)^2\xi_2^1 + \xi^r\partial_r\omega^2 = 0$$
$$\Omega^3 \equiv \omega^3(\xi_1^1 + \xi_2^2) + \omega^1\omega^3\xi_1^2 + \omega^2\omega^3\xi_2^1 + \xi^r\partial_r\omega^3 = 0$$
$$\Omega^4 \equiv \xi_{11}^1 + \omega^1\xi_{11}^2 + \omega^4\xi_1^1 + (2\omega^5 - \omega^1\omega^4)\xi_1^2 + \xi^r\partial_r\omega^3 = 0$$
$$\Omega^5 \equiv \xi_{12}^1 + \omega^1\xi_{12}^2 + \omega^4\xi_2^1 + (\omega^6 - \omega^1\omega^5)\xi_1^2 + \omega^5\xi_2^2 + \xi^r\partial_r\omega^5 + 0$$
$$\Omega^6 \equiv \xi_{22}^1 + \omega^1\xi_{22}^2 + 2\omega^6\xi_2^2 + 2\omega^5\xi_2^1 - \omega^6\xi_1^1 - \omega^1\omega^6\xi_1^2 + \xi^r\partial_r\omega^6 = 0$$
$$\Omega^7 \equiv \xi_{22}^2 + \omega^2\xi_{22}^1 + \omega^7\xi_2^2 + (2\omega^8 - \omega^2\omega^7)\xi_2^1 + \xi^r\partial_r\omega^7 = 0$$
$$\Omega^8 \equiv \xi_{12}^2 + \omega^2\xi_{12}^1 + \omega^7\xi_1^2 + (\omega^9 - \omega^2\omega^8)\xi_2^1 + \omega^8\xi_1^1 + \xi^r\partial_r\omega^8 = 0$$
$$\Omega^9 \equiv \xi_{11}^2 + \omega^2\xi_{11}^1 + 2\omega^9\xi_1^2 + 2\omega^8\xi_1^1 - \omega^8\xi_2^2 - \omega^2\omega^9\xi_2^1 + \xi^r\partial_r\omega^9 = 0$$

We check that this general system $R_2(\omega) \subset J_2(T)$ is invariant under the permutation $(1 \leftrightarrow 2)$.

Using the formal derivative $d_i\xi_\mu^k = \xi_{\mu+1_i}^k$, $d_ia(x) = \partial_ia(x)$ in order to study the induced map $R_3^0 \xrightarrow{\pi_2^3} R_2^0 \xrightarrow{\pi_1^2} R_1^0$, we obtain in particular:

$$d_2\Omega^3 - \omega^3\Omega^5 - \omega^3\Omega^7 - (\omega^5 + \omega^7)\Omega^3 =$$
$$(\omega^2(\partial_2\omega^3 - \omega^3(\omega^5 + \omega^7)) + \omega^3(\partial_2\omega^2 - \omega^8 + \omega^2\omega^7) + (\partial_1\omega^3 - \omega^3(\omega^4 + \omega^8)))\xi_2^1$$
$$+(\omega^1(\partial_2\omega^3 - \omega^3(\omega^5 + \omega^7)) + \omega^3(\partial_2\omega^1 - \omega^6 + \omega^1\omega^5))\xi_1^2$$
$$+(\partial_2\omega^3 - \omega^3(\omega^5 + \omega^7))\xi_2^2$$

whenever $\xi_1 \in R_1^0$ that is when we have $\xi^1 = 0, \xi^2 = 0$ and:

$$\xi_2^1 + \omega^1\xi_2^2 - \omega^1\xi_1^1 - (\omega^1)^2\xi_1^2 = 0$$

$$\xi_1^2 + \omega^2\xi_1^1 - \omega^2\xi_2^2 - (\omega^2)^2\xi_2^1 = 0$$

$$\xi_1^1 + \xi_2^2 + \omega^1\xi_1^2 + \omega^2\xi_2^1 = 0$$

leading in particular to:

$$\xi_1^1 + \xi_2^2 = 0, \quad \omega^1\xi_1^2 + \omega^2\xi_2^1 = 0$$

It follows that we have indeed:

$$d_2\Omega^3 - \omega^3\Omega^5 - \omega^3\Omega^7 - (\omega^5 + \omega^7)\Omega^3 \equiv \omega^3(\partial_2\omega^1 - \omega^6 + \omega^1\omega^5))\xi_1^2$$
$$+\omega^3(\partial_2\omega^2 - \omega^8 + \omega^2\omega^7)\xi_2^1$$
$$+(\partial_1\omega^3 - \omega^3(\omega^4 + \omega^8))\xi_2^1$$
$$+(\partial_2\omega^3 - \omega^3(\omega^5 + \omega^7))\xi_2^2$$

Using the permutation induced by $1 \leftrightarrow 2$, we may work out the 6 different linear equations which describe the fact that the 6 second order equations are coming from the derivatives of the first order equations. Accordingly, we obtain the 6 first order structure equations:

$$\partial_1\omega^1 - \omega^5 + \omega^1\omega^4 = 0 \qquad \partial_2\omega^1 - \omega^6 + \omega^1\omega^5 = 0$$
$$\partial_1\omega^2 - \omega^9 + \omega^2\omega^8 = 0 \qquad \partial_2\omega^2 - \omega^8 + \omega^2\omega^7 = 0$$
$$\partial_1\omega^3 - \omega^3(\omega^4 + \omega^8) = 0 \qquad \partial_2\omega^3 - \omega^3(\omega^5 + \omega^7) = 0$$

Replacing ω by Φ we check indeed that the linear part of the 6 equations with respect to $(\Phi^1, ..., \Phi^9)$ has coefficients only depending on (Φ^1, Φ^2, Φ^3) and determinant just equal to $\Phi^3(1 - \Phi^1\Phi^2)$ which cannot vanish, in a coherent way with the formulas already obtained for these invariants. Such a situation is absolutely similar to the one existing with the Christoffel symbols through the Levi-Civita isomorphism $j_1(\omega) \simeq (\omega, \gamma)$ in the case of a riemannian structure with $n = 2$.

It is quite more difficult to find the remaining 4 structure equations. For this, we notice that:

$$d_2\Omega^4 - d_1\Omega^5 \equiv (\partial_2\omega^1 - \omega^6 + \omega^1\omega^5)\xi_{11}^2 - (\partial_1\omega^1 - \omega^5 + \omega^1\omega^4)\xi_{12}^2$$
$$+(\partial_2(\omega^5 - \omega^1\omega^4) - \partial_1(\omega^6 - \omega^1\omega^5))\xi_1^2$$
$$(\partial_2\omega^4 - \partial_1\omega^5)(\xi_1^1 + \xi_2^2) + \xi^r\partial_r(\partial_2\omega^4 - \partial_1\omega^5)$$

Taking into acount the previous structure equations and the fact that $\partial_1(\partial_2\omega^1) - \partial_2(\partial_1\omega^1) = 0$, we finally get:

$$d_2\Omega^4 - d_1\Omega^5 \equiv (\partial_2\omega^4 - \partial_1\omega^5)(\xi_1^1 + \xi_2^2) + \xi^r\partial_r(\partial_2\omega^4 - \partial_1\omega^5) = 0$$

and a similar result obtained by permutation:

$$d_1\Omega^7 - d_2\Omega^8 \equiv (\partial_1\omega^7 - \partial_2\omega^8)(\xi_1^1 + \xi_2^2) + \xi^r\partial_r(\partial_1\omega^7 - \partial_2\omega^8) = 0$$

However, we have:

$$\Phi^3(1-\Phi^1\Phi^2) \equiv y_1^1 y_2^2 - y_2^1 y_1^2 = 1 \Rightarrow \omega^3(1-\omega^1\omega^2)(\xi_1^1+\xi_2^2)+\xi^r\partial_r(\omega^3(1-\omega^1\omega^2)) = 0$$

As we have already seen in the introduction while taking into account the permutation induced by $(1 \leftrightarrow 2)$, the only possibility is to have the 2 structure equations:

$$\partial_2\omega^4 - \partial_1\omega^5 = c_1\,\omega^3(1-\omega^1\omega^2), \quad \partial_1\omega^7 - \partial_2\omega^8 = c_2\,\omega^3(1-\omega^1\omega^2)$$

with the structure constants $c = (c_1, c_2)$.

It remains therefore to exhibit the two remaining structure equations that *must exist*. For this, we have to take into account that the system of CC for the involutive Lie equations *must also be involutive and thus formally integrable*, but also invariant by $(1 \leftrightarrow 2)$. For this, we have to consider the following 3 crossed derivatives that can be exhibited:

$$\partial_2(\omega^5 - \omega^1\omega^4) - \partial_1(\omega^6 - \omega^1\omega^5) = 0$$

$$\partial_1(\omega^8 - \omega^2\omega^7) - \partial_2(\omega^9 - \omega^2\omega^8) = 0$$

$$\partial_2(\omega^4 + \omega^8) - \partial_1(\omega^5 + \omega^7) = 0$$

The two first equations are the desired last two structure equations we were looking for while the third equation provides the unique generalized Jacobi condition $c_1 - c_2 = 0 \Rightarrow c_1 = c_2 = c$.

We let the reader check that the Lie pseudogroup:

$$\bar\Gamma = \{y^1 = \frac{ax^1 + b}{cx^1 + d}, y^2 = \frac{ax^2 + b}{cx^2 + d} \mid a, b, c, d = cst\}$$

is provided by the new specialization:

$$\bar\omega^1 = 0, \bar\omega^2 = 0, \bar\omega^3 = \frac{1}{(x^2 - x^1)^2},$$

$$\bar\omega^4 = \frac{2}{x^2 - x^1}, \bar\omega^5 = 0, \bar\omega^6 = 0, \bar\omega^7 = \frac{2}{x^1 - x^2}, \bar\omega^8 = 0, \bar\omega^9 = 0$$

giving $\bar c_1 = \bar c_2 = \bar c = -2$ while the previous specialization:

$$\omega^1 = 0, \omega_2 = 0, \omega^3 = 1, \omega^4 = 0, \omega^5 = 0, \omega_6 = 0, \omega^7 = 1, \omega^8 = 0, \omega^9 = 0$$

is giving $c_1 = c_2 = c = 0$.

It follows from these results that Γ is of codmension 1 in the connected component of the identity $\tilde{\Gamma} = \{y^1 = ax^1 + b, y^2 = cx^2 + d \mid a, b, c, d = cst\}$ of its normalizer which is also containing the transformation $(y^1 = x^2, y^2 = x^1)$ while $\tilde{\tilde{\Gamma}} = \bar{\Gamma}$ is its own normalizer with the same comment. We let the reader study similarly the case of a riemannian structure when $n = 2$ because all the dimensions of the corresponding systems and Janet bundles are the same but now one of the two structure constants must vanish ([30]). It is nevertheless important to notice that, in each case, if we keep only the differential invariants defining the corresponding first order system, there is only *one* integrability condition of order 2 with a *single* structure constant c which is the scalar cuvature in the riemannian case. In the present situation, it is sufficient to use the previous formulas giving ω^4 and ω^5 as rational functions of $\omega^1, \omega^2, \omega^3$ and their first derivatives before substituting them into $\partial_2\omega^4 - \partial_1\omega^5 = c\,\omega^3(1 - \omega^1\omega^2)$. This example perfectly explains why all these new concepts are still unknown.

The following definition only depends on R_q^0:

DEFINITION 5.46: A *truncated Lie algebra* is a couple (R_q^0, c) where $R_q^0 \subset J_q^0(T)$ is such that $[R_q^0, R_q^0] \subset R_q^0$ and $c \in Cart(R_q^0) = \{c \in A(R_q^0) \mid c$ is R_q^0-invariant, $\frac{1}{2}\{c, c\} = 0$ in $F_2\}$.

THEOREM 5.47: The infinitesimal deformation of $Cart(R_q) = \{c \in A(R_q^0) | c$ is R_q-invariant, $\frac{1}{2}\{c, c\} = 0\}$ in F_2 is just $Z_1(R_q)$.

Proof: If $c_t = c + tC + \ldots$ is a deformation of c and $\chi_q(t) = \chi_q + tX_q^0 + \ldots$is a representative of c_t, we must have $C \in F_1$ and $X_q^0 \in T^* \otimes J_q^0(T)$. Denoting simply by $\{c, C\} = 0$ the linearization of $\frac{1}{2}\{c, c\} = 0$, we must have $\{\chi_q, X_q^0\} \in \wedge^2 T^* \otimes R_q^0 + \delta(T^* \otimes S_{q+1}T^* \otimes T)$ and get successively with different lifts:

$$\{\chi_{q+1}(\xi), X_{q+1}^0(\eta)\} + \{X_{q+1}^0(\xi), \chi_{q+1}(\eta)\} - X_q^0(\{\chi_1(\xi), \chi_1(\eta)\})$$

$$-\chi_q(\{X_1^0(\xi), \chi_1(\eta)\} + \{\chi_1(\xi), X_1^0(\eta)\})$$

$$= L(\chi_q(\xi))X_q^0(\eta) - L(\chi_q(\eta))X_q^0(\xi)$$
$$-i(\xi)DX_{q+1}^0(\eta) + i(\eta)DX_{q+1}^0(\xi) - X_q^0(\{\}) - \chi_q()$$
$$= i(\eta)L(\chi_q(\xi))X_q^0 - i(\xi)L(\chi_q(\eta))X_q^0 - DX_{q+1}^0(\xi, \eta) - \chi_q()$$

where we have used the relation:

$$i(\eta)L(\chi_q(\xi))X_q^0 = L(\chi_q(\xi))X_q^0(\eta) - X_q^0(L(\chi_1(\xi)\eta)$$

with $L(\chi_1(\xi))\eta = [\xi, \eta] + i(\eta)D\chi_1(\xi)$ and the relation:

$$i(\eta)DX_{q+1}^0(\xi) - i(\xi)DX_{q+1}^0(\eta) + X_q^0([\xi, \eta]) = DX_{q+1}^0(\xi, \eta)$$

that is finally:

$$\{\chi_q, X_q^0\}(\xi, \eta) + DX_{q+1}^0(\xi, \eta) = i(\eta)\nabla_\xi X_q^0 - i(\xi)\nabla_\eta X_q^0 - \chi_q()$$

Now $L(\xi_q)C = 0, \forall \xi_q \in R_q = R_q^0 \oplus \chi_q(T) \Rightarrow \nabla_\xi C = 0$ in $F_1, \forall \xi \in T \Rightarrow \nabla_\xi X_q^0 \in T^* \otimes R_q^0 + \delta(S_{q+1}T^* \otimes T)$ and thus: $\{\chi_q, X_q^0\} + DX_{q+1}^0 \in \wedge^2 T^* \otimes R_q$. Modifying if necessary the lift X_{q+1}^0, we need only $DX_{q+1}^0 \in \wedge^2 T^* \otimes R_q$, a result leading to $L(\xi_q)C = 0, \forall \xi_q \in R_q$, that is $C \in \Upsilon_1$ and $\mathcal{D}_2 C = 0$ in F_2.

$$\text{Q.E.D.}$$

As the F_r only depend on R_q^0 both with the sub-bundles $E_r \subset F_r$, we finally obtain ([37], p 721):

COROLLARY 5.48: The deformation cohomology $H_r(R_q)$ only depends on the truncated Lie algebra (R_q^0, c).

We now provide a few more comments on the differential geometry of infinitesimal Lie equations but the first point is the following purely algebraic result:

DEFINITION 5.49: A *ring* A is a non-empty set with two associative binary operations called *addition* and *multiplication*, respectively sending $a, b \in A$ to $a + b \in A$ and $ab \in A$ in such a way that A becomes an abelian group for the multiplication, so that A has a zero element denoted by 0, every $a \in A$ has an additive inverse denoted by $-a$ and the multiplication is distributive over the addition, that is to say $a(b + c) = ab + ac, (a + b)c = ac + bc, \forall a, b, c \in A$.

A ring A is said to be *unitary* if it has a (unique) element $1 \in A$ such that $1a = a1 = a, \forall a \in A$ and *commutative* if $ab = ba, \forall a, b \in A$.

A non-zero element $a \in A$ is called a *zero-divisor* if one can find a non-zero $b \in A$

such that $ab = 0$ and a ring is called an *integral domain* if it has no zero-divisor.

A ring K is called a *field* if every non-zero element $a \in K$ is a *unit*, that is one can find an element $b \in K$ such that $ab = 1 \in K$.

DEFINITION 5.50: A *module* M over a ring A or simply an *A-module* is a set of elements $x, y, z, \ldots$ which is an abelian group for an addition $(x, y) \to x + y$ with an action $A \times M \to M : (a, x) \to ax$ satisfying:

- $\quad a(x + y) = ax + ay, \forall a \in A, \forall x, y \in M$
- $\quad a(bx) = (ab)x, \forall a, b \in A, \forall x \in M$
- $\quad (a + b)x = ax + bx, \forall a, b \in A, \forall x \in M$
- $\quad 1x = x, \forall x \in M$

The set of modules over a ring A will be denoted by $mod(A)$. A module over a field is called a *vector space*.

DEFINITION 5.51: A map $f : M \to N$ between two A-modules is called a *homomorphism* over A if $f(x + y) = f(x) + f(y), \forall x, y \in M$ and $f(ax) = af(x), \forall a \in A, \forall x \in M$. We successively define:

- $\quad ker(f) = \{x \in M | f(x) = 0\}$
- $\quad coim(f) = M/ker(f)$
- $\quad im(f) = \{y \in N | \exists x \in M, f(x) = y\}$
- $\quad coker(f) = N/im(f)$

with an isomorphism $coim(f) \simeq im(f)$ induced by f.

DEFINITION 5.52: We say that a chain of modules and homomorphisms is a *sequence* if the composition of two successive such homomorphisms is zero. A sequence is said to be *exact* if the kernel of each map is equal to the image of the map preceding it. An injective homomorphism is called a *monomorphism*, a surjective homomorphism is called an *epimorphism* and a bijective homomorphism is called an *isomorphism*. A short exact sequence is an exact sequence made by a monomorphism followed by an epimorphism.

The following proposition will provide the reason for introducing connections:

PROPOSITION 5.53: If one has a short exact sequence:

$$0 \longrightarrow M' \xrightarrow{f} M \xrightarrow{g} M'' \longrightarrow 0$$

then the following conditions are equivalent:
- There exists an epimorphism $u : M \to M'$ such that $u \circ f = id_{M'}$.
- There exists a monomorphism $v : M'' \to M$ such that $g \circ v = id_{M''}$.
- There are maps $u : M \to M'$ and $v : M'' \to M$ such that $f \circ u + v \circ g = id_M$ and this relation provides an isomorphism $(u, g) : M \to M' \oplus M''$ with inverse $f + v : M' \oplus M'' \to M$.

Proof: Let us construct the morphism $h = id_M - f \circ u : M \to M$ and notice that $h \circ f = f - f = 0$. Hence, if $x'' \in M''$, we may find $x \in M$ with $g(x) = x''$ because g is an epimorphism. The map $v : x'' \to h(x) : M'' \to M$ is well defined because $g(x_1) = g(x_2) = x'' \Rightarrow \exists x' \in M', x_1 - x_2 = f(x') \Rightarrow g(x_1) = g(x_2)$ and we have $v(x'') = x - f \circ u(x) \Rightarrow g \circ v(x'') = g(x) = x''$ with $h(x) = x - f \circ u(x) = v(x') = v \circ g(x), \forall x \in M$. The study of v is similar.

Q.E.D.

DEFINITION 5.54: In the above situation, we say that the short exact sequence *splits* and $u(v)$ is called a *lift* for $f(g)$. The short exact sequence $0 \to \mathbb{Z} \to \mathbb{Q} \to \mathbb{Q}/\mathbb{Z} \to 0$ cannot split over $\mathbb{Z}$.

The second point is the following commutative and exact diagram allowing to define F_1 when $R_q \subset J_q(E)$ is a system of order q on E and $R_{q+1} = \rho_1(R_q) = J_1(R_q) \cap J_{q+1}(E) \subset J_1(J_q(E))$ is its first prolongation:

$$
\begin{array}{ccccccc}
0 & & 0 & & 0 & & 0 \\
\downarrow & & \downarrow & & \downarrow & & \downarrow \\
0 \to g_{q+1} & \longrightarrow & S_{q+1}T^* \otimes E & \xrightarrow{\sigma_1(\Phi)} & T^* \otimes F_0 & \longrightarrow & F_1 \to 0 \\
\downarrow & & \downarrow & & \downarrow & & \| \\
0 \to R_{q+1} & \longrightarrow & J_{q+1}(E) & \xrightarrow{\rho_1(\Phi)} & J_1(F_0) & \longrightarrow & F_1 \to 0 \\
\downarrow & & \downarrow \pi_q^{q+1} & & \downarrow \pi_0^1 & & \downarrow \\
0 \to R_q & \longrightarrow & J_q(E) & \xrightarrow{\Phi} & F_0 & \longrightarrow & 0 \to 0 \\
& & \downarrow & & \downarrow & & \\
& & 0 & & 0 & &
\end{array}
$$

PROPOSITION 5.55: There exist a long exact sequence:

$$0 \to g_{q+1} \longrightarrow R_{q+1} \longrightarrow R_q \xrightarrow{\kappa} F_1$$

where the map κ is called the *curvature* of R_q.

Proof: Picking a section $\xi_q \in R_q$, we can consider it as a section of $J_{q+1}(E)$ and lift it up to a section $\xi_{q+1} \in J_{q+1}(E)$ in order to obtain a section $\eta_1 = \rho_1(\Phi)(\xi_{q+1}) \in J_1(F_0)$ projecting to zero in F_0. Hence, we have obtained a section of $T^* \otimes F_0$ that we may project to a section of F_1 which is well defined as it will not depend on the previous lift. Indeed, if we choose another section $\bar{\xi}_{q+1}$ over ξ_q, then $\bar{\xi}_{q+1} - \xi_{q+1} \in S_{q+1}T^* \otimes E$ gives zero projection in $F_1 = coker(\sigma_1(\Phi))$. An elementary chase backwards which is left to the reader as an exercise allows to prove that $\kappa(\xi_q) = 0 \Leftrightarrow \exists \xi_{q+1} \in R_{q+1}$ with $\xi_q = \pi_q^{q+1}(\xi_{q+1})$.

Q.E.D.

Our purpose is to combine these two preceding comments when $E = T$ and $R_q \subset J_q(T)$ is a system of infinitesimal transitive Lie equations, that is $[R_q, R_q] \subset R_q]$ and $\chi_q \in id_T^{-1}(T^* \otimes R_q) \subset id_T^{-1}(T^* \otimes J_q(T))$ is a R_q-connection. Indeed, in this case, we have the split short exact sequence:

$$0 \to R_q^0 \longrightarrow R_q \xrightarrow{\chi_q} T \to 0$$

and obtain therefore an isomorphism $R_q \simeq R_q^0 \oplus T$, that is any section $\xi_q \in R_q$ can be written as $\xi_q = \xi_q^0 + \chi_q(\xi)$ where $\xi_q^0 \in R_q^0$ and $\xi = \pi_0^q(\xi_q) \in T$. It remains to establish a link between the connection χ_q and the curvature κ.

As a first step towards the solution of this problem, we have the following proposition which is not evident at all in actual practice and that we recall independently of Corollary 5.8:

PROPOSITION 5.56: When $R_q \subset J_q(T)$ is a system of infinitesimal Lie equations of order q, then $R_{q+1} \subset J_{q+1}(T)$ is a system of infinitesimal Lie equations of order $q + 1$, *even if R_q is not formally integrable.*

Proof: One just need to use the delicate technical Proposition 5.7 with any $\xi_{q+1}, \eta_{q+1} \in R_{q+1}$ because $D\xi_{q+1}, D\eta_{q+1} \in T^* \otimes R_q$ in order to conclude that $D[\xi_{q+1}, \eta_{q+1}] \in T^* \otimes R_q$ because $[\xi_{q+1}, \eta_{q+1}] \in J_{q+1}(T)$ projects onto $[\xi_q, \eta_q] \in R_q$.

$$Q.E.D.$$

EXAMPLE 5.57: (See Example 5.1) If $n = 2, q = 1$, let $R_1 = \{\xi_1 \subset J_1(T) \mid L(\xi_1)\alpha = 0\}$ where $\alpha = x^2 dx^1 \in T^*$ be defined by the first order equations $x^2\xi_1^1 + \xi^2 = 0, \xi_2^1 = 0$. Then $R_2 \subset J_2(T)$ is defined by *adding* the new equations $\xi_1^1 + \xi_2^2 = 0, \xi_{12}^1 = 0, \xi_{22}^1 = 0, x^2\xi_{11}^1 + \xi_1^2 = 0$ and $\pi_1^2 : R_2 \to R_1$ is not surjective. However, we have $[R_1, R_1] \subset R_1 \Rightarrow [R_2, R_2] \subset R_2$ and thus $[R_1^{(1)}, R_1^{(1)}] \subset R_1^{(1)}$ where $R_1^{1)} = \{\xi_1 \subset J_1(T) \mid L(\xi_1)\alpha = 0, L(\xi_1)\beta = 0\}$ with $\beta = dx^1 \wedge dx^2 = -d\alpha \in \wedge^2 T^*$.

The following commutative diagram is extending the diagram provided before Remark 3.6:

$$
\begin{array}{ccccc}
0 & 0 & 0 & 0 \\
\downarrow & \downarrow & \downarrow & \downarrow \\
0 \to g_{q+1} \to R^0_{q+1} \to & R^0_q & \xrightarrow{\kappa} & F_1 \\
\| & \downarrow & \downarrow & \| \\
0 \to g_{q+1} \to R_{q+1} \to & R_q & \xrightarrow{\kappa} & F_1 \\
\downarrow & \downarrow & \downarrow\uparrow \chi_q & \downarrow \\
0 \to T = & T & \to & 0 \\
& & \downarrow & \\
& & 0 &
\end{array}
$$

and we have a map $\xi_q \in R_q \to \kappa(\xi_q) = L(\xi_q)c \in F_1$ with a map $\xi \in T \to L(\chi_q(\xi))c \in F_1$ which does not depend on the R_q-connection χ_q only if the induced map $\xi^0_q \in R^0_q \to L(\xi^0_q)c \in F_1$ is vanishing. Of course the first map vanishes if and only if the two other maps vanish.

Now, according to Proposition 5.52, we may associate to any section $\xi_q \in R_q$ the section $\xi^0_q = \xi_q - \chi_q(\xi) \in R^0_q$ and we get with any other section $\eta_q \in R_q$:

$$[\xi_q, \eta_q] = [\xi^0_q + \chi_q(\xi), \eta^0_q + \chi_q(\eta)] = [\xi^0_q, \eta^0_q] + [\xi^0_q, \chi_q(\eta)] + [\chi_q(\xi), \eta^0_q] + [\chi_q(\xi), \chi_q(\eta)]$$

Among these four terms, only the first is defined purely algebraically, that is to say fiber by fiber, or, equivalently, does not involve any differentiation. Hence, if $R^0_q \subset J^0_q(T)$ is given with $[R^0_q, R^0_q] \subset R^0_q$, the main problem is to *realize* R_q as a direct sum by choosing nicely the R_q-connection, in particular with the hope that R_q should be formally integrable or, at least that $\pi^{q+1}_q : R_{q+1} \longrightarrow R_q$ should be an epimorphism. For this, first of all, we notice that $[\xi^0_q, \chi_q(\eta)] \in R^0_q$ and $[\chi_q(\xi), \eta^0_q] \in R^0_q$ and thus we must have $[R^0_q, \chi_q(T)] \in R^0_q$. However, it folows from the Definition 5.4 and Proposition 5.10 that $J_q(T)$ is associated with $J_{q+1}(T)$ while $J^0_q(T)$ is associated with $J_q(T)$ and thus R^0_q is associated with $R_q = R^0_q \oplus \chi_q(T)$ with $L(\chi_q(\xi)\eta^0_q = [\chi_q(\xi), \eta^0_q] \in J^0_q(T)$. As for the final term, it is over $[\xi, \eta] \in T$ and we must have therefore $[\chi_q(\xi), \chi_q(\eta)] - \chi_q([\xi, \eta]) \in R^0_q$. We have the following technical lemma:

LEMMA 5.58: For any connection $\chi \in id_T^{-1}(T^* \otimes J_q(T))$, we have the following equality in $\wedge^2 T^* \otimes J^0_q(T)$:

82

$$[\chi_q(\xi), \chi_q(\eta)] - \chi_q([\xi, \eta]) = \{\chi_{q+1}(\xi), \chi_{q+1}(\eta)\} + D\chi_{q+1}(\xi, \eta) \quad, \forall \xi, \eta \in T$$

which is *bilinear* in $\xi, \eta \in T$ and *does not depend on the lift* $\chi_{q+1} \in id_T^{-1}(T^* \otimes J_{q+1}(T))$ of χ_q.

Proof: From the definition of the differential bracket, we have:

$$[\chi_q(\xi), \chi_q(\eta)] = \{\chi_{q+1}(\xi), \chi_{q+1}(\eta)\} + i(\xi)D(\chi_{q+1}(\eta)) - i(\eta)D(\chi_{q+1}(\xi))$$

and the right member of the equality does not depend on the lift of χ_q. However, we have:

$$\xi^r(\partial_r(\chi_{\mu,s}^k \eta^s) - \chi_{\mu+1_r,s}^k \eta^s) = \chi_{\mu,s}^k \xi^r \partial_r \eta^s + (\partial_r \chi_{\mu,s}^k - \chi_{\mu+1_r,s}^k)\xi^r \eta^s$$

and the last term on the right exactly describes the extension $D : T^* \otimes J_{q+1}(T) \longrightarrow \wedge^2 T^* \otimes J_q(T)$ of the Spencer operator $D : J_{q+1}(T) \longrightarrow T^* \otimes J_q(T)$. The Lemma follows by skewsymetrization and substraction.

Q.E.D.

DEFINITION 5.59: This bilinear form is called the *curvature* κ_q of the connection χ_q and comes from the Frobenius type system considered in Lemma 5.21. However, in the present situation, we have $\kappa_q(\xi, \eta) \in R_q^0$, that is $\kappa_q \in \wedge^2 T^* \times R_q^0$ and the curvature may theferore not vanish as a section of $\wedge^2 T^* \otimes R_q^0$. However, when the Lie equation $R_q \subset J_q(T)$ is given, using a section a of the finite Lie form $\mathcal{R}_q \subset \Pi_q$ over Y (care), we obtain a map $T(a) : T(Y) \to V(\mathcal{R}_q)$ that we can compose with the inverse of the isomorphism $\sharp : R_q(Y) \to V(\mathcal{R}_q)$ in order to obtain a R_q-connection over the target. It thus follows from §4 of Section 4 and the Frobenius theorem that this connection has zero curvature over the target. The link existing between the previous formula and the second non-linear Spencer operator has been fully described in ([39], Remark 12, p 226) but it is another story. Collecting all the previou results, we obtain:

THEOREM 5.60: In order that $R_q = R_q^0 \oplus \chi_q(T)$ become a system of infinitesimal Lie equations, we need:

1) $[R_q^0, R_q^0] \subset R_q^0$.

2) R_q^0 must be invariant by $\chi_q(T)$, that is $L(\chi_q(T))R_q^0 \subset R_q^0$.

3) The curvature κ_q of χ_q must belong to $\wedge^2 T^* \otimes R_q^0$ but need not vanish.

It finally remains to study a last but delicate problem, namely to compare the deformation cohomology of R_q to that of R_{q+1}. First of all, as R_q is supposed to be involutive in order to construct the corresponding Janet sequence with Janet bundles F_r, then R_{q+1} is of course involutive too and we may construct the corresponding Janet sequence with Janet bundles F_r'. The link between the two Janet sequences is described by the next theorem.

THEOREM 5.61: There is the following commutative diagram with exact columns and locally exact top row:

$$
\begin{array}{ccccccccccc}
 & & & & 0 & & 0 & & & & 0 \\
 & & & & \downarrow & & \downarrow & & & & \downarrow \\
 & & & 0 \to & K_0 & \to & K_1 & \to \ldots \to & & K_n & \to 0 \\
 & & & & \downarrow & & \downarrow & & & & \downarrow \\
0 \to & \Theta & \to & T \xrightarrow{\mathcal{D}'} & F_0' & \xrightarrow{\mathcal{D}_1'} & F_1' & \xrightarrow{\mathcal{D}_2'} \ldots \xrightarrow{\mathcal{D}_n'} & & F_n' & \to 0 \\
 & \| & & \| & \downarrow \Psi_0 & & \downarrow \Psi_1 & & & \downarrow \Psi_n & \\
0 \to & \Theta & \to & T \xrightarrow{\mathcal{D}} & F_0 & \xrightarrow{\mathcal{D}_1} & F_1 & \xrightarrow{\mathcal{D}_2} \ldots \xrightarrow{\mathcal{D}_n} & & F_n & \to 0 \\
 & & & & \downarrow & & \downarrow & & & & \downarrow \\
 & & & & 0 & & 0 & & & & 0 \\
\end{array}
$$

Proof: We may use the following commutative and exact diagram in order to construct successively the epimorphisms $\Psi_r : F_r' \to F_r$ with kernels K_r by using an induction on r starting with $r = 0$.

$$
\begin{array}{ccccccc}
0 & & 0 & & 0 & & 0 \\
\downarrow & & \downarrow & & \downarrow & & \downarrow \\
0 \to\ g_{q+r+1} & \to & S_{q+r+1}T^* \otimes T & \to & J_r(K_0) & \to \ldots \to & K_r & \to 0 \\
\downarrow & & \downarrow & & \downarrow & & \downarrow \\
0 \to\ R_{q+r+1} & \to & J_{q+r+1}(T) & \to & J_r(F_0') & \to \ldots \to & F_r' & \to 0 \\
\downarrow & & \downarrow \pi_{q+r}^{q+r+1} & & \downarrow J_r(\Psi_0) & & \downarrow \Psi_r \\
0 \to\ R_{q+r} & \to & J_{q+r}(T) & \to & J_r(F_0) & \to \ldots \to & F_r & \to 0 \\
\downarrow & & \downarrow & & \downarrow & & \downarrow \\
0 & & 0 & & 0 & & 0
\end{array}
$$

It follows that *the Janet sequence for $\mathcal{D}'$ projects onto the Janet sequence for $\mathcal{D}$* and we just need to prove that the kernel sequence, made up by first order operators only, is exact. For this, introducing the short exact sequences :

$$0 \to g_{q+r+1} \to S_{q+r+1}T^* \otimes T \to h_{r+1} \to 0$$

with $h_{r+1} \subset S_{r+1}T^* \otimes F_0$ the r-prolongation of the symbol $h_1 \subset T^* \otimes F_0$ of the system $B_1 \subset J_1(F_0)$, image of the first prolongation $J_{q+1}(T) \to J_1(F_0)$ of the epimorphism $J_q(T) \to F_0$ used in order to define $\mathcal{D}$. Here h_1 is identified with the reciprocal image of the symbol of $\mathcal{B}_1 \subset J_1(\mathcal{F})$ by ω. Using the Spencer operator D and its various extensions, we obtain the following commutative diagram where it is known that the vertical D-sequences are locally exact ([36],[49]).

$$
\begin{array}{ccccccccc}
& & 0 & & & & 0 & & \\
& & \downarrow & & & & \downarrow & & \\
0 & \to & K_0 & \to \ldots \to & & K_n & & \to 0 \\
\downarrow & & \downarrow j_{r+n} & & & & \downarrow j_r & & \\
0 \to & h_{r+n+1} & \to & J_{r+n}(K_0) & \to \ldots \to & J_r(K_n) & & \to 0 \\
& \downarrow -\delta & & \downarrow D & & \downarrow D & & \\
0 \to & T^* \otimes h_{r+n} & \to & T^* \otimes J_{r+n-1}(K_0) & \to \ldots \to & T^* \otimes J_{r-1}(K_n) & \to 0 \\
& \downarrow -\delta & & \downarrow D & & \downarrow D & & \\
& \vdots & & \vdots & & \vdots & & \\
& \downarrow -\delta & & \downarrow D & & & & \\
0 \to & \wedge^n T^* \otimes h_{r+1} & \to & \wedge^n T^* \otimes J_r(K_0) & \to \ldots & & & \\
& \downarrow & & \downarrow & & & & \\
& 0 & & 0 & & & &
\end{array}
$$

Now the left vertical sequence is exact because, applying the δ-sequence to the previous short exact sequence, then h_1 is involutive whenever g_q is involutive. The exactness of the top row finally follows from a diagonal chase in the last diagram because the other rows are exact by induction according to the previous diagram.

Q.E.D.

In order to use the last theorem in a natural way, we shall, for simplicity, restrict to the use of the classical Lie derivative by using the formulas:

$$\mathcal{L}(\xi)\eta_q = [j_q(\xi), \eta_q] \qquad \forall \xi \in T, \forall \eta_q \in J_q(T)$$

$$\mathcal{L}(\xi)\{\eta_q, \zeta_q\} = \{\mathcal{L}(\xi)\eta_q, \zeta_q\} + \{\eta_q, \mathcal{L}(\xi)\zeta_q\}$$

$$\mathcal{L}(\xi)[\eta_q, \zeta_q] = [\mathcal{L}(\xi)\eta_q, \zeta_q] + [\eta_q, \mathcal{L}(\xi)\zeta_q]$$

$$D\mathcal{L}(\xi) = \mathcal{L}(\xi)D$$

which are direct consequences of the fact that the algebraic bracket, the differential bracket and D only contain natural operations. Accordingly, when restricting to an involutive system $R_q \subset J_q(T)$ of infinitesimal Lie equations, we have to preserve the section ω of the natural bundle $\mathcal{F}$ in order to construct the Janet sequence and we obtain at once:

PROPOSITION 5.62: We have the formulas:
$$\mathcal{L}(\xi)\mathcal{D} = \mathcal{D}\mathcal{L}(\xi), \qquad \mathcal{L}(\xi)\mathcal{D}_r = \mathcal{D}_r\mathcal{L}(\xi), \qquad \forall \xi \in \Theta, \forall r = 0, ..., n.$$

EXAMPLE 5.63: Whenever $\mathcal{D}$ is a Lie operator already defined by the Lie derivative of ω with respect to a vector field, we get:

$$\begin{aligned}(\mathcal{L}(\xi)\mathcal{D} - \mathcal{D}\mathcal{L}(\xi))\eta = \mathcal{L}(\xi)\mathcal{L}(\eta)\omega - \mathcal{D}[\xi,\eta] &= (\mathcal{L}(\xi)\mathcal{L}(\eta) - \mathcal{L}([\xi,\eta]))\omega \\ &= \mathcal{L}(\eta)\mathcal{L}(\xi)\omega = \mathcal{L}(\eta)\mathcal{D}\xi = 0\end{aligned}$$

Such a result may be applied at once to all the known structures. The case of a symplectic structure with $n = 2p, \omega \in \wedge^2 T^*, det(\omega) \neq 0, d\omega = 0$ is particularly simple because a part of the corresponding Janet sequence is made by a part of the Poincaré sequence and it is well known that the Lie derivative $\mathcal{L}(\xi)$ commutes with the exterior derivative d (See [37], p 682 for more details).

The two following propositions will be obtained from the diagram of the last theorem by means of unusual chases in the tridimensional diagram obtained by applying $\mathcal{L}(\xi)$ for $\xi \in \Theta$ to the diagram of the last theorem. However, before providing them, it is essential to notice that a short exact sequence $0 \to F \to F' \to F'' \to 0$ of bundles associated with R_q may only provide in general the exact sequence $0 \to \Upsilon \to \Upsilon' \to \Upsilon''$ by letting $\mathcal{L}(\xi)$ acting on each bundle whenever $\xi \in \Theta$, unless one can split this sequence by using a natural map from F'' to F'. For example, in the case of the short exact sequence $0 \to J_q^0(T) \to J_q(T) \to T \to 0$ one can use $j_q : T \to J_q(T)$. We also recall that an isomorphism $Z'/B' = H' \simeq H = Z/B$ induced by maps $Z' \to Z$ and $B' \to B$ does not necessarily imply any property of these maps which may be neither monomorphisms nor epimorphisms.

PROPOSITION 5.64: There is an isomorphism $H_0' = H_0(R_{q+1}) \simeq H_0(R_q) = H_0$.

Proof: We shall provide two different proofs:
1) First of all, this result is a direct consequence of the short exact sequence provided by Proposition 5.32.

2) Let us prove that there is a monomorphism $0 \to H'_0 \to H_0$. First, if $a' \in \Upsilon'_0 \subset F'_0$ is killed by $\mathcal{D}'_1$ and such that $\Psi_0(a') = a = \mathcal{D}\eta$ for a certain $\eta \in \Upsilon(T) \subset T$, then $a = \Psi_0(a') = \Psi_0(\mathcal{D}'\eta)$ and thus $\Psi_0(a' - \mathcal{D}'\eta) = 0$, that is $a' - \mathcal{D}'\eta = b' \in K_0$ with $\mathcal{D}'_1 b' = \mathcal{D}'_1 a' - \mathcal{D}'_1 \mathcal{D}'\eta = 0$, that is $b' = 0 \Rightarrow a' = \mathcal{D}'\eta$.

Let us then prove that there is an epimorphism $H'_0 \to H_0 \to 0$. For this, if $a \in \Upsilon_0 \subset F_0$ is killed by $\mathcal{D}_1$, we may find $a' \in F'_0$ such that $a = \Psi_0(a')$ and $\mathcal{L}(\xi)a' \in K_0$. It follows that $c' = \mathcal{D}'_1 a' \in K_1$ with $\mathcal{D}'_2 c' = 0$, that is $c' = \mathcal{D}'_1 b'$ for a certain $b' \in K_0$. Accordingly, modifying a' if necessary by replacing it by $a' - b'$, we may suppose that $\mathcal{D}'_1 a' = 0$ with $\Psi_0(a') = a$ and we just need to prove that $\mathcal{L}(\xi)a' = 0$. Indeed, we have $\Psi_0(\mathcal{L}(\xi)a') = \mathcal{L}(\xi)a = 0 \Rightarrow \mathcal{L}(\xi)a' \in K_0$ and thus $\mathcal{D}'_1 \mathcal{L}(\xi)a' = \mathcal{L}(\xi)\mathcal{D}'_1 a' = 0 \Rightarrow \mathcal{L}(\xi)a' = 0$ because the restriction of $\mathcal{D}'_1$ to K_0 is a monomorphism.

$$\text{Q.E.D.}$$

PROPOSOTION 5.65: If $R_q \subset J_q(T)$ is involutive, the normalizer $N(\Theta)$ of Θ in T is defined by the involutive system $\tilde{R}_{q+1} \subset J_{q+1}(T)$ with involutive symbol $\tilde{g}_{q+1} = g_{q+1} \subset S_{q+1}T^* \otimes T$, defined by the purely algebraic condition $\{R_{q+1}, \tilde{R}_{q+1}\} \subset R_q$. Moreover, $Z_0(R_{q+1}) \simeq Z_0(R_q) \simeq \tilde{R}_{q+1}/R_{q+1}$.

Proof: With $Z_0 = Z_0(R_q)$ and $Z'_0 = Z_0(R_{q+1})$ as before, let us prove that there is an isomorphism $0 \to Z'_0 \to Z_0 \to 0$. For this, if $a' \in Z'_0$ is such that $\Psi_0(a') = 0$, then $a' \in K_0$ with $\mathcal{D}'_1 a' = 0 \Rightarrow a' = 0$ because the restriction of $\mathcal{D}'_1$ to K_0 is injective and we get a monomorphism $0 \to Z'_0 \to Z_0 \to 0$.

Similarly, if $a \in Z_0$, using the previous proposition, we can find $a' \in F'_0$ with $\mathcal{D}'_1 a' = 0$ and $\Psi_0(a') = a$. It follows that $\mathcal{L}(\xi)a' = d' \in K_0$ with $\mathcal{D}'_1 d' = \mathcal{D}'_1 \mathcal{L}(\xi)a' = \mathcal{L}(\xi)\mathcal{D}'_1 a' = 0 \Rightarrow d' = 0 \Rightarrow \mathcal{L}(\xi)a' = 0 \Rightarrow a' \in Z'_0$ and we get an epimorphism $Z'_0 \to Z_0 \to 0$, that is $Z(R_{q+1}) \simeq Z_0(R_q)$.

Now we have $\mathcal{L}(\xi_{q+1})\eta_q = \{\xi_{q+1}, \eta_{q+1}\} + i(\xi)D\eta_{q+1} \in R_q$ for *any lift* $\eta_{q+1} \in J_{q+1}(T)$ of $\eta_q \in J_q(T)$. However, cocycles in Z_0 are also killed by $\mathcal{D}_1$ and the representative $\eta_q \in J_q(T)$ must be such that $D\eta_{q+1} \in T^* \otimes R_q + \delta(S_{q+1}T^* \otimes T)$. Therefore , modifying η_{q+1} if necessary without changing η_q, we may suppose that $D\eta_{q+1} \in T^* \otimes R_q$, a result leading to the condition $\{\xi_{q+1}, \eta_{q+1}\} \in R_q \Rightarrow \eta_{q+1} \in \tilde{R}_{q+1}$. As we already know that $\tilde{g}_{q+1} = g_{q+1}$, introducing the projection $\tilde{R}_q \subset J_q(T)$

of $\tilde{R}_{q+1} \subset J_{q+1}(T)$, we obtain $Z_0(R_q) \simeq \tilde{R}_q/R_q \simeq \tilde{R}_{q+1}/R_{q+1}$ and similarly $Z_0(R_{q+1}) \simeq \tilde{R}_{q+2}/\tilde{R}_{q+2}$ in the following commutative and exact diagram:

$$
\begin{array}{ccccccc}
 & & 0 & & 0 & & \\
 & & \downarrow & & \downarrow & & \\
0 \to & g_{q+2} & \to & \tilde{g}_{q+2} & \to & 0 & \\
 & \downarrow & & \downarrow & & \downarrow & \\
0 \to & R_{q+2} & \to & \tilde{R}_{q+2} & \to & Z_0(R_{q+1}) & \to 0 \\
 & \downarrow & & \downarrow & & \downarrow & \\
0 \to & R_{q+1} & \to & \tilde{R}_{q+1} & \to & Z_0(R_q) & \to 0 \\
 & \downarrow & & \downarrow & & \downarrow & \\
 & 0 & & 0 & &
\end{array}
$$

Chasing in this diagram or counting the dimensions in order to get:

$$ dim(\tilde{R}_{q+2}) - dim(\tilde{R}_{q+1}) = dim(R_{q+2}) - dim(R_{q+1}) = dim(g_{q+2}) = dim(\tilde{g}_{q+2}) $$

it follows that $\pi_{q+1}^{q+2} : \tilde{R}_{q+2} \to \tilde{R}_{q+1}$ is an epimorphism.
Next, applying the formula of Proposition 5.6, we get:

$$
\begin{aligned}
\{\xi_{q+1}, i(\zeta)D\eta_{q+2}\} &= i(\zeta)D\{\xi_{q+2}, \eta_{q+2}\} - \{i(\zeta)D\xi_{q+2}, \eta_{q+1}\} \in R_q \\
&\Rightarrow D\eta_{q+2} \in T^* \otimes \tilde{R}_{q+1}
\end{aligned}
$$

It follows that $\tilde{R}_{q+2} = \rho_1(\tilde{R}_{q+1})$ because both systems project onto $\tilde{R}_{q+1}$ and have the same symbol $\tilde{g}_{q+2} = g_{q+2} = \rho_1(g_{q+1}) = \rho_1(\tilde{g}_{q+1})$.
Finally, as $\tilde{g}_{q+1} = g_{q+1}$ is an involutive symbol when g_q is involutive and $\pi_{q+1}^{q+2} :$ $\tilde{R}_{q+2} \to \tilde{R}_{q+1}$ is surjective, then $\tilde{R}_{q+1}$ is involutive because of the Janet/Goldschmidt/ Spencer criterion of formal integrability ([23],[36],[39],[49]).

$$ \text{Q.E.D.} $$

PROPOSITION 5.66: There is a monomorphism $0 \to H_1' \to H_1$. Accordingly, $H_1(R_{q+1}) = 0$ whenever $H_1(R_q) = 0$ and thus R_{q+1} is rigid whenever R_q is rigid and we may say that Θ is rigid.

Proof: If $a' \in \Upsilon_1' \subset F_1'$ is killed by $\mathcal{D}_2'$ and such that $\Psi_1(a') = a = \mathcal{D}_1 c$ for a certain $c \in \Upsilon_0 \subset F_0$, we may find $c' \in F_0'$ such that $\Psi_0(c') = c$ and thus

$a = \Psi_1(a') = \mathcal{D}_1\Psi_0 c' = \Psi_1\mathcal{D}'_1 c' \Rightarrow \Psi_1(a' - \mathcal{D}'_1 c') = 0 \Rightarrow a' - \mathcal{D}'_1 c' = b' \in K_1$.
As before, we get $\mathcal{D}'_2 b' = \mathcal{D}'_2 a' - \mathcal{D}'_2\mathcal{D}'_1 c' = 0 \Rightarrow \exists d' \in K_0$ with $\mathcal{D}'_1 d' = b'$ and thus $a' = \mathcal{D}\mathcal{D}'_1(c' + d')$. Hence, modifying c' if necessary, we may find $c' \in F'_0$ such that $\Psi_0(c') = c$ with $a' = \mathcal{D}'_1 c'$ and it only remains to prove that $c' \in \Upsilon'_0$. However, we have $\mathcal{L}(\xi)c' \in K_0$ because $\mathcal{L}(\xi)c = 0$ and $\mathcal{D}'_1\mathcal{L}(\xi)c' = \mathcal{L}(\xi)\mathcal{D}'_1 c' = \mathcal{L}(\xi)a' = 0$.
As the restriction of $\mathcal{D}'_1$ to K_0 is injective, we finally obtain $\mathcal{L}(\xi)c' = 0$ as we wanted.

Q.E.D.

Despite many attempts we have not been able to find additional results, in particular to compare $H_2(R_q)$ and $H_2(R_{q+1})$. However, integrating the Vessiot structure equations with structure constants c_t instead of c, we may exhibit a deformation ω_t of ω as a new section of $\mathcal{F}$. As $\mathcal{D}$ depends on $j_1(\omega)$, the $\mathcal{D}_r$ will also only depend on ω and various jets. Replacing ω by ω_t, we shall obtain operators $\mathcal{D}(t)$ and $\mathcal{D}_r(t)$ for $0 \leq r \leq n$. Therefore we may apply all the techniques and results of section 2 with only slight changes.

6 EXPLICIT COMPUTATIONS:

EXAMPLE 6.1: (1.1 *revisited*) This is a good example for understanding that the Janet bundles are only defined up to an isomorphism, contrary to the natural bundles. In this example, we have of course the basic natural bundle $\mathcal{F} = T^* \otimes V = T^* \times_X ... \times_X T^*$ (n-times) with $V = \mathbb{R}^n$. This is a vector bundle (even a tensor bundle) and we may therefore identity $\mathcal{F}$ with $\mathcal{F}_0 = V(\mathcal{F})$ and F_0. However, in this case, $g_1 = R_1^0 = 0$ and we get $F_0 = J_1^0(T)/R_1^0 = T^* \otimes T$. Similarly, we should get $F_r = \wedge^r T^* \otimes T \otimes T / \delta(\wedge^{r-1} T^* \otimes S_2 T^* \otimes T) = \wedge^{r+1} T^* \otimes T$ because we have the exact sequence :

$$\wedge^{r-1} T^* \otimes S_2 T^* \xrightarrow{\delta} \wedge^r T^* \otimes T^* \xrightarrow{\delta} \wedge^{r+1} T^* \longrightarrow 0$$

Using geometric objects, we should obtain:

$$(\mathcal{D}\xi)_i^\tau \equiv \omega_r^\tau(x)\partial_i \xi^r + \xi^r \partial_r \omega_i^\tau(x) = \Omega_i^\tau$$

$$\Rightarrow (\mathcal{D}_1\Omega)_{ij}^\tau \equiv \partial_i \Omega_j^\tau - \partial_j \Omega_i^\tau - c_{\rho\sigma}^\tau(\omega_i^\rho \Omega_j^\sigma + \omega_j^\sigma \Omega_i^\rho) = 0$$

and thus $F_1 = \wedge^2 T^* \otimes V$. On the contrary, using the solved form $\xi \to (\nabla\xi)_i^k \equiv \partial_i \xi^k + \xi^r a_r^k(x)\partial_r \omega_i^\tau(x)$ providing at once a unique R_1-connexion, a covariant derivative ∇ and thus a well defined ∇-sequence leading to $F_r = \wedge^{r+1} T^* \otimes T$ and thus $F_1 = \wedge^2 T^* \otimes T$. In fact, *the explanation is a confusion between the Janet sequence and the Spencer sequence* because we have indeed $C_0 = R_1 = T \Rightarrow C_{r+1} \simeq F_r \simeq \wedge^{r+1} T^* \otimes T$.

The sections of $\wedge^r T^* \otimes V$ invariant by R_1 are of the form $A_{\sigma_1...\sigma_r}^\tau \omega^{\sigma_1} \wedge ... \wedge \omega^{\sigma_r}$ while the sections of $\wedge^r T^* \otimes T$ invariant by R_1 are of the form $\alpha_\tau^k A_{\sigma_1...\sigma_r}^\tau \omega^{\sigma_1} \wedge ... \wedge \omega^{\sigma_r}$ and are thus different though depending on the same *constants* $A_{\sigma_1...\sigma_r}^\tau$, providing therefore a unique cohomology. From the Maurer-Cartan equations $d\omega^\tau = -c_{\rho\sigma}^\tau \omega^\rho \wedge \omega^\sigma$ where we change the sign of the structure constants fo convenience, we obtain terms of the form Ac which are *exactly* the first terms in the definition of the Chevalley-Eilenberg operator. Now, looking at the terms containing α, we obtain (care to the minus sign):

$$(\partial_i \alpha_\tau^k + \alpha_\tau^r a_\sigma^k \partial_r \omega_i^\sigma) A^\tau dx^i = \alpha_\tau^r a_\sigma^k (\partial_r \omega_i^\sigma - \partial_i \omega_r^\sigma) A^\tau dx^i = -\alpha_\tau^k c_{\rho\sigma}^\tau A^\rho \omega^\sigma$$

and recover the second terms of this operator.

It is however essential to notice that the two approaches are totally different. In particular, elements in $A(R^0_q)$ are 1-forms with value in some vector bundles and *it is therefore a pure chance that they could become 2-forms in this example* (See electromagnetism and gravitation in [43]).

We may also compare the concept of "change of basis" of the Lie algebra with the "label transformations". Indeed, using the Lie form, we get (care again to the minus sign):

$$\bar{\omega} \sim \omega \;\Leftrightarrow\; \bar{\alpha}^k_\tau \partial_r \bar{\omega}^\tau_i = \alpha^k_\rho \partial_r \omega^\rho_i = -\omega^\rho_i \partial_r \alpha^k_\rho$$
$$\Leftrightarrow\; \alpha^i_\rho \partial_r \bar{\omega}^\tau_i + \bar{\omega}^\tau_k \partial_r \alpha^k_\rho = 0$$
$$\Leftrightarrow\; \partial_r(\alpha^i_\rho \bar{\omega}^\tau_i) = 0$$
$$\Leftrightarrow\; \bar{\omega}^\tau_i = a^\tau_\sigma \omega^\sigma_i \,, a = cst$$

Accordingly, the sections of $\Upsilon_0 \subset T^* \otimes V$ are of the form $A^\tau_\sigma \omega^\sigma_i$ with $A = cst$ and it is rather extraordinary that two such different approaches can provide the same result.

Finally, we study $\Theta, Z(\Theta), C(\Theta)$ and $N(\Theta)$ in this framework. First of all, Θ is described by a linear combination with constant coefficients of the infinitesimal generators $\{\theta_\tau = \theta^i_\tau \partial_i\}$. Then we must have $-\xi^k_i \eta^i + \xi^r \partial_r \eta^k = 0$, whenever $\xi^k_i + \xi^r \alpha^k_\tau \partial_r \omega^\tau_i = 0$, that is $\alpha^k_\tau \partial_r \omega^\tau_i \eta^i + \partial_r \eta^k = 0$ or $\partial_r(\omega^\tau_i \eta^i) = 0$ and thus $\eta^i = \lambda^\tau \alpha^i_\tau$ with $\lambda = cst$ (*reciprocal distribution*). We already know that $[\alpha_\rho, \alpha_\sigma] = c^\tau_{\rho\sigma}\alpha_\sigma$ with our choice of sign. The isomorphism between Θ and $C(\Theta)$ is therefore a pure chance. As for $N(\Theta)$, it follows from Corollary 5.27 that the defining equations are $(\mathcal{L}(\xi)\omega)^\tau_i \equiv \omega^\tau_r(x)\partial_i \xi^r + \xi^r \partial_r \omega^\tau_i(x) = A^\tau_\sigma \omega^\sigma_i(x)$ where $A = cst$ is a derivation of the Lie algebra $\mathcal{G}$ with the same structure constants c.

EXAMPLE 6.2: (*1.2 revisited*) Let us study the equivalence relation $\bar{\omega} \sim \omega$ by asking first that $\bar{R}^0_1 = R^0_1$. We must have $\bar{\omega}_{rj}\xi^r_i + \bar{\omega}_{ir}\xi^r_j = 0 \Leftrightarrow \omega_{rj}\xi^r_i + \omega_{ir}\xi^r_j = 0$. Setting $\xi_{i,j} = \omega_{rj}\xi^r_i$, it amounts to check that $\bar{\omega}_{rj}\omega^{rt}\xi_{t,i} + \bar{\omega}_{ir}\omega^{rt}\xi_{t,j} = 0 \Leftrightarrow \xi_{j,i} + \xi_{i,j} = 0$ and we must have therefore:

$$\bar{\omega}_{rj}\omega^{rt}\delta^s_i + \bar{\omega}_{ir}\omega^{rt}\delta^s_j - \bar{\omega}_{rj}\omega^{rs}\delta^t_i - \bar{\omega}_{ir}\omega^{rs}\delta^t_j = 0$$

Contracting in s and j, we get $n\bar{\omega}_{ir}\omega^{rt} - \bar{\omega}_{rs}\omega^{rs}\delta^t_i = 0$, that is $\bar{\omega}_{ir}\omega^{rt} = a(x)\delta^t_i \Rightarrow \bar{\omega}_{ij} = a(x)\omega_{ij}$ whenever $\bar{\omega} \sim \omega$. Substituting in $\bar{R}_1 = R_1$, we finally get $\bar{\omega} = a\omega, a = cst$ and the group of label transformations is just the multiplicative group.

It then follows that $\bar{\gamma} = \gamma$ but such a property can be checked directly from the solved form of the second order equations $\mathcal{L}(\xi)\gamma = 0$. It follows that $\bar{\rho} = \rho$ and thus $\bar{c} = (1/a)c$. In any case, $N(\Theta)$ is defined by the infinitesimal Lie equations $\mathcal{L}(\xi)\omega = A\omega$ with $cA = 0$ as there is no Jacobi condition on the single structure constant c. Accordingly, $N(\Theta) = \Theta$ if $c \neq 0$ and $dim(N(\Theta)/\Theta) = 1$ if $c = 0$. This is the reason for which the Weyl group is the normalizer of the Poincaré group, obtained by adding the generator $x^i\partial_i$ of the dilatation on space-time. It is important to notice that the Galilée group is of codimension 2 in its normalizer obtained by dilatating *separately* space and time, another reason for which the Poincaré group cannot be obtained from the Galilée group by a continuous deformation. As for $C(\Theta)$, using again R_1^0, we have to solve $\eta^i\omega^{kj}\xi_{i,j} = 0 \Leftrightarrow \eta^i\xi_{i,j} = 0, \forall \xi_{i,j} + \xi_{j,i} = 0$ and thus $C(\Theta) = 0 \Rightarrow Z(\Theta) = 0$. Finally, looking at the linear equations $\hat{R}_1^0$ defined by $\omega_{rj}\xi_i^r + \omega_{ir}\xi_j^r - \frac{2}{n}\omega_{ij}\xi_r^r = 0$, we let the reader ckeck as a much more delicate exercise that $\bar{\omega} \sim \omega \Leftrightarrow \bar{\omega} = a\omega$ leading to $\hat{\bar{\omega}} = \hat{\omega}$ for the geometric object $\hat{\omega} = \omega/|det(\omega)|^{\frac{1}{n}}$ of the conformal pseudogroup which is therefore its own normalizer (See [36], p 346-348 and [39], p 434 for more details).

EXAMPLE 6.3: (1.3 *revisited*) This example is by far the most difficult to treat in dimension $n = 2p + 1$ ([37], p 684). In the present case when $n = 3$, this is one of the best examples where the Lie equations obtained by eliminating ρ, namely $\xi_3^1 - x^3\xi_3^2 = 0, \xi_2^1 - x^3\xi_2^2 + x^3\xi_1^1 - (x^3)^2\xi_1^2 - \xi^3 = 0$ could be used without even knowing about the underlying geometric object. However, *in this case R_1 is not involutive* and we must start afresh with the involutive system $R_1^{(1)} \subset R_1$ for constructing the canonical Janet sequence. We first notice that there is only one CC for the new involutive system $R_1 \subset J_1(T)$ of infinitesimal Lie equations and thus surely no Jacobi condition for the only structure constant c. It thus remains to study the inclusions $Z(\Theta) \subseteq \Theta \subseteq N(\Theta)$. Using $R_1^0 \subset J_1^0(T)$ for $C(\Theta)$, we have to solve $\eta^s\xi_s^k = 0$ whenever $\omega_r\xi_i^r - \frac{1}{2}\omega_i\xi_r^r = 0$ and thus $\eta^s\omega_s\xi_r^r = 0 \Rightarrow \eta^s\omega_s = 0$. As $\omega \neq 0$, changing coordinates if necessary, we may suppose that $\omega_1 \neq 0$. It follows from the involutive assumption that at least one jet coordinate of each class is parametric and in fact, changing coordinates if necessary, we have in all the examples presented $pri = \{\xi_3^3, \xi_3^1, \xi_2^1\} \Rightarrow par = \{\xi_1^1, \xi_1^2, \xi_1^3, \xi_2^2, \xi_2^3, \xi_3^2\}$. Hence, choosing successively $\xi_1^2, \xi_2^2, \xi_3^2$ as unique non-zero parametric jet, we obtain $\eta^1 = 0, \eta^2 = 0, \eta^3 = 0 \Rightarrow \eta = 0$ and thus $C(\Theta) = 0 \Rightarrow Z(\Theta) = 0$. As for $N(\Theta)$, we have to study the equivalence

$\bar{\omega} \sim \omega$, that is to study when we have $\bar{\omega}_r \xi_i^r - \frac{1}{2}\bar{\omega}_i \xi_r^r = 0 \Leftrightarrow \omega_r \xi_i^r - \frac{1}{2}\omega_i \xi_r^r$. Though it looks like to be a simple algebraic problem, one needs an explicit computation or computer algebra and we prefer to use another more powerful technique ([37], p 688). Introducing the completely skewsymmetrical symbol $\epsilon = (\epsilon^{i_1 i_2 i_3})$ where $\epsilon^{i_1 i_2 i_3} = 1$ if $(i_1 i_2 i_3)$ is an even permutation of (123) or -1 if it is an odd permutation and 0 otherwise, let us introduce the skewsymmetrical 2-contravariant density $\omega^{ij} = \epsilon^{ijk}\omega_k$. Then one can rewrite the lie equations R_1 as :

$$-\omega^{rj}(x)\xi_r^i - \omega^{ir}(x)\xi_r^j - \frac{1}{2}\omega^{ij}(x)\xi_r^r + \xi^r \partial_r \omega^{ij}(x) = 0$$

and we may exhibit a section $\xi_r^i = \omega^{is} A_{rs}$ with $A_{rs} = A_{sr}$ and thus $\xi_r^r = 0$. It is important to notice that $det(\omega) = 0$ when $n = 2p + 1$, contrary to the Riemann or symplectic case and ω cannot therefore be used in order to raise or lower indices. As we must have $\bar{R}_1^0 = R_1^0$, the same section must satisfy $(\bar{\omega}^{rj}\omega^{is} + \bar{\omega}^{ir}\omega^{js})A_{rs} = 0, \forall A_{rs} = A_{sr}$, and we must have $(\bar{\omega}^{rj}\omega^{is} + \bar{\omega}^{ir}\omega^{js}) + (\bar{\omega}^{sj}\omega^{ir} + \bar{\omega}^{is}\omega^{jr}) = 0$. Setting $s = j$, we get $\bar{\omega}^{rj}\omega^{ij} = \bar{\omega}^{ij}\omega^{rj} \Rightarrow \bar{\omega}^{ij}(x) = a(x)\omega^{ij}(x)$. Substituting and substracting, we get $\omega^{ij}(x)\xi^r \partial_r a(x) = 0 \Rightarrow a(x) = a = cst$ or similarly $\bar{\omega}_i(x) = a(x)\omega_i(x) \Rightarrow \omega_i(x)\xi^r \partial_r a(x) = 0 \Rightarrow a(x) = a = cst$ because $\omega \neq 0$ and one of the components at least must be nonzero. It follows at once that one has $N(\Gamma) = \{f \in aut(X) \mid j_1(f)^{-1}(\omega) = a\omega, a^2 c = c\}$. Accordingly, $N(\Theta) = \{\xi \in T \mid \mathcal{L}(\xi)\omega = A\omega, Ac = 0\}$ and Θ is of codimension 1 in its normalizer if $c = 0$ or $N(\Theta) = \Theta$ if $c \neq 0$. For example, in the case of a contact structure with $c = 1$, we have $N(\Theta) = \Theta$ but, when $\omega = (1, 0, 0) \Rightarrow c = 0$, we have to eliminate the constant A among the equations $\partial_3 \xi^3 + \partial_2 \xi^2 - \partial_1 \xi^1 = -2A, \partial_3 \xi^1 = 0, \partial_2 \xi^1 = 0$ and we may add the infinitesimal generator $x^i \partial_i$ of a dilatation providing $A = -\frac{1}{2}$.

EXAMPLE 6.4: (1.4 *revisited*) This is by far the most interesting example. First of all, contrary to all the previous examples, we have at once $\mathcal{F} = \mathcal{F}_0 = T^* \times_X \wedge^2 T^*$, $\mathcal{F}_1 = \wedge^2 T^* \times_X \wedge^3 T^*$, $\mathcal{F}_2 = \wedge^3 T^*$ and we may identify F_r with $\mathcal{F}_r$ for $r = 0, 1, 2, 3$ in order to obtain the Janet sequence:

$$0 \longrightarrow \Theta \longrightarrow T \xrightarrow{\mathcal{D}} F_0 \xrightarrow{\mathcal{D}_1} F_1 \xrightarrow{\mathcal{D}_2} F_2 \longrightarrow 0$$

with $dim(T) = 3, dim(F_0) = 6, dim(F_1) = 4, dim(F_2) = 1$ and $3 - 6 + 4 - 1 = 0$. Using local coordinates (u^1, u^2, u^3) for T^* and (u^4, u^5, u^6) for $\wedge^2 T^*$ in F_0, we obtain

$W = \{W_1 = u^1\frac{\partial}{\partial u^1} + u^2\frac{\partial}{\partial u^2} + u^3\frac{\partial}{\partial u^3}, W_2 = u^4\frac{\partial}{\partial u^4} + u^5\frac{\partial}{\partial u^5} + u^6\frac{\partial}{\partial u^6}\}$. With $\omega = (\alpha, \beta)$, we have $\mathcal{D}\xi = (\mathcal{L}(\xi)\alpha, \mathcal{L}(\xi)\beta)$ and recall the IC made by the Vessiot structure equations $d\alpha = c'\beta, d\beta = c''\alpha \wedge \beta$ with $\gamma = \alpha \wedge \beta \neq 0$. As before, we may easily find the label transformations $(\bar{\alpha} = a\alpha, \bar{\beta} = b\beta) \Rightarrow (\bar{c}' = \frac{a}{b}c', \bar{c}'' = \frac{1}{a}c'')$ and obtain $\Upsilon_0 = (A\alpha, B\beta) \subset F_0$ with $A, B = cst$. Similarly, we get $\Upsilon_1 = (C'\beta, C''\gamma) \subset F_1$ with $C', C'' = cst$ and the induced map $\mathcal{D}_1 : \Upsilon_0 \to \Upsilon_1 : (A, B) \to (c'(A - B) = C', -c''A = C'')$ in a coherent way with the linearization of the Vessiot structure equations:

$$Ad\alpha - c'B\beta = c'(A - B)\beta, Bd\beta - c''A\alpha \wedge \beta - c''B\alpha \wedge \beta = -c''A\gamma$$

We obtain therefore $N(\Theta) = \{\xi \in T \mid \mathcal{L}(\xi)\alpha = A\alpha, \mathcal{L}(\xi)\beta = B\beta, c'(A - B) = 0, c''A = 0\}$ and thus: $c = (0,0) \Rightarrow dim(N(\Theta)/\Theta) = 2, c = (1,0) \Rightarrow A = B \Rightarrow dim(N(\Theta)/\Theta) = 1, c = (0,1) \Rightarrow A = 0 \Rightarrow dim(N(\Theta)/\Theta) = 1$. The only Jacobi condition $c'c'' = 0$ provides $c''C' + c'C'' = 0$ in any case. For example, in the unimodular contact case, we must add $x^1\partial_1 + x^2\partial_2$ to Θ in order to obtain $N(\Theta)$.

As for the centralizer $C(\Theta)$, we must look first for $\eta \in T$ such that $\eta^r\xi_r^k = 0, \forall \xi_1^0 \in R_1^0$. For this, multiplying $\beta_{rj}\xi_i^r + \beta_{ir}\xi_j^r = 0$ by η_i and contracting on i, we get $(\eta^i\beta_{ir})\xi_j^r = 0$ whenever $\alpha_r\xi_i^r = 0$. Accordingly, we obtain $\eta^i\beta_{ir} = L(x)\alpha_r$ as 1-forms. Now, as $n = 3$, we may introduce the pseudo-vector $(\beta^1 = \beta_{23}, \beta^2 = \beta_{31}, \beta^3 = \beta_{12})$ transforming like a vector up to a division by the Jacobian Δ and thus, using the volume form $\gamma = \alpha \wedge \beta$, it follows that $\tilde{\beta} = (\beta^1/\gamma, \beta^2/\gamma, \beta^3/\gamma)$ is a true vector field. As $\beta_{ir}\tilde{\beta}^r = 0$ because $n = 3$, we obtain by contraction $L(x)\alpha_r\tilde{\beta}^r = L(x)(\alpha \wedge \beta/\gamma) = L(x) = 0 \Rightarrow \eta^i\beta_{ir} = 0 \to \eta^k = K(x)\tilde{\beta}^k$. But α, β, γ and thus $\tilde{\beta}$ are invariant by any $\xi \in \Theta$ and thus $K(x) = K = cst$, that is $C(\Theta) = \{\theta \in T | \eta = K\tilde{\beta}, K = cst\}$. In all the three special sections considered, we have simply $C(\Theta) = \{(K, 0, 0) \in T | K = cst\}$.

We obtain finally the folowing recapitulating picture:

$$
\begin{array}{ccccccccc}
(K) & \longrightarrow & (A,B) & \longrightarrow & (C',C'') & \longrightarrow & (D) & \to 0 \\
(K\tilde{\beta}) & \longrightarrow & (A\alpha, B\beta) & \longrightarrow & (C'\beta, C''\gamma) & \longrightarrow & (D\gamma) & \to 0 \\
0 \to \Theta \to \quad T & \xrightarrow{\;\mathcal{D}\;} & T^* \times_X \wedge^2 T^* & \xrightarrow{\;\mathcal{D}_1\;} & \wedge^2 T^* \times_X \wedge^3 T^* & \xrightarrow{\;\mathcal{D}_2\;} & \wedge^3 T^* & \to 0 \\
\xi & \longrightarrow & (\Phi, \Psi) & \longrightarrow & (U,V) & \longrightarrow & (W) & \to 0
\end{array}
$$

The *purely differential lower part* is describing the operators involved in the Janet sequence, namely and successively:

$$
\begin{aligned}
\mathcal{D} \quad & \qquad \xi \longrightarrow (\mathcal{L}(\xi)\alpha = \Phi, \mathcal{L}(\xi)\beta = \Psi) \\
\mathcal{D}_1 \quad & (\Phi, \Psi) \longrightarrow (d\Phi - c'\Psi = U, d\Psi - c''(\Phi \wedge \beta + \alpha \wedge \Psi) = V) \\
\mathcal{D}_2 \quad & \qquad (U,V) \longrightarrow (dU + c'V = W)
\end{aligned}
$$

The *purely algebraic upper part* is induced by these operators acting on the second line which is describing the invariant sections of the respective Janet bundles and we obtain the linear maps:

$$
\begin{aligned}
\mathcal{D} \quad & \qquad (K) \longrightarrow (0, c''K) \\
\mathcal{D}_1 \quad & (A,B) \longrightarrow (c'(A-B) = C', -c''A = C'') \\
\mathcal{D}_2 \quad & (C',C'') \longrightarrow (c''C' + c'C'' = D)
\end{aligned}
$$

only depending on the structure constants $c = (c', c'')$. Of course, we finally check that the final composition is $c'c''(A-B) - c'c''A = -c'c''B = 0$.

EXAMPLE 6.5: (1.5 *revisited*) This is the simplest example showing out the usefulness of the methods for constructing the normalizer. If $\omega = (\alpha, \beta)$ with $\alpha \in T^*, \beta \in \wedge^2 T^*$ is the geometric object involved, we know that the only Vessiot structure equation is $d\alpha = c\beta$ with $c = cst$. Now, using the previous examples, we have $\bar{\omega} \sim \omega \Leftrightarrow \bar{\alpha} = a\alpha, \bar{\beta} = b\beta$ with nonzero $a, b = cst$. Hence we obtain $d\bar{\alpha} = \bar{c}\bar{\beta} \Rightarrow \bar{c} = (a/b)c$, a result leading to two possibilities for studying the system $j_1(f)^{-1}(\omega) = \bar{\omega}$ with $\bar{c} = c$.

• $c \neq 0$: We must have $a = b$ and the pseudogroup is of codimension 1 in its normalizer. For example, if $\alpha = x^2 dx^1, \beta = dx^1 \wedge dx^2 \Rightarrow c = -1$, then $\Gamma = \{y^1 = f(x^1), y^2 = x^2/(\partial f(x^1)/\partial x^1)/\partial x^1)\}$ with $c = cst$ and we have to solve the system $y^2 y_1^1 = ax^2, y_2^1 = 0 \Rightarrow y_1^1 y_2^2 = a$ with solutions $N(\Gamma) = \{y^1 = f(x^1), y^2 = ax^2/(\partial f(x^1)\partial x^1)\}$. On the infinitesimal level, cocycles are made by arbitrary $C = cst$ while coboundaries are of the form $c(A - B) = C$. Hence, $C = 0 \Rightarrow A = B$ and we have to solve the system $\mathcal{L}(\xi)\omega = A\omega$ in order to define $N(\Theta)$.

• $c = 0$: In this case, we must also have $\bar{c} = 0$ too but the two non-zero constants (group parameters) a and b are arbitrary and the pseudogroup is of codimension 2 in its normalizer. For example, if $\alpha = dx^1, \beta = dx^1 \wedge dx^2 \Rightarrow c = 0$, then $\Gamma = \{y^1 = x^1 + c, y^2 = x^2 + g(x^1)\}$ with $c = cst$ and we have to solve the system $y_1^1 = a, y_2^1 = 0, y_1^1 y_2^2 = b$ with solutions $N(\Gamma) = \{y^1 = ax^1 + c, y^2 = (b/a)x^2 + g(x^1)\}$. On the infinitesimal level, cocycles are made by arbitrary $C = cst$ while coboundaries are of the form $c(A - B) = 0$. Hence, A and B may now be arbitrary and we have to solve the system $\mathcal{L}(\xi)\alpha = A\alpha, \mathcal{L}(\xi)\beta = B\beta$ in order to define $N(\Theta)$.

EXAMPLE 6.6: (1.6 *revisited*) Comparing the Lie group of transformations $y = x + b$ and $y = ax$, we notice that $exp(x + b) = exp(b)exp(x)$ and the trnsformation $x \rightarrow exp(x)$ should transform $(\alpha = 1, \gamma = 0$ to $\alpha = 1/x, \gamma = 0$ even though $c = 0$ in the first case but $c = -1$ in the second, a result that seems to contradict all what we have already said many times. However, from the Medolahi equations, we deduce at once that $(\bar{\alpha}, \bar{\gamma}) \sim (\alpha, \gamma) \Leftrightarrow \bar{\alpha} = a\alpha, \bar{\gamma} = \gamma + b\alpha$ by substraction, in such a way that we have now $\partial_x \bar{\alpha} - \bar{\gamma}\bar{\alpha} = \bar{c}\bar{\alpha}^2$ and thus $\bar{c} = (1/a)c - (b/a)$. Hence, it is just sufficient, in order to pass from $c = 0$ to $\bar{c} = -1$, to set $b = a, \forall a \neq 0$ (care to thenotations). It follows that $y = x + b$ is of codimension 1 in its normalizer $y = ax + b$. A similar procedure may be followed with the geometric object (α, β, γ).

7 CONCLUSION:

The work of E. Vessiot, motivated by the application of the local theory of Lie pseudogroups ([51]) to the *differential Galois theory*, namely the Galois theory for systems of partial differential equations (See references in [37]), has been deliberately ignored by E. Cartan and followers ([8],[26]). As a byproduct, the *Vessiot structure equations*, introduced as early as in 1903, are still unknown today. Similarly and twenty years later but for other reasons related to his work on general relativity ([6]), in particular his correspondence with A. Einstein, Cartan did not acknowledge the work of M. Janet on the formal theory of systems of partial differential equations (Compare [7] and [23]). Accordingly, the combination of natural bundles and geometric objects with differential sequences, in particular the *linear and nonlinear Janet sequences* has been presented for the first time in a rather self-contained manner through this paper. The main idea is to induce from the linear Janet sequence, considered as a linear differential sequence with vector bundles and linear differential operators, a purely algebraic *deformation sequence* with finite dimensional vector spaces and linear maps which may not be exact. We have shown that the equivalence problem for structures on manifolds has to do with the local exactness of the Janet sequence at F_0 while the deformation problem of the corresponding algebraic structures has to do with the exactness of the deformation sequence at $\Upsilon_1 = \Upsilon(F_1)$, the space of invariant sections of F_1, that is *one step further on in the sequence*. This result explains why the many tentatives done in order to link the *deformation of algebraic structures* like Lie algebras with the *deformation of geometric structures* on manifolds have not been successful. Meanwhile, we have emphasized the part that could be played by computer algebra in any effective computation and hope to have opened a new field of research for the future. Finally, ending with a wink, the reader must not forget that one of the first aplications of computer agebra in 1970 has been done in the deformation theory of Lie algebras where a few counterexamples could only be found in dimension greater than 10, that is with more than 500 structure constants !.

8 BIBLIOGRAPHY:

[1] U. AMALDI: Congrés de la Société Italienne pour le Progrés des Sciences, Parma, Italy, 1907 (in italian) (Estratto dagli Annali di Matematica, XV, III, 293-327).

[2] J.M. ANCOCHEA BERMUDEZ, R. CAMPOAMOR-STURSBERG, L. GARCIA VERGNOLLE, M. GOZE: Algèbres de Lie résolubles réelles algébriquement rigides, Monatsh. Math., 152, 2007, 187-195.

[3] M. BARAKAT: Jets. A MAPLE-package for formal differential geometry, Computer algebra in scientific computing, (EACA Konstanz), Springer, Berlin, 2001, 1-12.

http://wwwb.rwth-aachen.de/~barakat/jets/

[4] E. CARTAN: Sur une généralisation de la notion de courbure de Riemann et les espaces à torsion, C. R. Académie des Sciences Paris, 174, 1922, 437-439, 593-595, 734-737, 857-860.

[5] E. CARTAN: Sur les variétés à connexion affine et la théorie de la relativité généralisée, Ann. Ec. Norm. Sup., 40, 1923, 325-412; 41, 1924, 1-25; 42, 1925, 17-88.

[6] E. CARTAN, A. EINSTEIN: Letters on Absolute Parallelism 1929-1932 (Original letters with english translation), Princeton University Press and Académie Royale de Belgique, R. Debever Ed., Princeton University Press, 1979.

[7] E. CARTAN: Sur la théorie des systèmes en involution et ses applications à la relativité, Bull. Soc. Math. France, 59, 193, 88-118.

[8] E. CARTAN: La théorie de Galois et ses diverses généralisations, Oeuvres Complètes 1938.

[9] C. CHEVALLEY, S. EILENBERG: Cohomology theory of Lie groups and Lie algebras, Transactions of the American Mathematical Society, 63, (1), 1948, 85-124.

[10] L.P. EISENHART: Riemannian Geometry, Princeton University Press, Princeton, 1926.

[11] M. FLATO: Deformation view of physical theories, Czech. J. Phys., B, 32, 1982, 472-475.

[12] J. GASQUI, H. GOLDSCHMIDT: Déformations Infinitésimales des Structures Conformes Plates, Birkhauser, 1984.

[13] M. GERSTENHABER: On the deformation of rings and algebras, Ann. of

Math., 79, 1964, (I) 59-103; 84, 1966, (II) 1-99; 88, 1968, (III) 1-34; 99, 1974, (IV) 257-267.

[14] H. GOLDSCHMIDT: Sur la structure des équations de Lie, J. Differential Geometry, 6, 1972, (I) 357-373 + 7, 1972, (II) 67-95 + 11, 1976, (III) 167-223.

[15] H. GOLDSCHMIDT, D.C. SPENCER: On the nonlinear cohomology of Lie equations, Acta. Math., 136, 1973, (I) 103-170, (II) 171-239, (III) J. Diff. Geometry, 13, 1978, 409-453.

[15] R.E. GREENE, S.G. KRANTZ: Deformation of complex structures, Adv. Math. 43, 1982, 1-86.

[16] V. GUILLEMIN, S. STERNBERG: Deformation Theory of Pseudogroup Structures, Memo 66, Am. of Math. Soc., 1966.

[17] V. GUILLEMIN, S. STERNBERG: An algebraic model of transitive differential geometry, Bull. Amer. Math. Soc., 70, 1964, 16-47.

[18] V. GUILLEMIN, S. STERNBERG: The Lewy counterexample and the local equivalence problem for G-structures, J. Diff. Geometry, 1, 1967, 127-131.

[19] I. HAYASHI: Embedding and existence theorems of infinite Lie algebras, J. Math. Soc. Japan, 22, 1970, 1-14.

[20] M. HAZEWINKEL, M. GERSTENHABER: Deformation Theory of Algebras and Structures and Applications, Kluwer, NATO proceedings, 1988.

[21] G. HOCHSCHILD: On the cohomology groups of an associative algebra, Ann. of Math., 46, 1945, 58-67.

[22] E. INONU, E.P. WIGNER: On the contraction of Lie groups and Lie algebras, Proc. Nat. Acad. Sci. USA, 39, 1953, 510.

[23] M. JANET: Les Systèmes d'Equations aux Dérivées partielles, Journal de Mathématiques Pures et Appliquées, 3, 1920, 65-151.

[24] K. KODAIRA, L. NIRENBERG: On the existence of deformations of complex analytic structures, Ann. of Math., 68, 1958, 450-459.

[25] K. KODAIRA, D.C. SPENCER: On the deformation of complex analytic structures, I, II, Ann. of Math., 67, 1968, 328-466.

[26] A. KUMPERA, D.C. SPENCER: Lie Equations, Ann. Math. Studies 73, Princeton University Press, Princeton, 1972.

[27] M. KURANISHI: On the locally complete families of complex analytic structures, Ann. of Math., 75, 1962, 536-577.

[28] M. LEVY-NAHAS: Deformation and contraction of Lie algebras, J. Math.

Phys., 8, 1967, 1211-1222.

[29] M. LEVY-NAHAS: Two simple applications of the deformations of Lie algebras, Ann. Inst. H. Poincaré, A, 13, 1970, 221-227.

[30] A. LORENZ: Jet Groupoids, Natural Bundles and the Vessiot Equivalence Method, Ph.D. Thesis (published electronically), Department of Mathematics, RWTH Aachen-University, march 18, 2009.See also: On local integrability conditions of jet groupoids, Acta Appl. Math., 101, 2008, 205-213 at :
http://dx.doi.org/10.1007/s10440-008-9193-7

[31] A. NIJENHUIS, R.W. RICHARDSON: Cohomology and deformations of algebraic structures, Bull. Am. Math. Soc., 79, 1964, 406-411.

[32] V. OUGAROV: Théorie de la Relativité Restreinte, MIR, Moscow, 1969; french translation, 1979.

[33] W. PAULI: Theory of Relativity, Pergamon Press, London, 1958.

[34] W.S. PIPER: Algebraic deformation theory, J. Diff. Geometry, 1, 1967, 133-168.

[35] W.S. PIPER: Algebras of matrices under deformations, J. Diff. Geometry, 5, 1971, 437-449.

[36] J.-F. POMMARET: Systems of Partial Differential Equations and Lie Pseudogroups, Gordon and Breach, New York, 1978; Russian translation: MIR, Moscow, 1983.

[37] J.-F. POMMARET: Differential Galois Theory, Gordon and Breach, New York, 1983.

[38] J.-F. POMMARET: Lie Pseudogroups and Mechanics, Gordon and Breach, New York, 1988.

[39] J.-F. POMMARET: Partial Differential Equations and Group Theory: New Perspectives for Applications, Kluwer, Dordrecht, 1994.

[40] J.-F. POMMARET: Partial Differential Control Theory, Kluwer, Dordrecht, 2001.

[41] J.-F. POMMARET: Parametrization of Cosserat equations, Acta Mechanica, 215, 2010, 43-55.

[42] J.-F. POMMARET: Macaulay inverse systems revisited, Journal of Symbolic Computation, 46, 2011, 1049-1069.

[43] J.-F. POMMARET: Spencer Operator and Applications: From Continuum Mechanics to Mathematical Physics, in "Continuum Mechanics-Progress in Fundamen-

tals and Engineering Applications", Dr. Yong Gan (Ed.), ISBN: 978-953-51-0447–6, InTech, 2012, Available from:
http://www.intechopen.com/books/continuum-mechanics-progress-in-fundamentals-and-engineering-applications

[44] D.S. RIM: Deformations of transitive Lie algebras, Ann. of Math., 83, 1966, 349-357.

[45] A.N. RUDAKOV: Deformations of simple Lie algebras, Izv. Akad. Nauk. SSSR, Ser. Math., 35, 1971, 5.

[46] W. M. SEILER, Involution: The Formal Theory of Differential Equations and its Applications to Computer Algebra, Springer, 2009, 660 pp.
(See also doi:10.3842/SIGMA.2009.092, in particular sections 3 and 4).

[47] D.C. SPENCER: Some remarks on homological analysis and structures, Proc. Symp. Pure Math., 3, 1961, 56-86.

[48] D.C. SPENCER: deformation of structures on manifolds defined by transitive continuous pseudogroups, Ann. of Math., 76, 1962, (I) 306-445; 81, 1965, (II, III) 389-450.

[49] D. C. SPENCER: Overdetermined Systems of Partial Differential Equations, Bull. Am. Math. Soc., 75, 1965, 1-114.

[50] M. VERGNE: Cohomologie des algébres de Lie nilpotentes, Bull. Soc. Math. France, 98, 1970, 81-116.

[51] E. VESSIOT: Sur la théorie des groupes infinis, Ann. Ec. Norm. Sup., 20, 1903, 411-451.

[52] H. WEYL: Space, Time, Matter, Springer, 1918, 1958; Dover, 1952.

APPENDIX 1

ABSTRACT

The purpose of this Appendix 1, which is written in a rather self-contained way, is to revisit the Bianchi identities existing for the Riemann and Weyl tensors in the combined framework of the *formal theory* of systems of partial differential equations (Spencer cohomology, differential systems, formal integrability) and *Algebraic Analysis* (homological algebra, differential modules, duality). In particular, we prove that the $n^2(n^2 - 1)(n - 2)/24$ generating Bianchi identities for the Riemann tensor are *first order* and can be easily described by means of the Spencer cohomology of the first order Killing symbol in arbitrary dimension $n \geq 2$. Similarly, the $n(n^2-1)(n+2)(n-4)/24$ generating Bianchi identities for the Weyl tensor are *first order* and can be easily described by means of the Spencer cohomology of the first order conformal Killing symbol in arbitrary dimension $n \geq 5$. As a *most surprising result*, the 9 generating Bianchi identities for the Weyl tensor are of *second order* in dimension $n = 4$ while the analogue of the Weyl tensor has 5 components of *third order* in the metric with 3 *first order* generating Bianchi identities in dimension $n = 3$. The above results, *which could not be obtained otherwise*, are valid for any non-degenerate metric of constant riemannian curvature and do not depend on any conformal factor. They are checked in Appendix 2 by means of a computer algebra package produced by Alban Quadrat (INRIA, Lille). We finally explain why the work of Lanczos and followers is not coherent with these results and must therefore be also revisited.

KEY WORDS: Riemann tensor; Weyl tensor; Bianchi identities, Spencer cohomology, Vessiot structure equations; Poincaré sequence, Differential sequence; Differential modules; Compatibility conditions; Lanczos tensor.

1) INTRODUCTION

The language of *differential modules* has been recently introduced in applications as a way to understand the *structural properties* of systems of partial differential equations and the Poincaré duality between geometry and physics by using *adjoint operators* or variational calculus with differential constraints ([2],[23],[38]). In order to explain briefly the ideas of Lanczos as a way to justify the title of this paper, let us revisit briefly the foundation of n-dimensional elasticity theory as it can be found today in any textbook. If $x = (x^1, ..., x^n)$ is a point in space and $\xi = (\xi^1(x), ..., \xi^n(x))$ is the displacement vector, lowering the indices by means of the Euclidean metric, we may introduce the "small" deformation tensor $\epsilon = (\epsilon_{ij} = \epsilon_{ji} = (1/2)(\partial_i \xi_j + \partial_j \xi_i))$ with $n(n+1)/2$ (independent) *components* $(\epsilon_{i \leq j})$. If we study a part of a deformed body by means of a variational principle, we may introduce the local density of free energy $\varphi(\epsilon) = \varphi(\epsilon_{ij} | i \leq j)$ and vary the total free energy $\Phi = \int \varphi(\epsilon) dx$ with $dx = dx^1 \wedge ... \wedge dx^n$ by introducing $\sigma^{ij} = \partial \varphi / \partial \epsilon_{ij}$ for $i \leq j$ and "*deciding*" to define the stress tensor σ by a symmetric matrix with $\sigma^{ij} = \sigma^{ji}$ in a purely artificial way within such a variational principle. Indeed, the usual Cauchy Tetrahedron device (1828) assumes that each element of a boundary surface is acted on by a surface density of force $\vec{\sigma}$ with a linear dependence $\vec{\sigma} = (\sigma^{ir}(x) n_r)$ on the outward normal unit vector $\vec{n} = (n_r)$ and does not make any assumption on the stress tensor. It is only by an equilibrium of forces and couples, namely the well known *phenomenological static torsor equilibrium*, that one can *prove* the symmetry of σ. However, *if we assume this symmetry*, we may now consider the summation $\delta \Phi = \int \sigma^{ij} \delta \epsilon_{ij} dx = \int \sigma^{ir} \partial_r \delta \xi_i dx$. An integration by parts and a change of sign produce the integral $\int (\partial_r \sigma^{ir}) \delta \xi_i dx$ leading to the stress equations $\partial_r \sigma^{ir} = 0$. This classical approach to elasticity theory, based on invariant theory with respect to the group of rigid motions, cannot therefore describe equilibrium of torsors by means of a variational principle where the proper torsor concept is totally lacking. It is however widely used through the technique of "*finite elements*" where it can also be applied to electromagnetism (EM) with similar quadratic (piezoelectricity) or cubic (photoelasticity) lagrangian integrals, as will be shown in Appendix 3. In this situation, the 4-potential A of EM is used in place of ξ while the EM field $dA = F = (\vec{B}, \vec{E})$ is used in place of ϵ.

However, there exists another equivalent procedure dealing with a *variational calculus with constraint*. Indeed, as we shall see later on, the deformation tensor is not any symmetric tensor as it must satisfy $n^2(n^2 - 1)/12$ *Riemann* compatibility conditions (CC), that is the only condition $\partial_{22}\epsilon_{11} + \partial_{11}\epsilon_{22} - 2\partial_{12}\epsilon_{12} = 0$ when $n = 2$. In this case, introducing the *Lagrange multiplier* λ, *we have to vary the new integral* $\int[\varphi(\epsilon) + \lambda(\partial_{22}\epsilon_{11} + \partial_{11}\epsilon_{22} - 2\partial_{12}\epsilon_{12})]dx$ *for an arbitrary* ϵ. Setting $\lambda = -\phi$, a double integration by parts now provides the parametrization $\sigma^{11} = \partial_{22}\phi, \sigma^{12} = \sigma^{21} = -\partial_{12}\phi, \sigma^{22} = \partial_{11}\phi$ of the stress equations by means of the Airy function ϕ and the *formal adjoint* of the Riemann CC ([1],[26]). The same variational calculus with constraint may thus also be used in order to avoid the introduction of the EM potential A by using the Maxwell equations $dF = 0$ in place of the Riemann CC for ϵ but, in all these situations, *we have to eliminate the Lagrange multipliers or use them as potentials*.

In arbitrary dimension, the above compatibility conditions are nothing else but the linearized Riemann tensor in Riemannian geometry, a crucial mathematical tool in the theory of general relativity and a good reason for studying the work of Cornelius Lanczos (1893-1974) as it can be found in ([14],[15]) or in a few modern references ([5],[6],[7],[18],[36]). The starting point of Lanczos has been to take EM as a model in order to introduce a Lagrangian that should be quadratic in the Riemann tensor $(\rho_{l,ij}^k \Rightarrow \rho_{ij} = \rho_{i,rj}^r = \rho_{ji} \Rightarrow \rho = \omega^{ij}\rho_{ij})$ while considering it independently of its expression through the second order derivatives of a metric (ω_{ij}) with inverse (ω^{ij}) or the first order derivatives of the corresponding Christoffel symbols (γ_{ij}^k). According to the previous paragraph, the corresponding variational calculus *must* involve PD constraints made by the Bianchi identities and *the new lagrangian to vary must therefore contain as many Lagrange multipliers as the number of generating Bianchi identities* that can be written under the form:

$$\nabla_r\rho_{l,ij}^k + \nabla_i\rho_{l,jr}^k + \nabla_j\rho_{l,ri}^k = 0 \Rightarrow \nabla_r\rho_{l,ij}^r = \nabla_i\rho_{lj} - \nabla_j\rho_{li}$$

Meanwhile, Lanczos and followers have been looking for a kind of *"parametrization"* by using the corresponding *"Lanczos potential"*, exactly like the Lagrange multiplier has been used as an Airy potential for the stress equations. However, we shall prove that the definition of a *Riemann candidate* cannot be done without the knowledge of the Spencer cohomology. Moreover, we have pointed out the existence of well known couplings between elasticity and electromagnetism, namely piezoelectricity and photoelasticity, which are showing that, in the respective Lagrangians, the EM field is on equal footing with the deformation tensor and *not* with the Riemann tensor. The *shift by one step backwards* that must be used in the physical interpretation of the differential sequences involved cannot therefore be avoided. Meanwhile, the *ordinary derivatives* ∂_i can be used in place of the *covariant derivatives* ∇_i when dealing with the linearized framework as the Christoffel symbols vanish when Euclidean or Minkowskian metrics are used.

The next tentative of Lanczos has been to extend his approach to the Weyl tensor:

$$\tau^k_{l,ij} = \rho^k_{l,ij} - \frac{1}{(n-2)}(\delta^k_i \rho_{lj} - \delta^k_j \rho_{li} + \omega_{lj}\omega^{ks}\rho_{si} - \omega_{li}\omega^{ks}\rho_{sj}) + \frac{1}{(n-1)(n-2)}(\delta^k_i \omega_{lj} - \delta^k_j \omega_{li})\rho$$

The main problem is now that the Spencer cohomology of the symbols of the conformal Killing equations, in particular the 2-acyclicity, will be *absolutely needed* in order to provide the Weyl tensor and its relation with the Riemann tensor. It will follow that the CC for the Weyl tensor may not be first order contrary to the CC for the Riemann tensor made by the Bianchi identities, another reason for justifying the shift by one step already quoted. In order to provide an idea of the difficulty involved, let us define the following tensors:

$$Schouten = (\sigma_{ij} = \rho_{ij} - \frac{1}{2(n-1)}\omega_{ij}\rho) \Rightarrow Cotton = (\sigma_{k,ij} = \nabla_i \sigma_{kj} - \nabla_j \sigma_{ki})$$

An elementary but tedious computation allows to prove the formula:

$$\nabla_r \tau^r_{k,ij} = \frac{(n-3)}{(n-2)}\sigma_{k,ij}$$

Then, of course, *if Einstein equations in vacuum are valid*, the Schouten and Cotton tensors vanish but the left member is by no way a differential identity for the Weyl tensor and *great care must be taken when mixing up mathematics with physics*.

The author thanks Prof. Lars Andersson (Einstein Institute, Postdam) for having suggested him to study the *Lanczos potential* within this new framework and Alban Quadrat (INRIA, Lille), a specialist of control theory and computer algebra, for having checked directly in Appendix 2 the striking results contained in this Appendix.

2) HOMOLOGICAL ALGEBRA

We now need a few definitions andresults from homological algebra. In the following two classical theorems, $A, B, C, D, K, L, M, N, Q, R, S, T$ will be modules over a ring A or vector spaces over a field k and the linear maps are making the diagrams commutative. We start recalling the well known Cramer's rule for linear systems through the exactness of the ker/coker sequence for modules. When $f : M \to N$ is a linear map (homomorphism), we introduce the so-called ker/coker long exact sequence:

$$0 \longrightarrow ker(f) \longrightarrow M \overset{f}{\longrightarrow} N \longrightarrow coker(f) \longrightarrow 0$$

In the case of vector spaces over a field k, we successively have $rk(f) = dim(im(f))$, $dim(ker(f)) = dim(M) - rk(f)$ and $dim(coker(f)) = dim(N) - rk(f)$ is the proper number of compatibility conditions. We obtain by substraction:

$$dim(ker(f)) - dim(M) + dim(N) - dim(coker(f)) = 0$$

In the case of modules, we may replace the dimension by the rank with $rk_A(M) = r$ when $F \simeq A^r$ is the greatest free submodule of M and obtain the same relations because of the additive property of the rank ([23],[24],[33]). The following theorems will be crucially used through the whole paper ([3],[9],[17],[23],[37]):

SNAKE THEOREM 2.1: When one has the following commutative diagram resulting from the two central vertical short exact sequences by exhibiting the three corresponding horizontal ker/coker exact sequences:

$$
\begin{array}{ccccccccc}
& & 0 & & 0 & & 0 & & \\
& & \downarrow & & \downarrow & & \downarrow & & \\
0 & \to & K & \to & A & \to & A' & \to Q \to & 0 \\
& & \downarrow & & \downarrow f & & \downarrow f' & \downarrow & \\
0 & \to & L & \to & B & \to & B' & \to R \to & 0 \\
& & \downarrow & & \downarrow g & & \downarrow g' & \downarrow & \\
0 & \to & M & \to & C & \to & C' & \to S \to & 0 \\
& & \downarrow & & \downarrow & & \downarrow & & \\
& & 0 & & 0 & & 0 & &
\end{array}
$$

then there exists a connecting map $M \longrightarrow Q$ both with a long exact sequence:

$$0 \longrightarrow K \longrightarrow L \longrightarrow M \longrightarrow Q \longrightarrow R \longrightarrow S \longrightarrow 0.$$

Proof: We construct the connecting map by using the following succession of elements:

$$
\begin{array}{ccccc}
a & \cdots & a' & \longrightarrow & q \\
\vdots & & \downarrow & & \\
b & \longrightarrow & b' & & \\
\downarrow & & \vdots & & \\
m & \longrightarrow & c & \cdots & 0
\end{array}
$$

Indeed, starting with $m \in M$, we may identify it with $c \in C$ in the kernel of the next horizontal map. As g is an epimorphism, we may find $b \in B$ such that $c = g(b)$ and apply the next horizontal map to get $b' \in B'$ in the kernel of g' by the commutativity of the lower square. Accordingly, there is a unique $a' \in A'$ such that $b' = f'(a')$ and we may finally project a' to $q \in Q$. The map is well defined because, if we take another lift for c in B, it will differ from b by the image under f of a certain $a \in A$ having zero image in Q by composition. The remaining of the proof is similar. The above explicit procedure is called " *chase* " and will not be repeated.

Q.E.D.

We may now introduce *cohomology theory* through the following definition:

DEFINITION 2.2: If one has a sequence $L \xrightarrow{f} M \xrightarrow{g} N$, that is if $g \circ f = 0$, then one may introduce the submodules *coboundary* $= B = im(f) \subseteq ker(g) = cocycle = Z \subseteq M$ and define the cohomology at M to be the quotient $H = Z/B$. The sequence is said to be *exact* at M if $im(f) = ker(g)$.

COHOMOLOGY THEOREM 2.3: The following commutative diagram where the two central vertical sequences are long exact sequences and the horizontal lines

108

are ker/coker exact sequences:

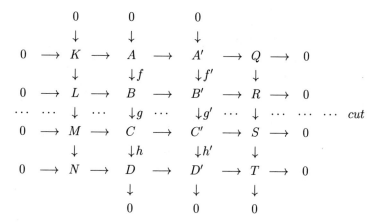

$$
\begin{array}{ccccccccc}
 & & 0 & & 0 & & 0 & & \\
 & & \downarrow & & \downarrow & & \downarrow & & \\
0 & \longrightarrow & K & \longrightarrow & A & \longrightarrow & A' & \longrightarrow Q \longrightarrow 0 \\
 & & \downarrow & & \downarrow f & & \downarrow f' & \quad \downarrow & \\
0 & \longrightarrow & L & \longrightarrow & B & \longrightarrow & B' & \longrightarrow R \longrightarrow 0 \\
\cdots & \cdots & \downarrow & \cdots & \downarrow g & \cdots & \downarrow g' \cdots \downarrow \cdots \cdots \cdots & & cut \\
0 & \longrightarrow & M & \longrightarrow & C & \longrightarrow & C' & \longrightarrow S \longrightarrow 0 \\
 & & \downarrow & & \downarrow h & & \downarrow h' & \quad \downarrow & \\
0 & \longrightarrow & N & \longrightarrow & D & \longrightarrow & D' & \longrightarrow T \longrightarrow 0 \\
 & & \downarrow & & \downarrow & & \downarrow & & \\
 & & 0 & & 0 & & 0 & & \\
\end{array}
$$

induces an isomorphism between the cohomology at M in the left vertical column and the kernel of the morphism $Q \to R$ in the right vertical column.

Proof. Let us "cut" the preceding diagram along the dotted line. We obtain the following two commutative and exact diagrams with $im(g) = ker(h), im(g') = ker(h')$:

$$
\begin{array}{ccccccccc}
 & & 0 & & 0 & & 0 & & \\
 & & \downarrow & & \downarrow & & \downarrow & & \\
0 \longrightarrow & & K & \longrightarrow & A & \longrightarrow & A' & \longrightarrow Q \longrightarrow 0 \\
 & & \downarrow & & \downarrow f & & \downarrow f' & \quad \downarrow & \\
0 \longrightarrow & & L & \longrightarrow & B & \longrightarrow & B' & \longrightarrow R \longrightarrow 0 \\
 & & \downarrow & & \downarrow g & & \downarrow g' & & \\
0 \longrightarrow & & cocycle & \longrightarrow & im(g) & \longrightarrow & im(g') & & \\
 & & & & \downarrow & & \downarrow & & \\
 & & & & 0 & & 0 & & \\
\end{array}
$$

$$
\begin{array}{ccccccc}
& & 0 & & 0 & & 0 \\
& & \downarrow & & \downarrow & & \downarrow \\
0 & \longrightarrow & cocycle & \longrightarrow & ker(h) & \longrightarrow & ker(h') \\
& & \downarrow & & \downarrow & & \downarrow \\
0 & \longrightarrow & M & \longrightarrow & C & \longrightarrow & C' \\
& & \downarrow & & \downarrow h & & \downarrow h' \\
0 & \longrightarrow & N & \longrightarrow & D & \longrightarrow & D' \\
& & & & \downarrow & & \downarrow \\
& & & & 0 & & 0
\end{array}
$$

Using the snake theorem, we successively obtain the following long exact sequences:

$$
\begin{aligned}
\Longrightarrow \quad \exists \qquad & 0 \longrightarrow K \longrightarrow L \xrightarrow{g} cocycle \longrightarrow Q \longrightarrow R \\
\Longrightarrow \quad \exists \qquad & 0 \longrightarrow coboundary \longrightarrow cocycle \longrightarrow ker\,(Q \longrightarrow R) \longrightarrow 0 \\
\Longrightarrow \qquad & cohomology \simeq ker\,(Q \longrightarrow R)
\end{aligned}
$$

Q.E.D.

We finally quote for a later use:

PROPOSITION 2.4: If one has a short exact sequence:

$$
0 \longrightarrow M' \xrightarrow{f} M \xrightarrow{g} M'' \longrightarrow 0
$$

then the following conditions are equivalent:
• There exists an epimorphism $u : M \to M'$ such that $u \circ f = id_{M'}$ (*left inverse* of f).
• There exists a monomorphism $v : M'' \to M$ such that $g \circ v = id_{M''}$ (*right inverse* of g).

DEFINITION 2.5: In the above situation, we say that the short exact sequence *splits*. The relation $f \circ u + v \circ g = id_M$ provides an isomorphism $(u, g) : M \to M' \oplus M''$ with inverse $f + v : M' \oplus M" \to M$. The short exact sequence $0 \to \mathbb{Z} \to \mathbb{Q} \to \mathbb{Q}/\mathbb{Z} \to 0$ cannot split over $\mathbb{Z}$.

3) DIFFERENTIAL SYSTEMS

If E is a vector bundle over the base manifold X with projection π and local co-ordinates $(x, y) = (x^i, y^k)$ projecting onto $x = (x^i)$ for $i = 1, ..., n$ and $k = 1, ..., m$, identifying a map with its graph, a (local) section $f : U \subset X \to E$ is such that $\pi \circ f = id$ on U and we write $y^k = f^k(x)$ or simply $y = f(x)$. For any change of local coordinates $(x, y) \to (\bar{x} = \varphi(x), \bar{y} = A(x)y)$ on E, the change of section is $y = f(x) \to \bar{y} = \bar{f}(\bar{x})$ such that $\bar{f}^l(\varphi(x)) \equiv A_k^l(x)f^k(x)$. The new vector bundle E^* obtained by changing the *transition matrix* A to its inverse A^{-1} is called the *dual vector bundle* of E. In particular, let T be the tangent vector bundle of vector fields on X, T^* be the cotangent vector bundle of 1-forms on X and $S_q T^*$ be the vector bundle of symmetric q-covariant tensors on X. Differentiating with respect to x^i and using new coordinates y_i^k in place of $\partial_i f^k(x)$, we obtain $\bar{y}_r^l \partial_i \varphi^r(x) = A_k^l(x)y_i^k + \partial_i A_k^l(x)y^k$. Introducing a multi-index $\mu = (\mu_1, ..., \mu_n)$ with length $| \mu |= \mu_1 + ... + \mu_n$ and prolonging the procedure up to order q, we may construct in this way, by patching coordinates, a vector bundle $J_q(E)$ over X, called the *jet bundle of order* q with local coordinates $(x, y_q) = (x^i, y_\mu^k)$ with $0 \leq | \mu |\leq q$ and $y_0^k = y^k$. We have therefore epimorphisms $\pi_q^{q+r} : J_{q+r}(E) \to J_q(E), \forall q, r \geq 0$ and the short exact sequences $0 \to S_q T^* \otimes E \to J_q(E) \xrightarrow{\pi_{q-1}^q} J_{q-1}(E) \to 0$. For a later use, we shall set $\mu + 1_i = (\mu_1, ..., \mu_{i-1}, \mu_i + 1, \mu_{i+1}, ..., \mu_n)$ and define the operator $j_q : E \to J_q(E) : f \to j_q(f)$ on sections by the local formula $j_q(f) : (x) \to (\partial_\mu f^k(x) \mid 0 \leq | \mu |\leq q, k = 1, ..., m)$. Moreover, a jet coordinate y_μ^k is said to be of *class i* if $\mu_1 = ... = \mu_{i-1} = 0, \mu_i \neq 0$. We finally inroduce the *Spencer operator* $D : J_{q+1}(E) \to T^* \otimes J_q(E) : f_{q+1} \to j_1(f_q) - f_{q+1}$ with $(Df_{q+1})_{\mu,i}^k = \partial_i f_\mu^k - f_{\mu+1_i}^k$.

DEFINITION 3.1: A *system* of PD equations of order q on E is a vector subbundle $R_q \subset J_q(E)$ locally defined by a constant rank system of linear equations for the jets of order q of the form $a_k^{\tau\mu}(x)y_\mu^k = 0$. Its *first prolongation* $R_{q+1} \subset J_{q+1}(E)$ will be defined by the equations $a_k^{\tau\mu}(x)y_\mu^k = 0, a_k^{\tau\mu}(x)y_{\mu+1_i}^k + \partial_i a_k^{\tau\mu}(x)y_\mu^k = 0$ which may not provide a system of constant rank as can easily be seen for $xy_x - y = 0 \Rightarrow xy_{xx} = 0$ where the rank drops at $x = 0$.

The next definition of *formal integrability* will be crucial for our purpose.

111

DEFINITION 3.2: A system R_q is said to be *formally integrable* if the R_{q+r} are vector bundles $\forall r \geq 0$ (regularity condition) and no new equation of order $q + r$ can be obtained by prolonging the given PD equations more than r times, $\forall r \geq 0$ or, equivalently, we have induced epimorphisms $\pi_{q+r}^{q+r+1} : R_{q+r+1} \to R_{q+r}, \forall r \geq 0$ allowing to compute " *step by step* " formal power series solutions.

A formal test first sketched by C. Riquier in 1910, has been improved by M. Janet in 1920 ([10],[19]) and by E. Cartan in 1945 ([4]), finally rediscovered in 1965, totally independently, by B. Buchberger who introduced Gröbner bases, using the name of his thesis advisor ([25]). However all these tentatives have been largely superseded and achieved in an intrinsic way, again totally independently of the previous approaches, by D.C. Spencer in 1965 ([19],[22],[39]).

DEFINITION 3.3: The family g_{q+r} of vector spaces over X defined by the purely linear equations $a_k^{\tau\mu}(x)v_{\mu+\nu}^k = 0$ for $\mid \mu \mid = q, \mid \nu \mid = r$ is called the *symbol* at order $q + r$ and only depends on g_q.

The following procedure, *where one may have to change linearly the independent variables if necessary*, is the key towards the next definition which is intrinsic even though it must be checked in a particular coordinate system called δ-*regular* (See [19],[22],[23] and [39] for more details):

• *Equations of class* n: Solve the maximum number β_q^n of equations with respect to the jets of order q and class n. Then call $(x^1, ..., x^n)$ *multiplicative variables.*

$- - - - - - - - - - - - - - - - -$

• *Equations of class* i: Solve the maximum number of *remaining* equations with respect to the jets of order q and class i. Then call $(x^1, ..., x^i)$ *multiplicative variables* and $(x^{i+1}, ..., x^n)$ *non-multiplicative variables.*

$- - - - - - - - - - - - - - - - - -$

• *Remaining equations equations of order* $\leq q-1$: Call $(x^1, ..., x^n)$ *non-multiplicative*

112

variables.

DEFINITION 3.4: The above multiplicative and non-multiplicative variables can be visualized respectively by integers and dots in the corresponding *Janet board*. A system of PD equations is said to be *involutive* if its first prolongation can be achieved by prolonging its equations only with respect to the corresponding multiplicative variables. The following numbers are called *characters*:

$$\alpha_q^i = m(q + n - i - 1)!/((q-1)!(n-i)!) - \beta_q^i, \quad \forall 1 \le i \le n \;\Rightarrow\; \alpha_q^1 \ge \dots \ge \alpha_q^n$$

For an involutive system, $(y^{\beta_q^n+1}, ..., y^m)$ can be given arbitrarily.

For an involutive system of order q in the above *solved form*, we shall use to denote by y_{pri} the *principal jet coordinates*, namely the leading terms of the solved equations in the sense of involution, and any formal derivative of a principal jet coordinate is again a principal jet coordinate. The remaining jet coordinates will be called *parametric jet coordinates* and denoted by y_{par}.

PROPOSITION 3.5: Using the Janet board and the definition of involutivity, we get:

$$dim(g_{q+r}) = \sum_{i=1}^{n} \frac{(r+i-1)!}{r!(i-1)!}\alpha_q^i \Rightarrow dim(R_{q+r}) = dim(R_{q-1}) + \sum_{i=1}^{n} \frac{(r+i)!}{r!i!}\alpha_q^i$$

Let now $\wedge^s T^*$ be the vector bundle of s-forms on X with usual bases $\{dx^I = dx^{i_1} \wedge \dots \wedge dx^{i_s}\}$ where we have set $I = (i_1 < \dots < i_s)$. . Moreover, if $\xi, \eta \in T$ are two vector fields on X, we may define their *bracket* $[\xi, \eta] \in T$ by the local formula $([\xi, \eta])^i(x) = \xi^r(x)\partial_r\eta^i(x) - \eta^s(x)\partial_s\xi^i(x)$ leading to the *Jacobi identity* $[\xi, [\eta, \zeta]] + [\eta, [\zeta, \xi]] + [\zeta, [\xi, \eta]] = 0, \forall \xi, \eta, \zeta \in T$. We may finally introduce the *exterior derivative* $d : \wedge^r T^* \to \wedge^{r+1} T^* : \omega = \omega_I dx^I \to d\omega = \partial_i\omega_I dx^i \wedge dx^I$ with $d^2 = d \circ d \equiv 0$ in the *Poincaré sequence*:

$$\wedge^0 T^* \xrightarrow{d} \wedge^1 T^* \xrightarrow{d} \wedge^2 T^* \xrightarrow{d} \dots \xrightarrow{d} \wedge^n T^* \longrightarrow 0$$

In a purely algebraic setting, one has ([19],[22],[23],[24],[39]):

113

PROPOSITION 3.6: There exists a map $\delta : \wedge^s T^* \otimes S_{q+1} T^* \otimes E \to \wedge^{s+1} T^* \otimes S_q T^* \otimes E$ which restricts to $\delta : \wedge^s T^* \otimes g_{q+1} \to \wedge^{s+1} T^* \otimes g_q$ and $\delta^2 = \delta \circ \delta = 0$.

Proof: Let us introduce the family of s-forms $\omega = \{\omega_\mu^k = v_{\mu,I}^k dx^I\}$ and set $(\delta\omega)_\mu^k = dx^i \wedge \omega_{\mu+1_i}^k$. We obtain at once $(\delta^2\omega)_\mu^k = dx^i \wedge dx^j \wedge \omega_{\mu+1_i+1_j}^k = 0$.

$$\text{Q.E.D.}$$

The kernel of each δ in the first case is equal to the image of the preceding δ but this may no longer be true in the restricted case and we set (See [22], p 85-88 for more details):

DEFINITION 3.7: Let $B_{q+r}^s(g_q) \subseteq Z_{q+r}^s(g_q)$ and $H_{q+r}^s(g_q) = Z_{q+r}^s(g_q)/B_{q+r}^s(g_q)$ be respectively the coboundary space, cocycle space and cohomology space at $\wedge^s T^* \otimes g_{q+r}$ of the restricted δ-sequence which only depend on g_q and may not be vector bundles. The symbol g_q is said to be *s-acyclic* if $H_{q+r}^1 = ... = H_{q+r}^s = 0, \forall r \geq 0$, *involutive* if it is n-acyclic and *finite type* if $g_{q+r} = 0$ becomes trivially involutive for r large enough. For a later use, we notice that a symbol g_q is involutive *and* of finite type if and only if $g_q = 0$. Finally, $S_q T^* \otimes E$ is involutive $\forall q \geq 0$ if we set $S_0 T^* \otimes E = E$. We shall prove later on that *any* symbol g_q is 1-acyclc.

CRITERION THEOREM 3.8: If $\pi_q^{q+1} : R_{q+1} \to R_q$ is an epimorphism of vector bundles and g_q is 2-acyclic (involutive), then R_q is formally integrable (involutive).

EXAMPLE 3.9: The system R_2 defined by the three PD equations

$$\Phi^3 \equiv y_{33} = 0, \quad \Phi^2 \equiv y_{23} - y_{11} = 0, \quad \Phi^1 \equiv y_{22} = 0$$

is homogeneous and thus automatically formally integrable but g_2 and g_3 are not involutive though finite type because $g_4 = 0$ and the sequence $0 \to \wedge^3 T^* \otimes g_3 \to 0$ is not exact. Elementary computations of ranks of matrices shows that the δ-map:

$$0 \to \wedge^2 T^* \otimes g_3 \xrightarrow{\delta} \wedge^3 T^* \otimes g_2 \to 0$$

is a 3×3 isomorphism and thus g_3 is 2-acyclic with $dim(g_3) = 1$, a *crucial intrinsic* property totally absent from any "old" work and quite more easy to handle than its Koszul dual. We invite the reader to treat similarly the system $y_{33} - y_{11} = 0, y_{23} = 0, y_{22} - y_{11} = 0$ and compare.

The main use of involution is to construct differential sequences that are made up by successive *compatibility conditions* (CC) of order one. In particular, when R_q is involutive, the differential operator $\mathcal{D} : E \xrightarrow{j_q} J_q(E) \xrightarrow{\Phi} J_q(E)/R_q = F_0$ of order q with space of solutions $\Theta \subset E$ is said to be *involutive* and one has the canonical *linear Janet sequence* ([22], p 144):

$$0 \longrightarrow \Theta \longrightarrow E \xrightarrow{\mathcal{D}} F_0 \xrightarrow{\mathcal{D}_1} F_1 \xrightarrow{\mathcal{D}_2} ... \xrightarrow{\mathcal{D}_n} F_n \longrightarrow 0$$

where each other operator is first order involutive and generates the CC of the preceding one with the *Janet bundles* $F_r = \wedge^r T^* \otimes J_q(E)/(\wedge^r T^* \otimes R_q + \delta(\wedge^{r-1} T^* \otimes S_{q+1} T^* \otimes E))$. As the Janet sequence can be "cut at any place", that is can also be constructed anew from any intermediate operator, *the numbering of the Janet bundles has nothing to do with that of the Poincaré sequence for the exterior derivative*, contrary to what many physicists still believe ($n = 3$ with $\mathcal{D} = div$ provides the simplest example). Moreover, the fiber dimension of the Janet bundles can be computed at once inductively from the board of multiplicative and non-multiplicative variables that can be exhibited for $\mathcal{D}$ by working out the board for $\mathcal{D}_1$ and so on. For this, the number of rows of this new board is the number of dots appearing in the initial board while the number $nb(i)$ of dots in the column i just indicates the number of CC of class i for $i = 1, ..., n$ with $nb(i) < nb(j), \forall i < j$.

MAIN THEOREM 3.10: When $R_q \subset J_q(E)$ is *not* involutive but formally integrable and *its symbol g_q becomes 2-acyclic after exactly s prolongations*, the generating CC are of order $s + 1$ (See [22], Example 6, p 120 and previous Example).

Proof: We may introduce the canonical epimorphism $\Phi = J_q(E) \to J_q(E)/R_q = F_0$ and denote by $\mathcal{D} = \Phi \circ j_q : E \to F_0$ the corresponding differential operator. As before, we may write formally $\Phi^\tau(x, y_q) \equiv a_{\mu,k}^\tau(x)y_\mu^k = z^\tau$, obtain $d_i \Phi^\tau \equiv a_{\mu,k}^\tau(x)y_{\mu+1_i}^k + \partial_i a_{\mu,k}^\tau(x)y_\mu^k = z_i^\tau$ for the first prolongation $\rho_1(\Phi) : J_{q+1}(E) \to j_1(F_0)$ and so on with $\rho_r(\Phi) : J_{q+r}(E) \to J_r(F_0)$ defined by $d_\nu \Phi^\tau = z_\nu^\tau$ with $0 \leq |\nu| \leq r$. Setting $B_r = im(\rho_r(\Phi)) \subseteq J_r(F_0)$, we may introduce the canonical epimorphism $\Psi : J_{s+1}(F_0) \to J_{s+1}(F_0)/B_{s+1} = F_1$. Taking into account the formal integrability

115

of R_q (care), we obtain by composition of jets the following commutative *prolongation diagrams* $\forall r \geq 1$:

$$
\begin{array}{ccccccc}
0 & & 0 & & 0 & & 0 \\
\downarrow & & \downarrow & & \downarrow & & \downarrow \\
0 \to \quad g_{q+r+s+1} & \to & S_{q+r+s+1}T^* \otimes E & \xrightarrow{\sigma_{r+s+1}(\Phi)} & S_{r+s+1}T^* \otimes F_0 & \xrightarrow{\sigma_r(\Psi)} & S_r T^* \otimes F_1 \\
\downarrow & & \downarrow & & \downarrow & & \downarrow \\
0 \to \quad R_{q+r+s+1} & \to & J_{q+r+s+1}(E) & \xrightarrow{\rho_{r+s+1}(\Phi)} & J_{r+s+1}(F_0) & \xrightarrow{\rho_r(\Psi)} & J_r(F_1) \\
\downarrow & & \downarrow & & \downarrow & & \downarrow \\
0 \to \quad R_{q+r+s} & \to & J_{q+r+s}(E) & \xrightarrow{\rho_{r+s}(\Phi)} & J_{r+s}(F_0) & \xrightarrow{\rho_{r-1}(\Psi)} & J_{r-1}(F_1) \\
\downarrow & & \downarrow & & \downarrow & & \downarrow \\
0 & & 0 & & 0 & & 0
\end{array}
$$

and the only thing we know is that the bottom sequence is exact for $r = 1$ by construction and that the upper induced sequence is exact when $r = 0$ when $\sigma_0(\Psi) = \sigma(\Psi)$ is the restriction of Ψ to $S_{s+1}T^* \otimes F_0 \subset J_{s+1}(F_0)$ after a chase in the following commutative diagram:

$$
\begin{array}{ccccccc}
0 & & 0 & & 0 & & 0 \\
\downarrow & & \downarrow & & \downarrow & & \downarrow \\
0 \to \quad g_{q+s+1} & \to & S_{q+s+1}T^* \otimes E & \xrightarrow{\sigma_{s+1}(\Phi)} & S_{s+1}T^* \otimes F_0 & \xrightarrow{\sigma(\Psi)} & F_1 \\
\downarrow & & \downarrow & & \downarrow & & \downarrow \\
0 \to \quad R_{q+s+1} & \to & J_{q+s+1}(E) & \xrightarrow{\rho_{s+1}(\Phi)} & J_{s+1}(F_0) & \xrightarrow{\Psi} & F_1 \\
\downarrow & & \downarrow & & \downarrow & & \\
0 \to \quad R_{q+s} & \to & J_{q+s}(E) & \xrightarrow{\rho_s(\Phi)} & J_s(F_0) & & \\
\downarrow & & \downarrow & & \downarrow & & \\
0 & & 0 & & 0 & &
\end{array}
$$

because R_q is formally integrable (care). Appying now the δ-maps to the upper row of the previous prolongation diagram and proceeding by induction, starting from $r = 1$, we shall prove that the upper row is exact. Indeed, setting $h_{r+s} = im(\sigma_{r+s}(\Phi)) \subseteq S_{r+s}T^* \otimes F_0$, we may cut the full commutative diagram thus obtained as in the proof of the previous "cohomology theorem" into the following two commutative diagrams:

116

$$
\begin{array}{ccccccccc}
& & 0 & & 0 & & 0 & & \\
& & \downarrow & & \downarrow & & \downarrow & & \\
0 \to & & g_{q+r+s+1} & \to & S_{q+r+s+1}T^* \otimes E & \to & h_{r+s+1} & \to 0 & \\
& & \downarrow & & \downarrow & & \downarrow & & \\
0 \to & & T^* \otimes g_{q+r+s} & \to & T^* \otimes S_{q+r+s}T^* \otimes E & \to & T^* \otimes h_{r+s} & \to 0 & \\
& & \downarrow & & \downarrow & & \downarrow & & \\
0 \to & & \wedge^2 T^* \otimes g_{q+r+s-1} & \to & \wedge^2 T^* \otimes S_{q+r+s-1}T^* \otimes E & \to & \wedge^2 T^* \otimes h_{r+s-1} & \to 0 & \\
& & \downarrow & & \downarrow & & & & \\
0 \to & & \wedge^3 T^* \otimes S_{q+r+s-2}T^* \otimes E & = & \wedge^3 T^* \otimes S_{q+r+s-2}T^* \otimes E & & & &
\end{array}
$$

$$
\begin{array}{ccccccc}
& & 0 & & 0 & & 0 \\
& & \downarrow & & \downarrow & & \downarrow \\
0 \to & & h_{r+s+1} & \to & S_{r+s+1}T^* \otimes F_0 & \to & S_r T^* \otimes F_1 \\
& & \downarrow & & \downarrow & & \downarrow \\
0 \to & & T^* \otimes h_{r+s} & \to & T^* \otimes S_{r+s}T^* \otimes F_0 & \to & T^* \otimes S_{r-1}T^* \otimes F_1 \\
& & \downarrow & & \downarrow & & \\
0 \to & & \wedge^2 T^* \otimes h_{r+s-1} & \to & \wedge^2 T^* \otimes S_{r+s-1}T^* \otimes F_0 & &
\end{array}
$$

An easy chase is showing that g_q is always 1-acyclic and that we have an induced monomorphism $0 \to h_{r+s+1} \to T^* \otimes h_{r+s}$. The crucial result that no classical approach could provide is that, whenever g_{q+s} is 2-acyclic, then the full right column of the first diagram is also exact or, equivalently, $h_{r+s+1} \subseteq S_{r+s+1}T^* \otimes F_0$ is the r-prolongation of the symbol $h_{s+1} \subseteq S_{s+1}T^* \otimes F_0$. Using finally the second diagram, it follows by induction and a chase that the upper row is exact whenever the central row is exact, a result achieving the first part of the proof.

We may also use an inductive chase in the full diagram, showing directly that the cohomology at $S_{r+s+1}T^* \otimes F_0$ of the upper sequence:

$$
0 \to g_{q+r+s+1} \to S_{q+r+s+1}T^* \otimes E \xrightarrow{\sigma_{r+s+1}(\Phi)} S_{r+s+1}T^* \otimes F_0 \xrightarrow{\sigma_r(\Psi)} S_r T^* \otimes F_1
$$

is the same as the δ-cohomology of the left column at $\wedge^2 T^* \otimes g_{q+r+s-1}$ because all the other vertical δ-sequences are exact.

Finally, starting from the *ker/coker* long exact sequence allowing to define Ψ and ending with F_1 while taking into account that the upper symbol row of the prolongation diagram is exact, we deduce by an inductive chase that the central row is also exact. It follows that B_{r+s+1} is the r-prolongation of B_{s+1} which is formally integrable because a chase shows that B_{r+s+1} projects onto B_{r+s}, $\forall r \geq 1$.The case of an involutive symbol can be studied similarly by choosing $s = 0$ and explains why all the CC operators met in the Janet sequence are first order involutive operators.

Q.E.D.

As we shall see through explicit examples, in particular the conformal Killing system, there is no rule in general in order to decide about the minimum number $s' \geq 0$ such that $h_{s+s'+1}$ becomes 2-acyclic in order to repeat the above procedure. However, replacing r by $s' + 2$ and chasing in the first of the last two diagrams, we have:

COROLLARY 3.11: The symbol $h_{s+s'+1}$ *becomes* 2-acyclic whenever the symbol $g_{q+s+s'}$ *becomes* 3-acyclic.

DEFINITION 3.12: More generally, a differential sequence is said to be *formally exact* if each operator generates the CC of the operator preceding it.

EXAMPLE 3.13: ([16],§38, p 40) The second order system $y_{11} = 0, y_{13} - y_2 = 0$ is neither formally integrable nor involutive. Indeed, we get $d_3 y_{11} - d_1(y_{13} - y_2) = y_{12}$ and $d_3 y_{12} - d_2(y_{13} - y_2) = y_{22}$, that is to say *each first and second* prolongation does bring a new second order PD equation. Considering the new system $y_{22} = 0, y_{12} = 0, y_{13} - y_2 = 0, y_{11} = 0$, the (evident !) permutation of coordinates $(1, 2, 3) \rightarrow (3, 2, 1)$ provides the following involutive second order system with one equation of class 3, 2 equations of class 2 and 1 equation of clas 1:

$$
\begin{cases}
\Phi^4 \equiv y_{33} & = 0 \\
\Phi^3 \equiv y_{23} & = 0 \\
\Phi^2 \equiv y_{22} & = 0 \\
\Phi^1 \equiv y_{13} - y_2 & = 0
\end{cases}
\quad
\begin{array}{|ccc|}
\hline
1 & 2 & 3 \\
1 & 2 & \bullet \\
1 & 2 & \bullet \\
1 & \bullet & \bullet \\
\hline
\end{array}
$$

We have $\alpha_2^3 = 0, \alpha_2^2 = 0, \alpha_2^1 = 2$ and we get therefore the (formally exact) Janet sequence:

$$0 \longrightarrow \Theta \longrightarrow 1 \longrightarrow 4 \longrightarrow 4 \longrightarrow 1 \longrightarrow 0$$

However, keeping only Φ^1 and Φ^4 while using the fact that d_{33} commutes with $d_{13} - d_2$, we get the formally exact sequence $0 \to \Theta \to 1 \to 2 \to 1 \to 0$ *which is not a Janet sequence* though the Euler-Poincaré characteristics vanishes in both cases with $1 - 4 + 4 - 1 = 1 - 2 + 1 = 0$ ([22], p 159 and [23]).

EXAMPLE 3.14: When $n = 3$ and $E = X \times \mathbb{R}$, the second order system $R_2 \subset J_2(E)$ defined by the three PD equations $Py \equiv y_{33} = 0, Qy \equiv y_{23} - y_{11} = 0, Ry \equiv y_{22} = 0$ with corresponding operators P, Q, R in the commutative differential ring of operators $D = D\mathbb{Q}[d_1, d_2, d_3]$ is trivially formally integrable because it is homogeneous but is not involutive because its symbol g_2 with $dim(g_2) = 6 - 3 = 3$ is finite type with $dim(g_3) = 1$ and $g_{4+r} = 0, \forall r \geq 0$. Accordingly, we have $dim(R_2) = 1 + 3 + 3 = 7$ while $dim(R_{3+r}) = 8 = 2^n, \forall r \geq 0$ (For the rigin of this formula, see [16], p 79). As we are dealing with one unknown only and D is commutative, the 3 *second order* (care) CC are produced by the 3 commutation relations obtained by circular permutations from $[P, Q] = P \circ Q - Q \circ P = 0$. Similarly, the *second order again* (care) next CC is produced by the Jacobi identity $[P, [Q, R]] + [Q, [R, P]] + [R, [P, Q]] \equiv 0$.

With more details, coming back to Example 3.9 while intoducing the three second order operators $P = d_{22}, Q = d_{23} - d_{11}, R = d_{33}$ which are commuting between themselves, we have now $q = 2, s = 1$ and we obtain the second order CC:

$$\Psi^3 \equiv P\Phi^2 - Q\Phi^1 = 0, \quad \Psi^2 \equiv R\Phi^1 - P\Phi^3 = 0, \quad \Psi^1 = Q\Phi^3 - R\Phi^2 = 0$$

Exactly like in the Poincaré sequence, we finally get the new second order CC:

$$P\Psi^1 + Q\Psi^2 + R\Psi^3 = 0$$

Writing out only the number of respective equations, we obtain the formally exact differential sequence with vanishing Euler-Poincaré characteristics:

$$0 \to \Theta \to 1 \xrightarrow{\mathcal{D}_1} 3 \xrightarrow{\mathcal{D}_2} 3 \xrightarrow{\mathcal{D}_3} 1 \to 0$$

which is made by second order operators only and is thus very far from being a Janet sequence.

Reversing the arrows, we obtain at once the following finite free resolution for the corresponding differential module M:

119

$$0 \to D \xrightarrow{\mathcal{D}_3} D^3 \xrightarrow{\mathcal{D}_2} D^3 \xrightarrow{\mathcal{D}_1} D \to M \to 0$$

We let the reader prove as an exercise of linear algebra that g_3 is 2-acyclic by showing the exactness of the δ-sequence $0 \to \wedge^2 T^* \otimes g_3 \xrightarrow{\delta} \wedge^3 T^* \otimes g_2 \to 0$, that h_2 is *not* 2-acyclic but that h_3 is 2-acyclic. It follows that $s' = 1$ because $2 + 1 + 1 = 4$ and $g_4 = 0$ is trivially involutive. A similar situation will be met with the conformal Killing equations.

We may consider the first prolongation $R_3 \subset J_3(E)$ defined by the following 12 PD equations:

$$\begin{cases} \phi^1 \equiv y_{333} = 0 \\ \phi^2 \equiv y_{233} = 0, \phi^3 \equiv y_{223} = 0, \phi^4 \equiv y_{222} = 0 \\ \phi^5 \equiv y_{133} = 0, \phi^6 \equiv y_{123} - y_{111} = \phi^7 \equiv y_{122} = 0, \phi^8 \equiv y_{113} = 0, \phi^9 \equiv y_{112} = 0 \\ \phi^{10} \equiv y_{33} = 0, \phi^{11} \equiv y_{23} - y_{11} = 0, \phi^{12} \equiv y_{22} = 0 \end{cases}$$

1	2	3
1	2	•
1	•	•
•	•	•

In this particular situation, that is when g_3 is already 2-acyclic though NOT involutive, we have just proved that the generating *compatibility conditions* (CC) are first order (See also [23], p 120) and described by the following 21 PD equations:

$$\begin{cases} \psi^1 \equiv d_3\phi^2 - d_2\phi^1 = 0, ..., \psi^8 \equiv d_3\phi^9 - d_2\phi^8 = 0 \\ \psi^9 \equiv d_2\phi^5 - d_1\phi^2 = 0, ..., \psi^{12} \equiv d_2\phi^9 - d_1\phi^7 = 0 \\ \psi^{13} \equiv d_3\phi^{10} - \phi^1 = 0, ..., \psi^{15} \equiv d_3\phi^{12} - \phi^3 = 0 \\ \psi^{16} \equiv d_2\phi^{10} - \phi^2 = 0, ..., \psi^{18} \equiv d_2\phi^{12} - \phi^4 = 0 \\ \psi^{19} \equiv d_1\phi^{10} - \phi^5 = 0, ..., \psi^{21} \equiv d_1\phi^{12} - \phi^7 = 0 \end{cases}$$

1	2	3
1	2	•
1	2	3
1	2	•
1	•	•

where we had to *skip* $d_2\phi^8 - d_1\phi^6 = y_{1111}$.

In order to discover this number, we just need to compute the dimensions in the long exact sequence used as a central row in the following commutative and exact diagram where $dim(E) = 1$ and the upper row is the corresponding symbol sequence where $g_4 = 0$:

$$
\begin{array}{ccccccc}
0 & & 0 & & 0 & & 0 \\
\downarrow & & \downarrow & & \downarrow & & \downarrow \\
0 \to g_4 & \longrightarrow & S_4 T^* \otimes E & \longrightarrow & T^* \otimes F_0 & \longrightarrow & F_1 \to 0 \\
\downarrow & & \downarrow & & \downarrow & & \parallel \\
0 \to R_4 & \longrightarrow & J_4(E) & \longrightarrow & J_1(F_0) & \longrightarrow & F_1 \to 0 \\
\downarrow & & \downarrow & & \downarrow & & \downarrow \\
0 \to R_3 & \longrightarrow & J_3(E) & \longrightarrow & F_0 & \longrightarrow & 0 \\
\downarrow & & \downarrow & & \downarrow & & \\
0 & & 0 & & 0 & &
\end{array}
$$

Each dot in the last Janet board is producing one CC *apart from one* as we may verify the relation:

$$d_3 \psi^{12} - d_2 \psi^8 + d_1 \psi^6 \equiv d_{22}\phi^8 - d_{11}\phi^3$$

which is describing in some sense the *defect of involutivity* and we let the reader check that $h_1 \subset T^* \otimes F_0$ is not involutive as it is even not 2-acyclic, while its first prolongation $h_2 \subset S_2 T^* \otimes F_0$ is 2-acyclic and even involutive.

We check therefore the remaining $13 - 1 = 12$ first order CC:

$$
\left\{
\begin{array}{ll}
\theta^1 \equiv d_3 \psi^9 - d_2 \psi^4 + d_1 \psi^1 = 0, ..., \theta^3 \equiv d_3 \psi^{11} - d_2 \psi^6 + d_1 \psi^3 = 0 & \boxed{1\ \ 2\ \ 3} \\
\theta^4 \equiv d_3 \psi^{16} - d_2 \psi^{13} + \psi^1 = 0, ..., \theta^6 \equiv d_3 \psi^{18} - d_2 \psi^{15} + \psi^3 = 0 & \boxed{1\ \ 2\ \ 3} \\
\theta^7 \equiv d_3 \psi^{19} - d_1 \psi^{13} + \psi^4 = 0, ..., \theta^9 \equiv d_3 \psi^{21} - d_1 \psi^{15} + \psi^6 = 0 & \boxed{1\ \ 2\ \ 3} \\
\theta^{10} \equiv d_2 \psi^{19} - d_1 \psi^{16} + \psi^9 = 0, ..., \theta^{12} \equiv d_2 \psi^{21} - d_1 \psi^{18} + \psi^{11} = 0 & \boxed{1\ \ 2\ \ \bullet}
\end{array}
\right.
$$

It is quite a pure chance that this system is involutive with the 3 first order CC:

$$
\left\{
d_3 \theta^{10} - d_2 \theta^7 + d_1 \theta^4 - \theta^1 = 0, ..., d_3 \theta^{12} - d_2 \theta^9 + d_1 \theta^6 - \theta^3 = 0 \quad \boxed{1\ \ 2\ \ 3}
\right.
$$

The following absolutely nontrivial points will be crucial for understanding the structure of the conformal equations later on (Compare to [32]).

Indeed, starting from the short exact sequence $0 \to R_3 \to J_3(E) \xrightarrow{\Phi} F_0 \to 0$ with fiber dimensions $0 \to 8 \to 20 \to 12 \to 0$ and using 2 prolongations in order to "reach" F_3, we get the following jet sequence of vector bundles, in fact the same that should be produced by any symbolic package:

$$0 \to R_5 \to J_5(E) \longrightarrow J_2(F_0) \longrightarrow J_1(F_1) \longrightarrow F_2 \to 0$$

with respective fiber dimensions:

$$0 \to 8 \to 56 \to 120 \to 84 \to dim(F_2) \to 0$$

Accordingly, *if the sequence were exact*, using the Euler-Poincaré formula ([15], Lemma 2.2, p 206), we should get $dim(F_2) = 56 + 84 - 120 - 8 = 12$, a result showing that *the sequence is indeed exact*. However, the differential sequence thus obtained should not be formally exact because we know that g_3 is 2-acyclic but not 3-acyclic, that is involutive. We have the long exact sequence:

$$0 \to R_6 \to J_6(E) \longrightarrow J_3(F_0) \longrightarrow J_2(F_1) \longrightarrow J_1(F_2) \longrightarrow F_3 \to 0$$

with respective fiber dimensions:

$$0 \to 8 \to 84 \to 240 \to 210 \to 48 \to 3 \to 0$$

and check that $8 - 84 + 240 - 210 + 48 - 3 = -1 \neq 0$.

We may therefore introduce the corresponding symbol sequence as upper row of the following commutative diagram in order to understand why the previous sequence is not exact and what is the resulting cohomology at $J_2(F_1)$. Indeed, in this diagram obtained by induction, *all the rows are exact but perhaps the upper one* and *all the columns are exact but perhaps the left one*:

	0		0		0		0		0
	$\downarrow$		$\downarrow$		$\downarrow$		$\downarrow$		$\downarrow$
$0 \to$	g_6	$\to$	S_6T^*	$\to$	$S_3T^* \otimes F_0$	$\to$	$S_2T^* \otimes F_1$	$\to$	$T^* \otimes F_2$
	$\downarrow \delta$		$\downarrow \delta$		$\downarrow \delta$		$\downarrow \delta$		$\parallel$
$0 \to$	$T^* \otimes g_5$	$\to$	$T^* \otimes S_5T^*$	$\to$	$T^* \otimes S_2T^* \otimes F_0$	$\to$	$T^* \otimes T^* \otimes F_1$	$\to$	$T^* \otimes F_2$
	$\downarrow \delta$		$\downarrow \delta$		$\downarrow \delta$		$\downarrow \delta$		$\downarrow$
$0 \to$	$\wedge^2 T^* \otimes g_4$	$\to$	$\wedge^2 T^* \otimes S_4T^*$	$\to$	$\wedge^2 T^* \otimes T^* \otimes F_0$	$\to$	$\wedge^2 T^* \otimes F_1$	$\to$	0
	$\downarrow \delta$		$\downarrow \delta$		$\downarrow \delta$		$\downarrow$		
$0 \to$	$\wedge^3 T^* \otimes g_3$	$\to$	$\wedge^3 T^* \otimes S_3T^*$	$\to$	$\wedge^3 T^* \otimes F_0$		0		
	$\downarrow$		$\downarrow$		$\downarrow$				
	0		0		0				

As $g_4 = g_5 = g_6 = 0$ and $dim(\wedge^3 T^* \otimes g_3) = dim(g_3) = 1$, a chase using the standard *snake lemma* of homological algebra ([32], p 174) proves that the upper seqence is not exact at $S_2T^* \otimes F_1$ with cohomology of dimension $dim(\wedge^3 T^* \otimes g_3) =$

$dim(g_3) = 1$. Hence, the previous sequence is not exact at $J_2(F_1)$, that is with $dim(im(J_3(F_0) \to J_2(F_1))) = 240 - 84 + 8 = 164$ while $dim(ker(J_2(F_1) \to J_1(F_2))) = 164 + 1 = 165$ and we have indeed $48 - 210 + 165 = 3$.

In a coherent way with the main theorem obtained, it follows from this result that *the previous first order* CC *have second order generating* CC, *a fact highly not evident at first sight* according to the preceding computations. Hence *we must define the correct F_2* by the long exact sequence:

$$0 \to R_6 \longrightarrow J_6(E) \longrightarrow J_3(F_0) \longrightarrow J_2(F_1) \longrightarrow F_2 \to 0$$

with respective fiber dimensions:

$$0 \to 8 \longrightarrow 84 \longrightarrow 240 \longrightarrow 210 \longrightarrow 46 \to 0$$

However, we must find back the $12 + 3 \times 12$ equations and their first derivatives but we have to take into account that they are related by 3 identities, that is we should find $12 + 36 - 3 = 45$ and *there is therefore one additional* CC *of second order* which is not a differential consequence of the first order ones already obtained (!). We ask the reader to make his mind for a few hours (... or days !) in order to discover it by himself and he will understand the kind of difficulty met for constructing explicitly differential sequences or resolutions while dealing with very big matrices.

For this, differentiating once all the $\psi^1, ..., \psi^{12}$ we obtain successively 8 equations of order 2 and class 3, $12 + 1 = 13$ equations of order 2 and class 2 including the one we already found for the defect of involutivity and 12 equations of order 2 and class 1. We obtain thus a total of $36 - 3 = 33$ linearly independent second order equations among which we select the 4 following ones that we differentiate conveniently as indicated below.

$$
\begin{array}{llll}
class\ 2 & d_{23}\phi^8 - d_{12}\phi^5 = d_2\psi^7 & \leftarrow & -d_2 \\
class\ 2 & d_{22}\phi^5 - d_{12}\phi^2 = d_2\psi^9 & \leftarrow & -d_1 \\
class\ 2 & d_{22}\phi^8 - d_{11}\phi^3 = d_3\psi^{12} - d_2\psi^8 + d_1\psi^6 & \leftarrow & d_3 \\
class\ 1 & d_{13}\phi^3 - d_{12}\phi^2 = d_1\psi^2 & \leftarrow & d_1
\end{array}
$$

We finally obtain the only additional second order homogeneous identity that we were looking for, namely:

$$d_{33}\psi^{12} - d_{23}\psi^8 - d_{22}\psi^7 + d_{13}\psi^6 - d_{12}\psi^9 + d_{11}\psi^2 \equiv 0$$

and check at once that it *cannot* be obtained by prolonging the previous system of 12 first order CC because its leading term $d_{33}\psi^{12}$ is different from $d_{33}\psi^9, d_{33}\psi^{10}, d_{33}\psi^{11}$.

Going one step further on, we obtain the long exact sequence:

$$0 \to R_7 \longrightarrow J_7(E) \longrightarrow J_4(F_0) \longrightarrow J_3(F_1) \longrightarrow J_1(F_2) \longrightarrow F_3 \to 0$$

with respective fiber dimensions:

$$0 \to 8 \longrightarrow 120 \longrightarrow 420 \longrightarrow 420 \longrightarrow 184 \to 72 \to 0$$

We let the reader compute himself the last *two steps* in order to find the formally exact differential sequence:

$$0 \to \Theta \to E \xrightarrow{\mathcal{D}_1} F_0 \xrightarrow{\mathcal{D}_2} F_1 \xrightarrow{\mathcal{D}_3} F_2 \xrightarrow{\mathcal{D}_4} F_3 \xrightarrow{\mathcal{D}_5} F_4 \xrightarrow{\mathcal{D}_6} F_5 \to 0$$

with operators of successive orders $3, 1, 2, 1, 1, 1$ and respective fiber dimensions:

$$0 \to \Theta \to 1 \xrightarrow{\mathcal{D}_1} 12 \xrightarrow{\mathcal{D}_2} 21 \xrightarrow{\mathcal{D}_3} 46 \xrightarrow{\mathcal{D}_4} 72 \xrightarrow{\mathcal{D}_5} 48 \xrightarrow{\mathcal{D}_6} 12 \to 0$$

We check that the Euler-Poincaré characteristic $1 - 12 + 21 - 46 + 72 - 48 + 12 = 0$ vanishes.

The reason for which we have to construct so many differential operators *necessarily*, is that the second order operator $\mathcal{D}_6$ is involutive but with 12 first order equations already exhibited and $33 + 1 = 34$ second order equations along with the new following modified commutative and exact diagram:

$$
\begin{array}{ccccccccccc}
& & 0 & & 0 & & 0 & & 0 & & 0 \\
& & \downarrow & & \downarrow & & \downarrow & & \downarrow & & \downarrow \\
0 \to & g_6 & \to & S_6 T^* \otimes E & \to & S_3 T^* \otimes F_0 & \to & S_2 T^* \otimes F_1 & \to & 34 & \to 0 \\
& \downarrow & & \downarrow & & \downarrow & & \downarrow & & \searrow \downarrow \\
0 \to & R_6 & \to & J_6(E) & \to & J_3(F_0) & \to & J_2(F_1) & \to & F_2 & \to 0 \\
& \downarrow & & \downarrow & & \downarrow & & \downarrow & & \downarrow \\
0 \to & R_5 & \to & J_5(E) & \to & J_2(F_0) & \to & J_1(F_1) & \to & 12 & \to 0 \\
& \downarrow & & \downarrow & & \downarrow & & \downarrow & & \downarrow \\
& 0 & & 0 & & 0 & & 0 & & 0
\end{array}
$$

where $dim(E) = 1$, $g_6 = 0$, $dim(R_6) = dim(R_5) \Rightarrow R_6 \simeq R_5$. As F_2 is the cokernel of the last map on the right in the central row and the bottom row is exact by induction, the central row is exact because of the next commutative diagram which is different from the preceding similar one.

$$
\begin{array}{ccccccccc}
& 0 & & 0 & & 0 & & 0 \\
& \downarrow & & \downarrow & & \downarrow & & \downarrow \\
0 \to & g_6 & \to & S_6 T^* & \to & S_3 T^* \otimes F_0 & \to & S_2 T^* \otimes F_1 & \to F_2 \\
& \downarrow \delta & & \downarrow \delta & & \downarrow \delta & & \downarrow \delta \\
0 \to & T^* \otimes g_5 & \to & T^* \otimes S_5 T^* & \to & T^* \otimes S_2 T^* \otimes F_0 & \to & T^* \otimes T^* \otimes F_1 \\
& \downarrow \delta & & \downarrow \delta & & \downarrow \delta & & \downarrow \delta \\
0 \to & \wedge^2 T^* \otimes g_4 & \to & \wedge^2 T^* \otimes S_4 T^* & \to & \wedge^2 T^* \otimes T^* \otimes F_0 & \to & \wedge^2 T^* \otimes F_1 & \to 0 \\
& \downarrow \delta & & \downarrow \delta & & \downarrow \delta & & \downarrow \\
0 \to & \wedge^3 T^* \otimes g_3 & \to & \wedge^3 T^* \otimes S_3 T^* & \to & \wedge^3 T^* \otimes F_0 & & 0 \\
& \downarrow & & \downarrow & & \downarrow & & \\
& 0 & & 0 & & 0
\end{array}
$$

but where we have also to take into account that $g_4 = g_5 = g_6 = 0$. Accordingly, the final part:

$$
F_1 \xrightarrow{\mathcal{D}_3} F_2 \xrightarrow{\mathcal{D}_4} F_3 \xrightarrow{\mathcal{D}_5} F_4 \xrightarrow{\mathcal{D}_6} F_5 \to 0
$$

is a Janet sequence for the second order operator $\mathcal{D}_3$ and thus ends with $\mathcal{D}_6$ *necessarily* because the maximum number of dots/non-multiplicative variables is equal to $n = 3$ as there are 12 first order equations involved, such a fact explaining why $dim(F_5) = 12$ *necessarily* (See [19], p 153,154, in particular Lemma 2.6, for more details). The reader trying to use computer algebra will have to deal with matrices as large as 441×460 in order to find this last operator (exercise).

The explanation of this tricky situation is not easy to grasp by somebody not familiar with homological algebra. As a byproduct, *one must only construct the Janet and Spencer sequences for involutive systems in order to connect them convenientely.* This result will be crucially used for many applications to mathematical physics in Appendix 3.

We may finally extend the restriction $D : R_{q+1} \to T^* \otimes R_q$ of the Spencer operator to:

$$D : \wedge^r T^* \otimes R_{q+1} \to \wedge^{r+1} T^* \otimes R_q : \alpha \otimes f_{q+1} \to d\alpha \otimes f_q + (-1)^r \alpha \wedge D f_{q+1} \Rightarrow D^2 = D \circ D \equiv 0$$

in order to construct the *first Spencer sequence* which is another resolution of Θ because the kernel of the first D is such that $f_{q+1} \in R_{q+1}, D f_{q+1} = 0 \Leftrightarrow f_{q+1} = j_{q+1}(f), f \in \Theta$ when q is large enough. This standard notation for the Spencer operator must not be confused with the same notation used in the next section for the ring of differential operators but the distinction will always be pointed out whenever a confusion could exist.

4) DIFFERENTIAL MODULES

Let K be a *differential field*, that is a field containing $\mathbb{Q}$ with n commuting *derivations* $\{\partial_1, ..., \partial_n\}$ with $\partial_i \partial_j = \partial_j \partial_i = \partial_{ij}, \forall i, j = 1, ..., n$ such that $\partial_i(a + b) = \partial_i a + \partial_i b$, $\partial_i(ab) = (\partial_i a)b + a\partial_i b, \forall a, b \in K$ and $\partial_i(1/a) = -(1/a^2)\partial_i a, \forall a \in K$. Using an implicit summation on multi-indices, we may introduce the (noncommutative) *ring of differential operators* $D = K[d_1, ..., d_n] = K[d]$ with elements $P = a^\mu d_\mu$ such that $| \mu | < \infty$ and $d_i a = a d_i + \partial_i a$. The highest value of $|\mu|$ with $a^\mu \neq 0$ is called the *order* of the *operator* P and the ring D with multiplication $(P, Q) \longrightarrow P \circ Q = PQ$ is filtered by the order q of the operators. We have the *filtration* $0 \subset K = D_0 \subset D_1 \subset ... \subset D_q \subset ... \subset D_\infty = D$. Moreover, it is clear that D, as an algebra, is generated by $K = D_0$ and $T = D_1/D_0$ with $D_1 = K \oplus T$ if we identify an element $\xi = \xi^i d_i \in T$ with the vector field $\xi = \xi^i(x)\partial_i$ of differential geometry, but with $\xi^i \in K$ now. It follows that $D = {}_D D_D$ is a *bimodule* over itself, being at the same time a left D-module ${}_D D$ by the composition $P \longrightarrow QP$ and a right D-module D_D by the composition $P \longrightarrow PQ$ with $D_r D_s = D_{r+s}, \forall r, s \geq 0$.

If we introduce *differential indeterminates* $y = (y^1, ..., y^m)$, we may extend $d_i y_\mu^k = y_{\mu+1_i}^k$ to $\Phi^\tau \equiv a_k^{\tau\mu} y_\mu^k \overset{d_i}{\longrightarrow} d_i \Phi^\tau \equiv a_k^{\tau\mu} y_{\mu+1_i}^k + \partial_i a_k^{\tau\mu} y_\mu^k$ for $\tau = 1, ..., p$. Therefore, setting $Dy^1 + ... + Dy^m = Dy \simeq D^m$ and calling $I = D\Phi \subset Dy$ the *differential module of equations*, we obtain by residue the *differential module* or *D-module* $M = Dy/D\Phi$, denoting the residue of y_μ^k by $\bar{y}_\mu^k$ when there can be a confusion. Introducing the two free differential modules $F_0 \simeq D^{m_0}, F_1 \simeq D^{m_1}$, we obtain equivalently the *free presentation* $F_1 \overset{d_1}{\longrightarrow} F_0 \to M \to 0$ of order q when $m_0 = m, m_1 = p$ and $d_1 = \mathcal{D} = \Phi \circ j_q$ with $(P_1, ..., P_p) \to (P_1, ..., P_p) \circ \mathcal{D} = (Q_1, ..., Q_m)$. We shall moreover assume that $\mathcal{D}$ provides a *strict morphism*, namely that the corresponding system R_q is formally integrable. It follows that M can be endowed with a *quotient filtration* obtained from that of D^m which is defined by the order of the jet coordinates y_q in $D_q y$. We have therefore the *inductive limit* $0 = M_{-1} \subseteq M_0 \subseteq M_1 \subseteq ... \subseteq M_q \subseteq ... \subseteq M_\infty = M$ with $d_i M_q \subseteq M_{q+1}$ but it is important to notice that $D_r D_q = D_{q+r} \Rightarrow D_r M_q = M_{q+r}, \forall q, r \geq 0 \Rightarrow M = DM_q, \forall q \geq 0$ *in this particular case*. It also follows from noetherian arguments and involution that $D_r I_q = I_{q+r}, \forall r \geq 0$ though we have in general only $D_r I_s \subseteq I_{r+s}, \forall r \geq 0, \forall s < q$. As $K \subset D$, we may introduce the *forgetful functor*

$for : mod(D) \to mod(K) : {}_D M \to {}_K M.$

More generally, introducing the successive CC as in the preceding section while changing slightly the numbering of the respective operators, we may finally obtain the *free resolution* of M, namely the exact sequence $\quad \dots \xrightarrow{d_3} F_2 \xrightarrow{d_2} F_1 \xrightarrow{d_1} F_0 \longrightarrow M \longrightarrow 0$. In actual practice, *one must never forget that $\mathcal{D} = \Phi \circ j_q$ acts on the left on column vectors in the operator case and on the right on row vectors in the module case*. Also, with a slight abuse of language, when $\mathcal{D} = \Phi \circ j_q$ is involutive as in section 3 and thus $R_q = ker(\Phi)$ is involutive, one should say that M has an *involutive presentation* of order q or that M_q is *involutive*.

DEFINITION 4.1: Setting $P = a^\mu d_\mu \in D \xleftrightarrow{ad} ad(P) = (-1)^{|\mu|} d_\mu a^\mu \in D$, we have $ad(ad(P)) = P$ and $ad(PQ) = ad(Q)ad(P), \forall P, Q \in D$. Such a definition can be extended to any matrix of operators by using the transposed matrix of adjoint operators and we get:

$$< \lambda, \mathcal{D}\xi >=< ad(\mathcal{D})\lambda, \xi > + div\,(\dots)$$

from integration by part, where λ is a row vector of test functions and $<>$ the usual contraction. We quote the useful formulas $[ad(\xi), ad(\eta)] = ad(\xi)ad(\eta) - ad(\eta)ad(\xi) = -ad([\xi, \eta]), \forall \xi, \eta \in T$ (*care about the minus sign*) and $rk_D(\mathcal{D}) = rk_D(ad(\mathcal{D}))$ as in ([23], p 610-612).

REMARK 4.2: As can be seen from the last two examples of Section 3, when $\mathcal{D}$ is involutive, then $ad(\mathcal{D})$ may not be involutive. In the differential framework, we may set $rk_D(\mathcal{D}) = m - \alpha_q^n = \beta_q^n$. Comparing to similar concepts used in *differential algebra*, this number is just the maximum number of differentially independent equations to be found in the differential module I of equations. Indeed, pointing out that differential indeterminates in differential algebra are nothing else than jet coordinates in differential geometry and using standard notations, we have $K\{y\} = lim_{q \to \infty} K[y_q]$. In that case, the differential ideal I *automatically* generates a prime differential ideal $\mathfrak{p} \subset K\{y\}$ providing a *differential extension* L/K with $L = Q(K\{y\}/\mathfrak{p})$ and *differential transcendence degree* $diff\,trd(L/K) = \alpha_q^n$, a result explaining the notations ([12],[22]). Now, from the dimension formulas of R_{q+r}, we obtain at once $rk_D(M) = \alpha_q^n$ and thus $rk_D(\mathcal{D}) = m - rk_D(M)$ in a coherent way with any free presentation of M starting with $\mathcal{D}$. However, $\mathcal{D}$ acts on the left in differential geometry but on the right in the theory of differential modules. For an operator of order zero, we recognize the fact that the rank of a matrix is eqal to the rank of the transposed matrix.

PROPOSITION 4.3: If $f \in aut(X)$ is a local diffeomorphisms on X, we may set $x = f^{-1}(y) = g(y)$ and we have the *identity*:

$$\frac{\partial}{\partial y^k}\left(\frac{1}{\Delta(g(y))}\partial_i f^k(g(y))\right) \equiv 0 \ \Rightarrow \ \frac{\partial}{\partial y^k}\left(\frac{1}{\Delta}\frac{\partial f^k}{\partial x^i}\mathcal{A}^i\right) = \frac{1}{\Delta}\frac{\partial f^k}{\partial x^i}\frac{\partial \mathcal{A}^i}{\partial y^k} = \frac{1}{\Delta}\partial_i \mathcal{A}^i$$

and the adjoint of the well defined intrinsic operator $\wedge^0 T^* \xrightarrow{d} \wedge^1 T^* = T^* : A \longrightarrow \partial_i A$ is (*minus*) the well defined intrinsic operator $\wedge^n T^* \xleftarrow{d} \wedge^n T^* \otimes T \simeq \wedge^{n-1} T^* :$ $\partial_i \mathcal{A}^i \longleftarrow \mathcal{A}^i$. Hence, if we have an operator $E \xrightarrow{D} F$, we obtain the *formal adjoint* operator $\wedge^n T^* \otimes E^* \xleftarrow{ad(D)} \wedge^n T^* \otimes F^*$.

Having in mind that D is a K-algebra, that K is a left D-module with the standard action $(D, K) \longrightarrow K : (P, a) \longrightarrow P(a) : (d_i, a) \longrightarrow \partial_i a$ and that D is a bimodule over itself, *we have only two possible constructions leading to the following two definitions*:

DEFINITION 4.4: We may define the *inverse system* $R = hom_K(M, K)$ of M and introduce $R_q = hom_K(M_q, K)$ as the *inverse system of order q*.

DEFINITION 4.5: We may define the right differential module $M^* = hom_D(M, D)$ by using the bimodule structure of $D = {}_D D_D$.

THEOREM 4.6: When M and N are left D-modules, then $hom_K(M, N)$ and $M \otimes_K N$ are left D-modules. In particular $R = hom_K(M, K)$ is also a left D-module for the *Spencer operator*.

Proof: For any $f \in hom_K(M, N)$, let us define:

$$(af)(m) = af(m) = f(am) \qquad \forall a \in K, \forall m \in M$$

$$(\xi f)(m) = \xi f(m) - f(\xi m) \qquad \forall \xi = \xi^i d_i \in T, \forall m \in M$$

It is easy to check that $\xi a = a\xi + \xi(a)$ in the operator sense and that $\xi\eta - \eta\xi = [\xi, \eta]$ is the standard bracket of vector fields. We have in particular with d in place of any d_i:

$$
\begin{aligned}
((da)f)(m) = (d(af))(m) = d(af(m)) - af(dm) &= (\partial a)f(m) + ad(f(m)) - af(dm) \\
&= (a(df))(m) + (\partial a)f(m) \\
&= ((ad + \partial a)f)(m)
\end{aligned}
$$

For any $m \otimes n \in M \otimes_K N$ with arbitrary $m \in M$ and $n \in N$, we may then define:

$$a(m \otimes n) = am \otimes n = m \otimes an \in M \otimes_K N$$

$$\xi(m \otimes n) = \xi m \otimes n + m \otimes \xi n \in M \otimes_K N$$

and conclude similarly with:

$$
\begin{aligned}
(da)(m \otimes n) = d(a(m \otimes n)) &= d(am \otimes n) \\
&= d(am) \otimes n + am \otimes dn \\
&= (\partial a)m \otimes n + a(dm) \otimes n + am \otimes dn \\
&= (ad + \partial a)(m \otimes n)
\end{aligned}
$$

Using K in place of N, we finally get $(d_i f)_\mu^k = (d_i f)(y_\mu^k) = \partial_i f_\mu^k - f_{\mu+1_i}^k$ that is *we recognize exactly the Spencer operator* with *now $Df = dx^i \otimes d_i f$* and thus:

$$(d_i(d_j f))_\mu^k = \partial_{ij} f_\mu^k - (\partial_i f_{\mu+1_j}^k + \partial_j f_{\mu+1_i}^k) + f_{\mu+1_i+1_j}^k \Rightarrow d_i(d_j f) = d_j(d_i f) = d_{ij} f$$

In fact, R is the *projective limit* of $\pi_q^{q+r} : R_{q+r} \to R_q$ in a coherent way with jet theory ([2],[27],[38]).

Q.E.D.

COROLLARY 4.7: If M and N are right D-modules, then $hom_K(M, N)$ is a left D-module. Moreover, if M is a left D-module and N is a right D-module, then $M \otimes_K N$ is a right D-module.

Proof: If M and N are right D-modules, we just need to set $(\xi f)(m) = f(m\xi) - f(m)\xi, \forall \xi \in T, \forall m \in M$ and conclude as before. Similarly, if M is a left D-module and N is a right D-module, we just need to set $(m \otimes n)\xi = m \otimes n\xi - \xi m \otimes n$.

Q.E.D.

REMARK 4.8: When $M = {}_DM \in mod(D)$ and $N = N_D$, , then $hom_K(N, M)$ cannot be endowed with any left or right differential structure. When $M = M_D$ and $N = N_D$, then $M \otimes_K N$ cannot be endowed with any left or right differential structure (See [2], p 24 for more details).

As $M^* = hom_D(M, D)$ is a right D-module, let us define the right D-module N_D by the ker/coker long exact sequence $0 \longleftarrow N_D \longleftarrow F_1^* \overset{D^*}{\longleftarrow} F_0^* \longleftarrow M^* \longleftarrow 0$.

COROLLARY 4.9: We have the *side changing* procedure $N_D \to N = {}_DN = hom_K(\wedge^n T^*, N_D)$ with inverse $M = {}_DM \to M_D = \wedge^n T^* \otimes_K M$ whenever $M, N \in mod(D)$.

Proof: According to the above Theorem, we just need to prove that $\wedge^n T^*$ has a natural right module structure over D. For this, if $\alpha = a dx^1 \wedge ... \wedge dx^n \in T^*$ is a volume form with coefficient $a \in K$, we may set $\alpha.P = ad(P)(a)dx^1 \wedge ... \wedge dx^n$ when $P \in D$. As D is generated by K and T, we just need to check that the above formula has an intrinsic meaning for any $\xi = \xi^i d_i \in T$. In that case, we check at once:

$$\alpha.\xi = -\partial_i(a\xi^i)dx^1 \wedge ... \wedge dx^n = -\mathcal{L}(\xi)\alpha$$

by introducing the Lie derivative of α with respect to ξ, along the intrinsic formula $\mathcal{L}(\xi) = i(\xi)d + di(\xi)$ where $i()$ is the interior multiplication and d is the exterior derivative of exterior forms. According to well known properties of the Lie derivative, we get :

$$\alpha.(a\xi) = (\alpha.\xi).a - \alpha.\xi(a), \quad \alpha.(\xi\eta - \eta\xi) = -[\mathcal{L}(\xi), \mathcal{L}(\eta)]\alpha = -\mathcal{L}([\xi, \eta])\alpha = \alpha.[\xi, \eta].$$

$$Q.E.D.$$

Collecting the previous results, if a differential operator $\mathcal{D}$ is given in the framework of differential geometry, we may keep the same notation $\mathcal{D}$ in the framework of differential modules which are *left* modules over the ring D of linear differential operators and apply duality, provided we use the notation $\mathcal{D}^*$ and deal with *right* differential modules or use the notation $ad(\mathcal{D})$ and deal again with *left* differential modules by using the *left* $\leftrightarrow$ *right conversion* procedure.

DEFINITION 4.10: If an operator $\xi \xrightarrow{\mathcal{D}} \eta$ is given, a *direct problem* is to look for (generating) *compatibility conditions* (CC) as an operator $\eta \xrightarrow{\mathcal{D}_1} \zeta$ such that $\mathcal{D}\xi = \eta \Rightarrow \mathcal{D}_1\eta = 0$. Conversely, given $\eta \xrightarrow{\mathcal{D}_1} \zeta$, the *inverse problem* will be to look for $\xi \xrightarrow{\mathcal{D}} \eta$ such that $\mathcal{D}_1$ generates the CC of $\mathcal{D}$ and we shall say that $\mathcal{D}_1$ is *parametrized by $\mathcal{D}$* if such an operator $\mathcal{D}$ is existing.

As $ad(ad(P)) = P, \forall P \in D$, any operator is the adjoint of a certain operator and we get:

DOUBLE DUALITY CRITERION 4.11: An operator $\mathcal{D}_1$ can be parametrized by an operator $\mathcal{D}$ if, whenever $ad(\mathcal{D})$ generates the CC of $ad(\mathcal{D}_1)$, then $\mathcal{D}_1$ generates the CC of $\mathcal{D}$. However, as shown in the example below, many other parametrizations may exist.

Reversing the arrows, we finally obtain:

TORSION-FREE CRITERION 4.12: A differential module M having a finite free presentation $F_1 \xrightarrow{\mathcal{D}_1} F_0 \to M \to 0$ is *torsion-free*, that is to say $t(M) = \{m \in M \mid \exists 0 \neq P \in D, Pm = 0\} = 0$, if and only if there exists a free differential module E and an exact sequence $F_1 \xrightarrow{\mathcal{D}_1} F_0 \xrightarrow{\mathcal{D}} E$ providing the *parametrization $M \subseteq E$*.

REMARK 4.13: Of course, solving the direct problem (Janet, Spencer) is *necessary* for solving the inverse problem. However, though the direct problem always has a solution, the inverse problem may not have a solution at all and the case of the Einstein operator is one of the best non-trivial PD counterexamples ([24],[30]). It is rather striking to discover that, in the case of OD operators, it took almost 50 years to understand that the possibility to solve the inverse problem was equivalent to the controllability of the corresponding control system ([24],[34]).

However, we have yet not proved the most difficult result that could not be obtained without homological algebra and the next three examples will explain this additional difficulty.

EXAMPLE 4.14: (*contact transformations*) With $n = 3$, $K = \mathbb{Q}(x^1, x^2, x^3)$, let us consider the Lie pseudogroup of transfomations preserving the first order geometric object ω like a 1-form but up to the square root of Δ. The infinitesimal transformations are among the solutions Θ of the *general* system:

$$\Omega_i \equiv (\mathcal{L}(\xi)\omega)_i \equiv \omega_r(x)\partial_i\xi^r - (1/2)\omega_i(x)\partial_r\xi^r + \xi^r\partial_r\omega_i(x) = 0$$

When $\omega = (1, -x^3, 0)$, we obtain the *special* involutive system:

$$\partial_3\xi^3 + \partial_2\xi^2 + 2x^3\partial_1\xi^2 - \partial_1\xi^1 = 0, \partial_3\xi^1 - x^3\partial_3\xi^2 = 0,$$
$$\partial_2\xi^1 - x^3\partial_2\xi^2 + x^3\partial_1\xi^1 - (x^3)^2\partial_1\xi^2 - \xi^3 = 0$$

with 2 equations of class 3, 1 equation of class 2 and thus only 1 first order CC for the second members coming from the linearization of the Vessiot structure equation:

$$\omega_1(\partial_2\omega_3 - \partial_3\omega_2) + \omega_2(\partial_3\omega_1 - \partial_1\omega_3) + \omega_3(\partial_1\omega_2 - \partial_2\omega_1) = c$$
$$\omega = (1, -x^3, 0) \Rightarrow c = 1 \quad , \quad \Omega_1 + \partial_2\Omega_3 - \partial_3\Omega_2 - x^3\partial_3\Omega_1 + x^3\partial_1\Omega_3 = 0$$

involving the only *structure constant* c. This system can be parametrized by a single potential ϕ:

$$-x^3\partial_3\phi + \phi = \xi^1, -\partial_3\phi = \xi^2, \partial_2\phi - x^3\partial_1\phi = \xi^3 \quad \Rightarrow \quad \xi^1 - x^3\xi^2 = \phi$$

We have the formally exact split differential sequence:

$$0 \to 1 \xrightarrow{D_{-1}} 3 \xrightarrow{D} 3 \xrightarrow{D_1} 1 \to 0$$

and the adjoint sequence is thus also a formally exact split differential sequence ([9],[17],[23],[37]). Finally, starting *anew* with the defining system:

$$\partial_3\xi^1 - x^3\partial_3\xi^2 = 0, \partial_2\xi^1 - x^3\partial_2\xi^2 + x^3\partial_1\xi^1 - (x^3)^2\partial_1\xi^2 - \xi^3 = 0$$

which is not formally integrable with no CC, the adjoint system has the only solution $\mu^1 = 0, \mu^2 = 0$.

EXAMPLE 4.15: With $n = 2$, $K = \mathbb{Q}(x^1, x^2)$, let us consider the Lie pseudogroup of transformaions preserving the geometric object $\omega = (\alpha, \beta)$ with $\alpha = x^2 dx^1 \in \wedge^1 T^*$, $\beta = dx^1 \wedge dx^2 \in \wedge^2 T^*$. The general system $\Omega \equiv \mathcal{D}\xi = \mathcal{L}(\xi)\omega = 0$ is involutive if and only if the Vessiot structure equation $d\alpha = c\beta$ is satisfied with $c = -1$ for the special case considered. The corresponding system of infinitesimal Lie equations is:

$$\Omega^1 \equiv x^2 \partial_1 \xi^1 + \xi^2 = 0, \Omega^2 \equiv x^2 \partial_2 \xi^1 = 0, \Omega^3 \equiv \partial_1 \xi^1 + \partial_2 \xi^2 = 0$$

$$\Rightarrow \mathcal{D}_1 \Omega \equiv \partial_1 \Omega^2 - \partial_2 \Omega^1 + \Omega^3 = 0$$

Multiplying the three first equations by test functions $\mu = (\mu^1, \mu^2, \mu^3)$ we get the adjoint system:

$$-\xi^1 \Rightarrow x^2 \partial_1 \mu^1 + x^2 \partial_2 \mu^2 + \partial_1 \mu^3 + \mu^2 = 0, \quad -\xi^2 \Rightarrow \partial_2 \mu^3 - \mu^1 = 0$$

Multiplying the last equation by the only test function λ, we get the adjoint system:

$$\Omega^1 \Rightarrow \partial_2 \lambda = \mu^1, \quad \Omega^2 \Rightarrow -\partial_1 \lambda = \mu^2, \quad \Omega^3 \Rightarrow \lambda = \mu^3$$

Substituting $\mu^1 = \partial_2 \mu^3$, we discover that $(x^2 \partial_2 + 1)(\partial_1 \mu^3 + \mu^2) = 0$ and $\partial_1 \mu^3 + \mu^2 \neq 0$ is a totally unexpected torsion element showing that $ad(\mathcal{D})$ does not generates the CC of $ad(\mathcal{D}_1)$. Finally, starting *anew* with the defining system $x^2 \partial_1 \xi^1 + \xi^2 = 0, \partial_2 \xi^1 = 0$ which is not formally integrable with no CC, the adjoint system becomes $\mu^1 = 0, \partial_2 \mu^2 = 0$ and $\mu^2 \neq 0$ is a torsion element.

EXAMPLE 4.16: With $\partial_{22}\xi = \eta^2, \partial_{12}\xi = \eta^1$ for $\mathcal{D}$, we get $\partial_1 \eta^2 - \partial_2 \eta^1 = \zeta$ for $\mathcal{D}_1$. Then $ad(\mathcal{D}_1)$ is defined by $\mu^2 = -\partial_1 \lambda, \mu^1 = \partial_2 \lambda$ while $ad(\mathcal{D})$ is defined by $\nu = \partial_{12}\mu^1 + \partial_{22}\mu^2$ but the CC of $ad(\mathcal{D}_1)$ are generated by $\nu' = \partial_1 \mu^1 + \partial_2 \mu^2$. In the operator framework, we have the differential sequences:

$$\xi \xrightarrow{\mathcal{D}} \eta \xrightarrow{\mathcal{D}_1} \zeta \rightarrow 0$$
$$0 \leftarrow \nu \xleftarrow{ad(\mathcal{D})} \mu \xleftarrow{ad(\mathcal{D}_1)} \lambda$$

where the upper sequence is formally exact at η but the lower sequence is not formally exact at μ.

Passing to the module framework, we obtain the sequences:

$$0 \to D \xrightarrow{\mathcal{D}_1} D^2 \xrightarrow{\mathcal{D}} D \to M \to 0$$
$$D \xleftarrow{ad(\mathcal{D}_1)} D^2 \xleftarrow{ad(\mathcal{D})} D \leftarrow 0$$

where the lower sequence is not exact at D^2.

Therefore, we have to find out situations in which $ad(\mathcal{D})$ generates the CC of $ad(\mathcal{D}_1)$ whenever $\mathcal{D}_1$ generates the CC of $\mathcal{D}$ and conversely. This problem will be studied in Section 5, Part C.

5) APPLICATIONS

Though the next pages will only be concerned with a study of the Lie pseudogroups of isometries (A) and conformal isometries (B), the reader must never forget that they can be used similarly for *any* arbitrary transitive Lie pseudogroup of transformations as above ([19],[21],[22],[29]).

A) RIEMANN TENSOR

Let $\omega = (\omega_{ij} = \omega_{ji}) \in S_2T*$be a non-degenerate metric with $det(\omega) \neq 0$. We shall apply the Main Theorem to the first order *Killing system* $R_1 \subset J_1(T)$ defined by the $n(n+1)/2$ linear equations $\Omega_{ij} \equiv \omega_{rj}\xi_i^r + \omega_{ir}\xi_j^r + \xi^r\partial_r\omega_{ij} = 0$ for any section $\xi_1 \in R_1$. Its symbol $g_1 \subset T^* \otimes T$ is defined by the $n(n+1)/2$ linear equations $\omega_{rj}\xi_i^r + \omega_{ir}\xi_j^r = 0$ and we obtain at once isomorphisms $g_1 \simeq \wedge^2T \simeq \wedge^2T^*$ by lowering or raising the indices by means of the metric, obtaining for example $\xi_{i,j} + \xi_{j,i} = 0$. As $det(\omega) \neq 0$, we may introduce the well known Chrisoffel symbols $\gamma = (\gamma_{ij}^k = \gamma_{ji}^k)$ through the standard Ricci/Levi-Civita isomorphism $j_1(\omega) \simeq (\omega, \gamma)$ and obtain by one prolongation the linear second order equations for any section $\xi_2 \in R_2$:

$$\Gamma_{ij}^k \equiv \xi_{ij}^k + \gamma_{rj}^k\xi_i^r + \gamma_{ir}^k - \gamma_{ij}^r\xi_r^k + \xi^r\partial_r\gamma_{ij}^k = 0$$

and we have $\Omega \in S_2T^* \Rightarrow \Gamma \in S_2T^* \otimes T$ for the respective linearization/variation of ω and γ. As we shall see that g_1 is *not* 2-acyclic and $g_2 = 0$ is defined by the $n^2(n+1)/2$ linear equations $\xi_{ij}^k = 0$, we may apply the Main Theorem with $q = 1, s = 1, E = T$, on the condition that R_1 should be formally integrable as it is finite type and cannot therefore be involutive. First of all, we have the following commutative and exact diagram allowing to define $F_0 = S_2T^*$:

$$
\begin{array}{ccccc}
& 0 & & 0 & & 0 \\
& \downarrow & & \downarrow & & \downarrow \\
0 \to & g_1 & \longrightarrow & T^* \otimes T & \xrightarrow{\sigma(\Phi)} & F_0 \to 0 \\
& \downarrow & & \downarrow & & \| \\
0 \to & R_1 & \longrightarrow & J_1(T) & \xrightarrow{\Phi} & F_0 \to 0 \\
& \downarrow & & \downarrow & & \downarrow \\
0 \to & T & = & T & \longrightarrow & 0 \\
& \downarrow & & \downarrow & & \\
& 0 & & 0 & &
\end{array}
$$

Now, $R_2 \xrightarrow{\pi_1^2} R_1$ is an isomorphism because $g_2 = 0$ and $dim(R_2) = dim(R_1) = n(n+1)/2$. Hence, $R_2 \subset J_2(T)$ is involutive if and only if $R_3 \xrightarrow{\pi_2^3} R_2$ is also an isomorphism too because $g_2 = 0 \Rightarrow g_{2+r} = 0, \forall r \geq 0$. Such a differential condition for ω has been shown by L.P. Eisenhart in ([8]) to be equivalent to the *Vessiot structure equation with one constant* called *constant riemannian curvature* $\rho_{l,ij}^k = c(\delta_i^k \omega_{lj} - \delta_j^k \omega_{li})$ (See [19],[22] and [29] for effective calculations still not acknowledged today). In this formula, c is an arbitrary constant and the *Riemann tensor* $(\rho_{l,ij}^k) \in \wedge^2 T^* \otimes T^* \otimes T$ satisfies the two types of purely algebraic relations:

$$
\omega_{rl} \rho_{k,ij}^r + \omega_{kr} \rho_{l,ij}^r = 0, \quad \rho_{l,ij}^k + \rho_{i,jl}^k + \rho_{j,li}^k = 0
$$

We shall suppose that ω is the Euclidean metric if $n = 2, 3$ and the Minkowskian metric if $n = 4$ but any other compatible choice should be convenient. As a next step, we know from the Main Theorem that the generating CC for the operator $Killing = \Phi \circ j_1 : T \to F_0$ are made by an operator $Riemann = \Psi \circ j_2 : F_0 \to F_1$ of order $s + 1 = 2$. We shall define F_1 by setting $q = 1, r = 0, s = 1$ in the corresponding diagram in order to get the following commutative diagram:

$$
\begin{array}{ccccccccc}
& & 0 & & 0 & & 0 & & \\
& & \downarrow & & \downarrow & & \downarrow & & \\
0 \to & & g_3 & \to & S_3 T^* \otimes T & \to & S_2 T^* \otimes F_0 & \to F_1 & \to 0 \\
& & \downarrow \delta & & \downarrow \delta & & \downarrow \delta & & \\
0 \to & T^* \otimes g_2 & \to & T^* \otimes S_2 T^* \otimes T & \to & T^* \otimes T^* \otimes F_0 & \to & 0 & \\
& \downarrow \delta & & \downarrow \delta & & \downarrow \delta & & & \\
0 \to & \wedge^2 T^* \otimes g_1 & \to & \underline{\wedge^2 T^* \otimes T^* \otimes T} & \to & \wedge^2 T^* \otimes F_0 & \to & 0 & \\
& \downarrow \delta & & \downarrow \delta & & \downarrow & & & \\
0 \to & \wedge^3 T^* \otimes T & = & \wedge^3 T^* \otimes T & \to & 0 & & & \\
& \downarrow & & \downarrow & & & & & \\
& 0 & & 0 & & & & &
\end{array}
$$

where all the rows are exact and all the columns are also exact but the first at $\wedge^2 T^* \otimes g_1$ with $g_2 = 0 \Rightarrow g_3 = 0$. We shall denote by $B^2(g_1)$ the *coboundary* as the image of the central δ, by $Z^2(g_1)$ the *cocycle* as the kernel of the lower δ and by $H^2(g_1) = Z^2(g_1)/B^2(g_1)$ the *Spencer δ-cohomology* at $\wedge^2 T^* \otimes g_1$ as the quotient. Chasing in the previous diagram, we discover that the *Riemann tensor* is a section of the bundle $F_1 = H^2(g_1) = Z^2(g_1)$ with $dim(F_1) = (n^2(n+1)^2/4) - (n^2(n+1)(n+2)/6) = (n^2(n-1)^2/4) - (n^2(n-1)(n-2)/6) = n^2(n^2-1)/12$ by using the top row or the left column. We discover at once the two properties of the (linearized) Riemann tensor through the chase involved, namely $(R^k_{l,ij}) \in \wedge^2 T^* \otimes T^* \otimes T$ is killed by both δ and $\sigma_0(\Phi)$. Similarly, going one step further, we get the (linearized) Bianchi identities by defining the first order operator $Bianchi : F_1 \to F_2$ where $F_2 = H^3(g_1) = Z^3(g_1)$ with $dim(F_2) = dim(\wedge^3 T^* \otimes g_1) - dim(\wedge^4 T^* \otimes T) = n^2(n-1)^2(n-2)/12 - n^2(n-1)(n-2)(n-3)/24 = n^2(n^2-1)(n-2)/24$ may be defined by the following commutative diagram:

$$
\begin{array}{ccccccccccc}
& & 0 & & 0 & & 0 & & 0 & & \\
& & \downarrow & & \downarrow & & \downarrow & & \downarrow & & \\
0 \to & & g_4 & \to & S_4 T^* \otimes T & \to & S_3 T^* \otimes F_0 & \to & T^* \otimes F_1 & \to & F_2 \to 0 \\
& & \downarrow & & \downarrow & & \downarrow & & \| & & \\
0 \to & & T^* \otimes g_3 & \to & T^* \otimes S_3 T^* \otimes T & \to & T^* \otimes S_2 T^* \otimes F_0 & \to & T^* \otimes F_1 & \to & 0 \\
& & \downarrow & & \downarrow & & \downarrow & & \downarrow & & \\
0 \to & & \wedge^2 T^* \otimes g_2 & \to & \wedge^2 T^* \otimes S_2 T^* \otimes T & \to & \wedge^2 T^* \otimes T^* \otimes F_0 & \to & 0 & & \\
& & \downarrow & & \downarrow & & \downarrow & & & & \\
0 \to & & \wedge^3 T^* \otimes g_1 & \to & \underline{\wedge^3 T^* \otimes T^* \otimes T} & \to & \wedge^3 T^* \otimes F_0 & \to & 0 & & \\
& & \downarrow & & \downarrow & & \downarrow & & & & \\
0 \to & & \wedge^4 T^* \otimes T & = & \wedge^4 T^* \otimes T & \to & 0 & & & & \\
& & \downarrow & & \downarrow & & & & & & \\
& & 0 & & 0 & & & & & &
\end{array}
$$

This approach is relating for the first time the concept of *Riemann tensor candidate*, introduced by Lanczos and others, to the Spencer δ-cohomology of the Killing symbols. We obtain therefore the formally exact sequence:

$$
0 \to \Theta \to n \overset{Killing}{\longrightarrow} n(n+1)/2 \overset{Riemann}{\longrightarrow} n^2(n^2-1)/12 \overset{Bianchi}{\longrightarrow} n^2(n^2-1)(n-2)/24 \to \dots
$$

with operators of successive orders $1, 2, 1, \dots$ and so on.

In the present situation, we have the (split) short exact sequences:

$$
0 \to F_1 \to \wedge^2 T^* \otimes g_1 \overset{\delta}{\longrightarrow} \wedge^3 T^* \otimes T \to 0, \quad 0 \to F_2 \to \wedge^3 T^* \otimes g_1 \overset{\delta}{\to} \wedge^4 T^* \otimes T \to 0
$$

and obtain the operator $ad(Bianchi) : \wedge^n T^* \otimes F_2^* \to \wedge^n T^* \otimes F_1^*$ with the short exact sequence:

$$
0 \leftarrow \wedge^n T^* \otimes F_2^* \leftarrow \wedge^{n-3} T^* \otimes g_1^* \leftarrow \wedge^{n-2} T^* \otimes T^* \leftarrow 0
$$

explaining at once why the *Lagrange multipliers* $\lambda \in \wedge^n T^* \otimes F_2^*$ can be represented by a section of $\wedge^{n-3} T^* \otimes \wedge^2 T^*$, that is by a Lanczos potential in $T \otimes \wedge^2 T^*$ when $n = 4$. We shall see in part C that $ad(Bianchi)$ is parametrizing $ad(Riemann)$ contrary to the claims of Lanczos. Moreover, we have already pointed out in many books ([21],[23]) or papers ([28],[32]) that continuum mechanics may be presented through a variational problem with a differential constraints which is *shifted by one step backwards in the previous differential sequence* because the infinitesimal deformation tensor $\epsilon = \frac{1}{2}\Omega \in S_2 T^*$ must be now killed by the operator $Riemann$

and the corresponding Lagrange multipliers $\lambda \in \wedge^n T^* \otimes F_1^*$ must be used because $ad(Riemann)$ is parametrizing $ad(Killing) = Cauchy$. Anybody using computations with finite elements also knows that a similar situation is held by electromagnetism too because the EM field is killed by $d : \wedge^2 T^* \to \wedge^3 T^* \Rightarrow ad(d) : \wedge^1 T^* \to \wedge^2 T^*$, another fact contradicting Lanczos claims.

Finally, the passage to differential modules can be achieved easily by using $K = \mathbb{Q}$ as will be done in the Appendix or $K = \mathbb{Q} < \omega >$ with standard notations because the Lie pseudogroup of isometries is an *algebraic Lie pseudogroup* as it can be defined by differential polynomials in the jets of order ≥ 1 (See [12],[20],[22] for details).

B) WEYL TENSOR

If the study of the Riemann tensor/operator has been related to many classical results, the study of the Weyl tensor/operator in this new framework is quite different because these new mathematical tools have not been available before 1975 and are still not acknowledged today by mathematical physicists. In particular, we may quote the link existing between acyclicity and formal integrability both with the possibility to use the *Vessiot structure equations* in order to combine in a unique framework the constant riemannian curvature condition needed for the Killing system, which only depends on one arbitrary constant, with the zero Weyl tensor condition needed for the conformal Killing system, which does not depend on any constant. For this reason, we shall follow as closely as possible the previous part A, putting a "*hat*" on the corresponding concepts.

The *conformal Killing system* $\hat{R}_1 \subset J_1(T)$, simply called *conformal* or *CKilling* in the sequel, is defined by eliminating the function $A(x)$ in the system $\mathcal{L}(\xi)\omega = A(x)\omega$. It is also a *Lie operator* $\hat{\mathcal{D}}$ with solutions $\hat{\Theta} \subset T$ satisfying $[\hat{\Theta}, \hat{\Theta}] \subset \hat{\Theta}$. Its symbol $\hat{g}_1$ is defined by the linear equations $\omega_{rj}\xi_i^r + \omega_{ir}\xi_j^r - \frac{2}{n}\omega_{ij}\xi_r^r = 0$ which do not depend on any conformal factor and is finite type because $\hat{g}_3 = 0$ when $n \geq 3$. We have ([19],[20],[32]):

LEMMA 5.1: $\hat{g}_2 \subset S_2 T^* \otimes T$ is *now* 2-acyclic *only when* $n \geq 4$ and 3-acyclic *only when* $n \geq 5$.

140

It is known that $\hat{R}_2$ and thus $\hat{R}_1$ too (by a chase) are formally integrable if and only if ω has zero *Weyl tensor*:

$$\tau^k_{l,ij} \equiv \rho^k_{l,ij} \ -\tfrac{1}{(n-2)}(\delta^k_i \rho_{lj} - \delta^k_j \rho_{li} + \omega_{lj}\omega^{ks}\rho_{si} - \omega_{li}\omega^{ks}\rho_{sj})$$
$$+\tfrac{1}{(n-1)(n-2)}(\delta^k_i \omega_{lj} - \delta^k_j \omega_{li})\rho = 0$$

If we use the formula $id_M - f \circ u = v \circ g$ of Proposition 2.4 in the *split short exact sequence* induced by the inclusions $g_1 \subset \hat{g}_1, 0 = g_2 \subset \hat{g}_2, g_3 = \hat{g}_3 = 0$ ([21],[22],[28]):

$$0 \longrightarrow Ricci \longrightarrow Riemann \longrightarrow Weyl \longrightarrow 0$$

according to the Vessiot structure equations, in particular if ω has constant Riemannian curvature and thus $\rho_{ij} = \rho^r_{i,rj} = c(n-1)\omega_{ij} \Rightarrow \rho = \omega^{ij}\rho_{ij} = cn(n-1)$ ([19],[21],[30],[31]). Using the same diagrams as before, we get $\hat{F}_0 = T^* \otimes T/\hat{g}_1$ with $dim(\hat{F}_0) = (n-1)(n+2)/2$ and $\hat{F}_1 = H^2(\hat{g}_1) \neq Z^2(\hat{g}_1)$ for defining any *Weyl tensor candidate*. As a byproduct, *we could believe* that the linearized operator $Weyl : \hat{F}_0 \to \hat{F}_1$ is of order 2 with a symbol $\hat{h}_2 \subset S_2 T^* \otimes \hat{F}_0$ which is *not* 2-acyclic by applying the δ-map to the short exact sequence:

$$0 \to \hat{g}_{3+r} \longrightarrow S_{3+r}T^* \otimes T \overset{\sigma_{2+r}(\Phi)}{\longrightarrow} \hat{h}_{2+r} \to 0$$

and chasing through the commutative diagram thus obtained with $r = 0, 1, 2$. As $\hat{h}_3$ becomes 2-acyclic after one prolongation of $\hat{h}_2$ only, it follows that *the generating CC for the Weyl operator are of order 2*, a result showing that the so-called Bianchi identities for the Weyl tensor are *not* CC in the strict sense of the definition as they do not involve only the Weyl tensor.

In fact, things are quite different and we have to distinguish three different cases:

- $n = 3$: According to the last Lemma, $\hat{g}_2$ is *not* 2-acyclic but $\hat{g}_3 = 0$ becomes trivially 2-ayclic and even involutive, that is $s = 2$. According to the Main Theorem, the operator $Weyl : \hat{F}_0 \to \hat{F}_1$ is *third order* because $s + 1 = 3$ (See Appendix 2) and $\hat{F}_1$ is defined by the short exact sequences:

$$0 \to S_4 T^* \otimes T \to S_3 T^* \otimes \hat{F}_0 \to \hat{F}_1 \to 0, \quad 0 \to \hat{F}_1 \to \wedge^2 T^* \otimes \hat{g}_2 \overset{\delta}{\longrightarrow} \wedge^3 T^* \otimes \hat{g}_1 \to 0$$

with $dim(\hat{F}_1) = 50 - 45 = 9 - 4 = 5$. As *now* $\hat{h}_3 \subset S_3 T^* \otimes \hat{F}_0$, applying the δ-map to the short exact sequence:

$$0 \to \hat{g}_6 \to S_6 T^* \otimes T \to \hat{h}_5 \to 0$$

and chasing, we discover that $\hat{h}_3$ is 2-acyclic because $\hat{g}_3 = 0$. Accordingly, the operator $Bianchi : \hat{F}_1 \to \hat{F}_2$ is *first order* and $\hat{F}_2$ is defined by the long exact sequence:

$$0 \to S_5 T^* \otimes T \to S_4 T^* \otimes \hat{F}_0 \to T^* \otimes \hat{F}_1 \to \hat{F}_2 \to 0$$

or by the isomorphism $0 \to \hat{F}_2 \to \wedge^3 T^* \otimes \hat{g}_2 \to 0$ giving $dim(\hat{F}_2) = 63 - 75 + 15 = 1 \times 3 = 3$.

Recapitulating, when $n = 3$ we have the formally exact differential sequence with $3 - 5 + 5 - 3 = 0$:

$$0 \to \hat{\Theta} \to 3 \xrightarrow{CKilling} 5 \xrightarrow{Weyl} 5 \xrightarrow{Bianchi} 3 \to 0$$

In actual practice, introducing the new geometric object $\hat{\omega}_{ij} = \omega_{ij}/|\det(\omega)|^{\frac{1}{n}}$ and the corresponding infinitesimal Lie equations:

$$\hat{\Omega}_{ij} \equiv \hat{\omega}_{rj}\xi_i^r + \hat{\omega}_{ir}\xi_j^r - \frac{2}{n}\hat{\omega}_{ij}\xi_r^r + \xi^r \partial_r \hat{\omega}_{ij} = 0 \quad \Rightarrow \quad \omega^{ij}\hat{\Omega}_{ij} = 0$$

we let the reader check that, *when $n = 3$*, we get for example (See Appendix 2):

$$-d_{123}\hat{\Omega}_{11} + (d_{223} + d_{333})\hat{\Omega}_{12} - (d_{222} + d_{233})\hat{\Omega}_{13} - 2d_{123}\hat{\Omega}_{22} + (d_{122} - d_{133})\hat{\Omega}_{23} \equiv 0$$

• $n = 4$: *This situation is even more striking* because $\hat{g}_2$ is 2 acyclic but *not* 3-acyclic and thus $s = 1$. As before, we have $dim(\hat{F}_0) = (n-1)(n+2)/2 = 9$ but, according to the Main Theorem, the operator $Weyl : \hat{F}_0 \to \hat{F}_1$ is of order $s + 1 = 2$ and $\hat{F}_1$ is defined by the short exact sequence:

$$0 \to S_3 T^* \otimes T \to S_2 T^* \otimes \hat{F}_0 \to \hat{F}_1 \to 0 \quad \Rightarrow \quad dim(\hat{F}_1) = 90 - 80 = 10$$

or by $\hat{F}_1 = H^2(\hat{g}_1) = Z^2(\hat{g}_1)/B^2(\hat{g}_1)$ with exact sequences:

$$0 \to T^* \otimes \hat{g}_2 \xrightarrow{\delta} B^2(\hat{g}_1) \to 0, \quad 0 \to Z^2(\hat{g}_1) \to \wedge^2 T^* \otimes \hat{g}_1 \xrightarrow{\delta} \wedge^3 T^* \otimes T \to 0$$

providing again $dim(\hat{F}_1) = 26 - 16 = 10$. The main problem is that, *now*, $\hat{g}_2$ is *not* 3-

142

acyclic and thus $\hat{h}_2$ is *not* 2-acyclic according to a chase in the commutative diagram:

$$
\begin{array}{ccccccc}
& 0 & & 0 & & 0 & \\
& \downarrow & & \downarrow & & \downarrow & \\
0 \to & \hat{g}_5 & \to & S_5 T^* \otimes T & \to & \hat{h}_4 & \to 0 \\
& \downarrow & & \downarrow & & \downarrow & \\
0 \to & T^* \otimes \hat{g}_4 & \to & T^* \otimes S_4 T^* \otimes T & \to & T^* \otimes \hat{h}_3 & \to 0 \\
& \downarrow & & \downarrow & & \downarrow & \\
0 \to & \wedge^2 T^* \otimes \hat{g}_3 & \to & \wedge^2 T^* \otimes S_3 T^* \otimes T & \to & \wedge^2 T^* \otimes \hat{h}_2 & \to 0 \\
& \downarrow & & \downarrow & & \downarrow & \\
0 \to & \wedge^3 T^* \otimes \hat{g}_2 & \to & \wedge^3 T^* \otimes S_2 T^* \otimes T & \to & \wedge^3 T^* \otimes T^* \otimes \hat{F}_0 & \to 0 \\
& \downarrow & & \downarrow & & \downarrow & \\
0 \to & \wedge^4 T^* \otimes \hat{g}_1 & \to & \wedge^4 T^* \otimes T^* \otimes T & \to & \wedge^4 T^* \otimes \hat{F}_0 & \to 0 \\
& \downarrow & & \downarrow & & \downarrow & \\
& 0 & & 0 & & 0 &
\end{array}
$$

but $\hat{h}_3$ *becomes* 2-acyclic by chasing in the next diagram:

$$
\begin{array}{ccccccc}
& 0 & & 0 & & 0 & \\
& \downarrow & & \downarrow & & \downarrow & \\
0 \to & \hat{g}_6 & \to & S_6 T^* \otimes T & \to & \hat{h}_5 & \to 0 \\
& \downarrow & & \downarrow & & \downarrow & \\
0 \to & T^* \otimes \hat{g}_5 & \to & T^* \otimes S_5 T^* \otimes T & \to & T^* \otimes \hat{h}_4 & \to 0 \\
& \downarrow & & \downarrow & & \downarrow & \\
0 \to & \wedge^2 T^* \otimes \hat{g}_4 & \to & \wedge^2 T^* \otimes S_4 T^* \otimes T & \to & \wedge^2 T^* \otimes \hat{h}_3 & \to 0 \\
& \downarrow & & \downarrow & & \downarrow & \\
0 \to & \wedge^3 T^* \otimes \hat{g}_3 & \to & \wedge^3 T^* \otimes S_3 T^* \otimes T & \to & \wedge^3 T^* \otimes \hat{h}_2 & \to 0 \\
& \downarrow & & \downarrow & & \downarrow & \\
0 \to & \wedge^4 T^* \otimes \hat{g}_2 & \to & \wedge^4 T^* \otimes S_2 T^* \otimes T & \to & \wedge^4 T^* \otimes T^* \otimes \hat{F}_0 & \to 0 \\
& \downarrow & & \downarrow & & \downarrow & \\
& 0 & & 0 & & 0 &
\end{array}
$$

and we have $s' = 1$. Accordingly, the operator $Bianchi : \hat{F}_1 \to \hat{F}_2$ is of order $s' + 1 = 2$ and $\hat{F}_2$ is defined by following commutative diagram where all the rows are exact and all the columns are exact but the first:

$$
\begin{array}{ccccccccc}
& 0 & & 0 & & 0 & & 0 & \\
& \downarrow & & \downarrow & & \downarrow & & \downarrow & \\
0 \to & \hat{g}_5 & \to & S_5 T^* \otimes T & \to & S_4 T^* \otimes \hat{F}_0 & \to & S_2 T^* \otimes \hat{F}_1 & \to \hat{F}_2 \to 0 \\
& \downarrow & & \downarrow & & \downarrow & & \downarrow & \\
0 \to & T^* \otimes \hat{g}_4 & \to & T^* \otimes S_4 T^* \otimes T & \to & T^* \otimes S_3 T^* \otimes \hat{F}_0 & \to & T^* \otimes T^* \otimes \hat{F}_1 & \to 0 \\
& \downarrow & & \downarrow & & \downarrow & & \downarrow & \\
0 \to & \wedge^2 T^* \otimes \hat{g}_3 & \to & \wedge^2 T^* \otimes S_3 T^* \otimes T & \to & \wedge^2 T^* \otimes S_2 T^* \otimes \hat{F}_0 & \to & \wedge^2 T^* \otimes \hat{F}_1 & \to 0 \\
& \downarrow & & \downarrow & & \downarrow & & \downarrow & \\
0 \to & \wedge^3 T^* \otimes \hat{g}_2 & \to & \wedge^3 T^* \otimes S_2 T^* \otimes T & \to & \wedge^3 T^* \otimes T^* \otimes \hat{F}_0 & \to & 0 & \\
& \downarrow & & \downarrow & & \downarrow & & & \\
0 \to & \wedge^4 T^* \otimes \hat{g}_1 & \to & \wedge^4 T^* \otimes T^* \otimes T & \to & \wedge^4 T^* \otimes \hat{F}_0 & \to & 0 & \\
& \downarrow & & \downarrow & & \downarrow & & & \\
& 0 & & 0 & & 0 & & & \\
\end{array}
$$

$$
\begin{array}{ccccccccc}
0 \to & \hat{g}_5 & \to & S_5 T^* \otimes T & \to & S_4 T^* \otimes \hat{F}_0 & \to & S_2 T^* \otimes \hat{F}_1 & \to & \hat{F}_2 & \to 0 \\
0 \to & 224 & \to & 315 & \to & 100 & \to & 9 & \to 0 \\
\end{array}
$$

$$
\begin{array}{ccccccc}
0 \to & \hat{g}_4 & \to & S_4 T^* \otimes T & \to & S_3 T^* \otimes \hat{F}_0 & \to & T^* \otimes \hat{F}_1 & \to 0 \\
0 \to & 140 & \to & 180 & \to & 40 & \to 0 \\
\end{array}
$$

We could define similarly the *first order* generating CC $\hat{F}_2 \to \hat{F}_3$ where $\hat{F}_3$ is defined by the following long exact sequence:

$$
\begin{array}{ccccccccc}
0 \to & \hat{g}_6 & \to & S_6 T^* \otimes T & \to & S_5 T^* \otimes \hat{F}_0 & \to & S_3 T^* \otimes \hat{F}_1 & \to & T^* \otimes \hat{F}_2 & \to & \hat{F}_3 & \to 0 \\
0 \to & 336 & \to & 504 & \to & 200 & \to & 36 & \to & 4 & \to 0 \\
\end{array}
$$

and obtain finally the formally exact differential sequence:

$$
0 \to \hat{\Theta} \to 4 \stackrel{CKilling}{\longrightarrow} 9 \stackrel{Weyl}{\longrightarrow} 10 \stackrel{Bianchi}{\longrightarrow} 9 \longrightarrow 4 \to 0
$$

with vanishing Euler-Poincaré characteristic $4 - 9 + 10 - 9 + 4 = 0$. We conclude the study of $n = 4$ by exhibiting the short exact sequence:

$$
\begin{array}{ccccccc}
0 \to & \wedge^2 T^* \otimes \hat{g}_2 & \to & \wedge^3 T^* \otimes \hat{g}_1 & \to & \wedge^4 T^* \otimes T & \to 0 \\
0 \to & 24 & \to & 28 & \to & 4 & \to 0 \\
\end{array}
$$

a result showing that $H^3(\hat{g}_1) = 0$ for $n = 4$, *contrary to what will happen when* $n \geq 5$.

- $n \geq 5$: We still have $dim(\hat{F}_0) = (n-1)(n+2)/2$ but now $\hat{g}_2$ is 2-acyclic *and* 3-acyclic with $s = 1$ again but *now* with $s' = 0$. Hence, according to the Main Theorem and its Corollary, the operator $Weyl$ is *second order* while the operator $Bianchi$ is *first order*. We may therefore use the same diagrams already introduced in part A but with a *"hat"* symbol and $n \geq 4$. In particular we get:

$$dim(Z^3(\hat{g}_1)) = dim(\wedge^3 T^* \otimes \hat{g}_1) - dim(\wedge^4 T^* \otimes T) = n(n-1)(n-2)(n^2+n+4)/24$$

$$dim(B^3(\hat{g}_1)) = n^2(n-1)/2 \quad \Rightarrow \quad dim(H^3(\hat{g}_1)) = n(n^2-1)(n+2)(n-4)/24$$

We find again $H^3(\hat{g}_1) = 0$ when $n = 4$ but $H^3(\hat{g}_1) \neq 0$ when $n \geq 5$, a key step that no classical technique can even imagine. We have therefore a *first order* operator $Bianchi : \hat{F}_1 \to \hat{F}_2$ with $\hat{F}_1 = H^2(\hat{g}_1)$ and $\hat{F}_2 = H^3(\hat{g}_1)$ when $n \geq 5$. We obtain therefore a formally exact differential sequence:

$$0 \to \hat{\Theta} \to n \xrightarrow{CKilling} (n-1)(n+2)/2 \xrightarrow{Weyl} n(n+1)(n+2)(n-3)/12 \xrightarrow{Bianchi} n(n^2-1)(n+2)(n-4)/24$$

In order to convince the reader about the powerfulness of the previous methods, let us prove that the generating CC of $Bianchi$ is a second order operator with 14 equations when $n = 5$. First, we ask the reader to prove, as an exercise, that $H^4(\hat{g}_1) = 0$ through the exact δ-sequence $0 \to \wedge^3 T^* \otimes \hat{g}_2 \to \wedge^4 T^* \otimes \hat{g}_1 \to \wedge^5 T^* \otimes T \to 0$ (Hint: $50 = 55 - 5$). Conclude that the generating CC cannot be of first order. Then, prove that $dim(H^4(\hat{g}_2)) = 25 - 11 = 14$. Finally, prove that the generating CC of these CC is first order with 5 equations (Hint: Use the vanishing of the Euler-Poincaré characteristic $5 - 14 + 35 - 35 + 14 - 5 = 0$) (See Appendix 2 for confirmation).

C) PARAMETRIZATIONS

Let E and F be two vector bundles with respective fiber dimensions $dim(E) = m$ and $dim(F) = p$. Starting with a differential operator $\mathcal{D} : E \to F$ of order q with solutions $\Theta \subset E$ and such that the corresponding system $R_q \subset J_q(E)$ is formally integrable, we have explained in Section 3 how to construct a formally exact differential sequence:

$$0 \to \Theta \to E \xrightarrow{\mathcal{D}} F_0 \xrightarrow{\mathcal{D}_1} F_1 \xrightarrow{\mathcal{D}_2} F_2 \xrightarrow{\mathcal{D}_3} \ldots$$

where $F_0 = F$ and each operator $\mathcal{D}_i$ of order q_i generates the CC of the previous one. In particular, if the starting operator $\mathcal{D}$ is involutive, then $q_1 = \ldots = q_n = 1$ and each $\mathcal{D}_i$ is involutive in the resulting Janet sequence finishing at F_n in the sense that $\mathcal{D}_n$ is formally surjective. Equivalently, it is possible to pass to the framework of differential module and look for a free resolution of a differential module M starting with a free finite presentation $D^p \xrightarrow{\mathcal{D}} D^m \to M \to 0$ with an operator acting on the right. In general, we have proved in the last two parts A and B that the succession of the orders $q, q_1, q_2, q_3, \ldots$ can be nevertheless quite strange. Meanwhile, we have proved through examples that many possible finite length such sequences can be exhibited and the purpose of *Homological Algebra* is to study formal properties that should not depend on the sequence used. At the end od Section 4, we have pointed out the fact that, whenever an operator $\mathcal{D}_1$ generates all the CC of an operator $\mathcal{D}$, this does not imply in general that the operator $ad(\mathcal{D})$ generates all the CC of $ad(\mathcal{D}_1)$. The following (quite difficult) theorem is a main result of homological algebra, adapted to differential systems and differential modules ([2],[3],[9],[17],[23],[37]):

THEOREM 5.2: The fact that a formally exact differential sequence considered as a resolution of Θ has the property that the adjoint sequence is also formally exact does not depend on the sequence but only on Θ.

COROLLARY 5.3: The fact that a free resolution of a differential module M has the property that the adjoint sequence is also a free resolution does not depend on the sequence but only on M.

Our problem in this part C is to describe a sufficiently general situation in such

146

a way that *all* the results of the parts A and B can fit together, in the sense that we shall no longer need to use a *hat* in order to distinguish them. For this, with $E = T$, let us say that an operator $\mathcal{D} = \Phi \circ j_q : T \to F$ is a *Lie operator* if $\mathcal{D}\xi = 0, \mathcal{D}\eta = 0 \Rightarrow \mathcal{D}[\xi, \eta] = 0$ that is to say $[\Theta, \Theta] \subseteq \Theta$. The corresponding system $R_q = ker(\Phi)$ is called a *system of infinitesimal Lie equations* and one can define a "*bracket on sections*" satisfying $[R_q, R_q] \subseteq R_q$ in order to check formally the previous definition ([19],[29],[30]). It has been found by E. Vessiot, as early as in ... 1903 ([19],[40]), that the condition of formal integrability of R_q can be described by the *Vessiot structure equations*, a set of (non-linear in general) differential conditions depending on a certain number of constants, for one or a family of *geometric objects* that can be vectors, forms, tensors or even higher order objects. The idea has been to look for "*general*" systems or symbols having the same dimensions as for a model object called "*special*", for example the euclidean metric when $n = 2, 3$ or the minkowskian metric when $n = 4$. The case of part A has been a metric ω with $det(\omega) \neq 0$ and constant riemannian curvature with one constant while the case of part B has been a metric density $\hat{\omega} = \omega/(\mid det(\omega) \mid^{\frac{1}{n}})$ with zero Weyl tensor and no constant involved. The following results will be local.

Let us suppose that we have a Lie group of transformations of X, namely a Lie group G and an *action* $X \times G \to X : (x, a) \to y = ax = f(x, a)$ or, better, its graph $X \times G \to X \times X : (x, a) \to (x, y = ax = f(x, a))$. Differentiating enough times, we may eliminate the parameters a among the equations $y_q = j_q(f)(x, a)$ for q large enough and get a (non-linear in general) *system of finite Lie equations*. Linearising this system for a close to the identity $e \in G$, that is for y close to x, provides the system $R_q \subset J_q(T)$ and the corresponding Lie operator of finite type. Equivalently, the three theorems of Sophus Lie assert that there exists a *finite number of infinitesimal generators* $\{\theta_\tau\}$ of the action that should be linearly independent over the constants and satisfy $[\theta_\rho, \theta_\sigma] = c_{\rho\sigma}^\tau \theta_\tau$ where the *structure constants* c define a Lie algebra $\mathcal{G} = T_e(G)$. We have therefore $\xi \in \Theta \Leftrightarrow \xi = \lambda^\tau \theta_\tau$ with $\lambda^\tau = cst$. Hence, we may replace locally the system of infinitesimal Lie equations by the system $\partial_i \lambda^\tau = 0$, getting therefore the differential sequence:

$$0 \to \Theta \to \wedge^0 T^* \otimes \mathcal{G} \xrightarrow{d} \wedge^1 T^* \otimes \mathcal{G} \xrightarrow{d} ... \xrightarrow{d} \wedge^n T^* \otimes \mathcal{G} \to 0$$

which is the tensor product of the Poincaré sequence by $\mathcal{G}$. Finally, we are in a position to apply the previous Theorem and Corollary because the Poincaré sequence is self adjoint (up to sign), that is $ad(d)$ generates the CC of $ad(d)$ at any position, exactly like d generates the CC of d at any position. We invite the reader to compare with the situation of the Maxwell equations in electromagnetisme. However, we have proved in ([21],[22],[30],[31],[32]) why neither the Janet sequence nor the Poincaré sequence can be used in physics and must be replaced by the *Spencer sequence* which is another resolution of Θ. We provide a few additional details on the motivations for such a procedure.

For this, if q is large enough in such a way that $g_q = 0$ and thus $R_{q+1} \simeq R_q$, let us define locally a section $\xi_{q+1} \in R_{q+1}$ by the formula $\xi^k_{\mu+1_i} = \lambda^\tau(x)\partial_{\mu+1_i}\theta^k_\tau(x)$ and apply the Spencer operator D. We obtain at once $(D\xi_{q+1})^k_{\mu,i} = \partial_i\xi^k_\mu - \xi^k_{\mu+1_i} = \partial_i\lambda^\tau\partial_\mu\theta^k_\tau$, a result proving that the previous sequence is (locally) isomorphic to the Spencer sequence:

$$0 \to \Theta \to \wedge^0 T^* \otimes R_q \xrightarrow{D} \wedge^1 T^* \otimes R_q \xrightarrow{D} ... \xrightarrow{D} \wedge^n T^* \otimes R_q \to 0$$

In the present paper, we had $dim(\mathcal{G}) = n(n+1)/2$ in part A and $dim(\hat{\mathcal{G}}) = (n+1)(n+2)/2$ in part B. Moreover, whatever is the part concerned, $ad(\mathcal{D})$ generates the CC of $ad(\mathcal{D}_1)$ because $\mathcal{D}_1$ generates the CC of $\mathcal{D}$ while, similarly, $ad(\mathcal{D}_1)$ generates the CC of $ad(\mathcal{D}_2)$ because $\mathcal{D}_2$ generates the CC of $\mathcal{D}_1$ and so on. We conclude with the following comments.

1) Coming back to the first differential sequence constructed in part A, the Riemann tensor is a section of F_1 which is in the image of the operator $Riemann$ or in the kernel of the operator $Bianchi$. Indeed, Lanczos has been considering the Riemann tensor as a section of F_1 killed by the operator $Bianchi$ considered as a differential constraint. Accordingly, he has used the action of $ad(Bianchi)$ on the corresponding Lagrange multipliers. However, this operator parametrizes $ad(Riemann)$ as we saw and *cannot be used in order to parametrize the Riemann tensor by means of the Lanczos potential.*

2) Coming back to the second differential sequence constructed (independently) in part B, the Weyl tensor is a section of $\hat{F}_1$ which is in the image of the operator $Weyl$ or in the kernel of the operator $Bianchi$. As most of the results presented are

148

unknown, in particular the fact that both operators $Weyl$ and $Bianchi$ are second order in dimension $n = 4$, we believe that even the proper concept of a Weyl tensor candidate must be revisited within this new framework.

3) In parts A and B, only linear differential operators have been used. However, it is known from *the formal theory of Lie pseudogroups* that non-linear differential sequences can be similarly constructed ([13],[19],[22]). *As a matter of fact*, if non-linear analogues of $\mathcal{D}$ and $\mathcal{D}_1$ may be exhibited, *this is not possible for $\mathcal{D}_2$*. Moreover, the only important problem is to compare the image of $\mathcal{D}$ with the kernel of $\mathcal{D}_1$ in the finite/infinitesimal *equivalence problem* (See [19], p 333 for a nice counterexample). We believe that *this shift by one step backwards in the interpretation of a differential sequence* will become important for future physics. It is done in engineering variational calculus where the deformation tensor and the EM field are on equal footing in the free energy.

6) CONCLUSION

In most textbooks, the Weyl tensor is always presented today by comparison with the Riemann tensor after eliminating a conformal factor and its derivatives. We have exhibited new methods in order to introduce both the Riemann and the Weyl tensor *independently* by using the formal theory of systems of partial differential equations (Spencer cohomology) in the study of arbitrary Lie pseudogroups while using the *Vessiot structure equations* for the Killing and conformal Killing systems *separately*. In particular, we have revisited, in both cases, the proper concept of Bianchi identities by means of homological algebra and diagram chasing, obtaining explicit numbers and orders for each dimension, *without the need of explicit indices*. These striking results are confirmed by means of computer algebra in the Appendix 2. They prove that the work of Lanczos and followers must be revisited within this new framework.

REFERENCES FOR THE APPENDIX 1

[1] G.B. AIRY: On the Strains in the Interior of Beams, Phil. Trans. Roy. Soc.London, 153, 1863, 49-80.

[2] J.E. BJORK: Analytic D-Modules and Applications, Kluwer, 1993.

[3] N. BOURBAKI: Eléments de Mathématiques, Algèbre, Ch. 10, Algèbre Homologique, Masson, Paris, 1980.

[4] E. CARTAN: Les Systèmes Différentiels Extérieurs et Leurs Applications Géométriques, Hermann, Paris, 1945.

[5] S.B. EDGAR: On Effective Constraints for the Riemann-Lanczos Systems of Equations. J. Math. Phys., 44, 2003, 5375-5385.
http://arxiv.org/abs/gr-qc/0302014

[6] S.B. EDGAR, A. HÖGLUND: The Lanczos potential for Weyl-Candidate Tensors Exists only in Four Dimension, General Relativity and Gravitation, 32, 12, 2000, 2307.
http://rspa.royalsocietypublishing.org/content/royprsa/453/1959/835.full.pdf

[7] S.B. EDGAR, J.M.M. SENOVILLA: A Local Potential for the Weyl tensor in all dimensions, Classical and Quantum Gravity, 21, 2004, L133.
http://arxiv.org/abs/gr-qc/0408071

[8] L.P. EISENHART: Riemannian Geometry, Princeton University Press, 1926.

[9] S.-T. HU: Introduction to Homological Algebra, Holden-Day, 1968.

[10] M. JANET: Les Systèmes d'Equations aux Dérivées Partielles, Journal de Math. Pures et Appliquées, 3,1920, 65-151.

[11] M. KASHIWARA: Algebraic Study of Systems of Partial Differential Equations, Mémoires de la Société Mathématique de France 63, 1995, (Transl. from Japanese of his 1970 Master's Thesis).

[12] E.R. KOLCHIN: Differential Algebra and Algebraic Groups, Academic Press, 1973.

[13] A. KUMPERA, D.C. SPENCER, Lie Equations, Ann. Math. Studies 73, Princeton University Press, Princeton, 1972.

[14] C. LANCZOS: Lagrange Multiplier and Riemannian Spaces, Reviews of Modern Physics, 21, 1949, 497-502.

[15] C. LANCZOS: The Splitting of the Riemann Tensor, Rev. Mod. Phys. 34, 1962, 379-389.

[16] F. S. MACAULAY, The Algebraic Theory of Modular Systems, Cambridge Tracts 19, Cambridge University Press, London, 1916; Reprinted by Stechert-Hafner Service Agency, New York, 1964.

[17] D.G. NORTHCOTT: An Introduction to Homological Algebra, Cambridge University Press, 1966.

[18] P. O'DONNELL, H. PYE: A Brief Historical Review of the Important Developments in Lanczos Potential Theory, EJTP, 24, 2010, 327-350.

[19] J.-F. POMMARET: Systems of Partial Differential Equations and Lie Pseudogroups, Gordon and Breach, New York, 1978 (Russian translation by MIR, Moscow, 1983)

[20] J.-F. POMMARET: Differential Galois Theory, Gordon and Breach, New York, 1983.

[21] J.-F. POMMARET: Lie Pseudogroups and Mechanics, Gordon and Breach, New York, 1988.

[22] J.-F. POMMARET: Partial Differential Equations and Group Theory,New Perspectives for Applications, Mathematics and its Applications 293, Kluwer, 1994. http://dx.doi.org/10.1007/978-94-017-2539-2

[23] J.-F. POMMARET: Partial Differential Control Theory, Kluwer, 2001, 957 pp.

[24] J.-F. POMMARET: Algebraic Analysis of Control Systems Defined by Par-

tial Differential Equations, in Advanced Topics in Control Systems Theory, Lecture Notes in Control and Information Sciences 311, Chapter 5, Springer, 2005, 155-223.

[25] J.-F. POMMARET: Gröbner Bases in Algebraic Analysis: New perspectives for applications, Radon Series Comp. Appl. Math 2, 1-21, de Gruyter, 2007.

[26] J.-F. POMMARET: Parametrization of Cosserat Equations, Acta Mechanica, 215, 2010, 43-55.

[27] J.-F. POMMARET: Macaulay Inverse Systems Revisited, Journal of Symbolic Computation, 46, 2011, 1049-1069.
http://dx.doi.org/10.1016/j.jsc.2011.05.007

[28] J.-F. POMMARET: Spencer Operator and Applications: From Continuum Mechanics to Mathematical Physics, in "Continuum Mechanics-Progress in Fundamentals and Engineering Applications", Dr. Yong Gan (Ed.), ISBN: 978-953-51-0447-6, InTech, 2012, Chapter 1, Available from:
http://www.intechopen.com/books/continuum-mechanics-progress-in-fundamentals-and-engineering-applications

[29] J.-F. POMMARET: Deformation Cohomology of Algebraic and Geometric Structures,
arXiv:1207.1964.
http://arxiv.org/abs/1207.1964

[30] J.-F. POMMARET: The Mathematical Foundations of General Relativity Revisited, Journal of Modern Physics, 4, 2013, 223-239.
http://dx.doi.org/10.4236/jmp.2013.48A022

[31] J.-F. POMMARET: The Mathematical Foundations of Gauge Theory Revisited, Journal of Modern Physics, 2014, 5, 157-170.
http://dx.doi.org/10.4236/jmp.2014.55026

[32] J.-F. POMMARET: From Thermodynamics to Gauge Theory: the Virial Theorem Revisited, in " Gauge Theories and Differential geometry ", NOVA Science Publishers, 2015, Chapter 1, 1-44.
http://arxiv.org/abs/1504.04118

[33] J.-F. POMMARET: Airy, Beltrami, Maxwell, Einstein and Lanczos potentials Revisited.
http://arxiv.org/abs/1512.05982

[34] J.-F. POMMARET, A. QUADRAT: Algebraic Analysis of Linear Multidimensional Control Systems, IMA Journal of Mathematical Control and Informations, 16,

1999, 275-297.

[35] A. QUADRAT, D. ROBERTZ: A Constructive Study of the Module Structure of Rings of Partial Differential Operators, Acta Applicandae Mathematicae, 133, 2014, 187-234.

http://hal-supelec.archives-ouvertes.fr/hal-00925533

[36] M.D. ROBERTS: The Physical Interpretation of the Lanczos Tensor, Il Nuovo Cimento, B110, 1996, 1165-1176.

http://arxiv.org/abs/gr-qc/9904006

[37] J.J. ROTMAN: An Introduction to Homological Algebra, Pure and Applied Mathematics, Academic Press, 1979.

[38] J.-P. SCHNEIDERS: An Introduction to D-Modules, Bull. Soc. Roy. Sci. Liège, 63, 1994, 223-295.

[39] D.C. SPENCER: Overdetermined Systems of Partial Differential Equations, Bull. Amer. Math. Soc., 75, 1965, 1-114.

[40] E. VESSIOT: Sur la Théorie des Groupes Continus, Annales Scientifiques de l'Ecole Normale Supérieure, Vol. 20, 1903, 411-451

(Can be obtained from   http://www.numdam.org).

APPENDIX 2

1 Weyl tensor

Using an Euclidean or Minkowskian metric, the system of conformal Killing equations can be written:

$$\omega_{rj}\, \xi_i^r + \omega_{ir}\, \xi_j^r - \frac{2}{n}\, \omega_{ij}\, \xi_r^r = 0. \tag{1}$$

but any other choice could be convenient too as we have explained in the paper.

1.1 Weyl tensor in dimension 2

Let us consider $n = 2$ and the euclidean metric $\omega_{ij} = \delta_j^i$ for $1 \leq i < j \leq 2$. If $D = \mathbb{Q}[d_1, d_2]$ denotes the commutative ring of PD operators in $d_1 = \frac{\partial}{\partial x_1}$ and $d_2 = \frac{\partial}{\partial x_2}$, then the system can be rewritten as $W_2\, \eta = 0$, where $\eta = (\xi^1 \quad \xi^2)^T$ and:

$$W_2 = \begin{pmatrix} d_1 & -d_2 \\ d_2 & d_1 \\ -d_1 & d_2 \end{pmatrix}$$

If we note $\lambda_2 = \begin{pmatrix} 1 & 0 & 1 \end{pmatrix}$, then we have $\lambda_2\, W_2 = 0$, which shows that the last row of W_2 is minus the first one. Hence, the D-module $\mathrm{im}_D(W_2) = \{\lambda\, W_2 \mid \lambda \in D^{1\times 2}\}$ generated by the rows of W_2 can be generated by the first two rows of W_2, i.e., we have $\mathrm{im}_D(W_2) = \mathrm{im}_D(R_2)$, where R_2 is the following matrix:

$$R_2 = \begin{pmatrix} d_1 & -d_2 \\ d_2 & d_1 \end{pmatrix}$$

We can check that the D-module $\ker_D(R_2) = \{\mu \in D^{1\times 2} \mid \mu\, R_2 = 0\}$, called the *second syzygy module* ([4]) of the D-module $M = D^2/(D^2\, R_2)$, is reduced to 0,

154

i.e., R_2 has full row rank. Hence, we obtain the following *finite free resolution* of M:

$$0 \to D^2 \xrightarrow{R_2} D^2 \to M \to 0$$

Similarly, indicating only the dimensions of the vector bundles involved and the order of the corresponding differential operators, we have the formally exact differential sequence with the same operator matrices but now acting on column vectors on the right:

$$0 \to \hat{\Theta} \to 2 \xrightarrow{1} 2 \to 0$$

In the present situation we just recognize the well known Cauchy-Riemann operator defining complex analytic functions.

1.2 Weyl tensor in dimension 3

Let us consider $n = 3$ and the euclidean metric $\omega_{ij} = \delta_j^i$ for $1 \leq i < j \leq 3$. If $D = \mathbb{Q}[d_1, d_2, d_3]$ denotes the commutative ring of PD operators in $d_i = \frac{\partial}{\partial x_i}$ for $i = 1, 2, 3$, then the system can be rewritten as $W_3 \, \eta = 0$, where $\eta = (\xi^1 \quad \xi^2 \quad \xi^3)^T$ and:

$$
W_3 = \begin{pmatrix}
\frac{4}{3} d_1 & -\frac{2}{3} d_2 & -\frac{2}{3} d_3 \\
d_2 & d_1 & 0 \\
d_3 & 0 & d_1 \\
-\frac{2}{3} d_1 & \frac{4}{3} d_2 & -\frac{2}{3} d_3 \\
0 & d_3 & d_2 \\
-\frac{2}{3} d_1 & -\frac{2}{3} d_2 & \frac{4}{3} d_3
\end{pmatrix}
$$

If we note $\lambda_3 = \begin{pmatrix} 1 & 0 & 0 & 1 & 0 & 1 \end{pmatrix} \in D^{1 \times 6}$, then we have $\lambda_3 W_3 = 0$ and shows that the last row of W_3 is a linear combination of the first and fourth rows of W_3. Hence, the D-module $\mathrm{im}_D(W_3)$, generated by the rows of W_3, can be generated by the first fifth rows of W_3, i.e., we have $\mathrm{im}_D(W_3) = \mathrm{im}_D(R_3)$, where R_3 is the following matrix:

$$
R_3 = \begin{pmatrix}
\frac{4}{3} d_1 & -\frac{2}{3} d_2 & -\frac{2}{3} d_3 \\
d_2 & d_1 & 0 \\
d_3 & 0 & d_1 \\
-\frac{2}{3} d_1 & \frac{4}{3} d_2 & -\frac{2}{3} d_3 \\
0 & d_3 & d_2
\end{pmatrix}
$$

Using the OREMODULES package, we can show that the left kernel of R_3, i.e., the D-module $\ker_D(R_3) = \{\lambda \in D^{1 \times 3} \mid \lambda R_3 = 0\}$, is generated by the rows of the following matrix S_3:

$$S_3 = \begin{pmatrix}
-d_2 d_3 d_1 & d_2^2 d_3 + d_3^3 & -d_2^3 - d_2 d_3^2 & -2 d_2 d_3 d_1 & d_1 d_2^2 - d_1 d_3^2 \\
-d_1^2 d_3 + 2 d_2^2 d_3 + d_3^3 & 0 & -2 d_1 d_2^2 - 2 d_1 d_3^2 & -2 d_1^2 d_3 + d_2^2 d_3 - d_3^3 & 2 d_1^2 d_2 + 2 d_2 d_3^2 \\
d_1^2 d_2 + 2 d_2^3 + d_2 d_3^2 & -2 d_1 d_2^2 - 2 d_1 d_3^2 & 0 & 2 d_1^2 d_2 + d_2^3 - d_2 d_3^2 & 2 d_1^2 d_3 + 2 d_2^2 d_3 \\
2 d_2 d_3 d_1 & -d_1^2 d_3 - d_3^3 & -d_1^2 d_2 + d_2 d_3^2 & d_2 d_3 d_1 & d_1^3 + d_1 d_3^2 \\
d_1^3 + 2 d_1 d_2^2 - d_1 d_3^2 & -2 d_1^2 d_2 - 2 d_2 d_3^2 & 2 d_1^2 d_3 + 2 d_2^2 d_3 & 2 d_2^3 + d_1 d_2^2 + d_1 d_3^2 & 0
\end{pmatrix}$$

i.e., we have $\ker_D(R_3) = \operatorname{im}_D(S_3) = \{\mu \in D^{1 \times 5} \mid \mu\, S_3\}$. The computation of S_3 takes 0.026 CPU seconds with Maple 18 on Mac OS 10.10.5 equipped with 2.8 GHz Intel Core i7 and 16 Go. For more details on the left kernel/syzygy computation, see

Algorithm 1 on page 330 of ([1]) (see also [3]). Similarly, the left kernel of T_3 is generated by the rows of the following matrix

$$T_3 = \begin{pmatrix} -2\,d_3 & 0 & d_1 & -2\,d_3 & -d_2 \\ 2\,d_1 & -d_2 & d_3 & 0 & 0 \\ 0 & d_1 & 0 & -2\,d_2 & d_3 \end{pmatrix}$$

i.e., we have $\ker_D(S_3) = \mathrm{im}_D(T_3)$. We can check that that the matrix T_3 has full row rank, i.e., $\ker_D(T_3) = 0$, or equivalently that the rows of T_3 are D-linearly independent. Finally, if we denote by $M = D^3/(D^5\,R_3) = D^3/(D^6\,W_3)$ the D-module *finitely presented* by R_3, associated with the system, then we obtain the following finite free resolution:

$$0 \to D^3 \xrightarrow{T_3} D^5 \xrightarrow{S_3} D^5 \xrightarrow{R_3} D^3 \to M \to 0$$

We have the corresponding formally exact *striking differential sequence*:

$$0 \to \hat{\Theta} \to 3 \xrightarrow{1} 5 \xrightarrow{3} 5 \xrightarrow{1} 3 \to 0$$

1.3 Weyl tensor in dimension 4

Let us consider $n = 4$ and the Minkowski metric $\omega = (1,\ 1,\ 1,\ -1)$. If $D = \mathbb{Q}[d_1, d_2, d_3, d_4]$ denotes the commutative ring of PD operators in $d_i = \frac{\partial}{\partial x_i}$ for $i = 1, \ldots, 4$, then the system can be rewritten as $W_4\, \eta = 0$, where $\eta = (\xi^1\ \ \xi^2\ \ \xi^3\ \ \xi^4)^T$ and:

$$
W_4 = \begin{pmatrix}
\frac{3}{2}d_1 & -\frac{1}{2}d_2 & -\frac{1}{2}d_3 & -\frac{1}{2}d_4 \\[4pt]
d_2 & d_1 & 0 & 0 \\[4pt]
d_3 & 0 & d_1 & 0 \\[4pt]
d_4 & 0 & 0 & -d_1 \\[4pt]
-\frac{1}{2}d_1 & \frac{3}{2}d_2 & -\frac{1}{2}d_3 & -\frac{1}{2}d_4 \\[4pt]
0 & d_3 & d_2 & 0 \\[4pt]
0 & d_4 & 0 & -d_2 \\[4pt]
-\frac{1}{2}d_1 & -\frac{1}{2}d_2 & \frac{3}{2}d_3 & -\frac{1}{2}d_4 \\[4pt]
0 & 0 & d_4 & -d_3 \\[4pt]
\frac{1}{2}d_1 & \frac{1}{2}d_2 & \frac{1}{2}d_3 & -\frac{3}{2}d_4
\end{pmatrix}
$$

If we note $\lambda_4 = \begin{pmatrix} 1 & 0 & 0 & 0 & 1 & 0 & 0 & 1 & 0 & -1 \end{pmatrix}$, then we have $\lambda_4 W_4 = 0$, which shows that the last row of W_4 is a linear combination of the first, fourth, and eighth rows. Hence, the D-module $\operatorname{im}_D(W_4)$, generated by the rows of W_4, can be generated by the first ninth rows of W_4, i.e., we have $\operatorname{im}_D(W_4) = \operatorname{im}_D(R_4)$, where R_4 is the following matrix:

$$
R_4 = \begin{pmatrix}
\frac{3}{2}d_1 & -\frac{1}{2}d_2 & -\frac{1}{2}d_3 & -\frac{1}{2}d_4 \\[4pt]
d_2 & d_1 & 0 & 0 \\[4pt]
d_3 & 0 & d_1 & 0 \\[4pt]
d_4 & 0 & 0 & -d_1 \\[4pt]
-\frac{1}{2}d_1 & \frac{3}{2}d_2 & -\frac{1}{2}d_3 & -\frac{1}{2}d_4 \\[4pt]
0 & d_3 & d_2 & 0 \\[4pt]
0 & d_4 & 0 & -d_2 \\[4pt]
-\frac{1}{2}d_1 & -\frac{1}{2}d_2 & \frac{3}{2}d_3 & -\frac{1}{2}d_4 \\[4pt]
0 & 0 & d_4 & -d_3
\end{pmatrix}
$$

We can show that the left kernel of R_4, i.e., the D-module $\ker_D(R_4)$, is generated by the rows of the following matrix S_4:

$$
S_4 = \begin{bmatrix}
d_4 d_1 & -d_1 d_3 & 0 & d_1 d_2 & -2 d_1 d_3 & d_3 d_4 & -d_2^2 - d_4^2 & d_2 d_3 & -d_1 d_3 \\[2pt]
0 & 2 d_1 d_2 & -d_4 d_1 & -d_1 d_3 & d_1 d_2 & -d_4 d_2 & -d_2 d_3 & d_3^2 + d_4^2 & d_1 d_2 \\[2pt]
d_1^2 - d_2^2 & 0 & d_2 d_3 & d_4 d_2 & -d_3 d_4 & -d_1 d_3 & -d_4 d_1 & 0 & d_3 d_4 \\[2pt]
d_2 d_3 & -d_4 d_2 & d_1^2 - d_3^2 & d_3 d_4 & 0 & -d_1 d_2 & 0 & -d_4 d_1 & d_4 d_2 \\[2pt]
-d_4 d_2 & d_2 d_3 & -d_3 d_4 & d_1^2 + d_4^2 & d_2 d_3 & 0 & -d_1 d_2 & -d_1 d_3 & 2 d_2 d_3 \\[2pt]
2 d_3 d_4 & -d_1^2 + d_2^2 - d_3^2 - d_4^2 & -2 d_4 d_2 & 0 & d_1^2 + d_2^2 - d_3^2 + d_4^2 & 0 & 2 d_1 d_3 & -2 d_1 d_2 & 2 d_2^2 - 2 d_3^2 \\[2pt]
-2 d_3 d_4 & 3 d_1^2 - 3 d_2^2 + d_3^2 + d_4^2 & 4 d_4 d_2 & 2 d_2 d_3 & -2 d_2^2 - 2 d_4^2 & -2 d_4 d_1 & -4 d_1 d_3 & 2 d_1 d_2 & d_1^2 - 3 d_1^2 + 3 d_3^2 + d_4^2 \\[2pt]
d_1 d_2 & 0 & 0 & -d_4 d_1 & 0 & -d_2 d_3 & 0 & d_3 d_4 & 0 \\[2pt]
-d_1 d_3 & d_4 d_1 & d_1 d_2 & 0 & -d_4 d_1 & -d_2^2 + d_3^2 & -d_3 d_4 & d_4 d_2 & 0 \\[2pt]
0 & 0 & d_1 d_3 & -d_4 d_1 & 0 & -d_2 d_3 & d_4 d_2 & 0 & 0
\end{bmatrix}
$$

i.e., we have $\ker_D(R_4) = \mathrm{im}_D(S_4)$. In fact, using the OREMODULES package, we can prove that $\ker_D(R_4) = \mathrm{im}_D(\overline{S}_4)$, where $\overline{S}_4 = (S_4^T \quad S_4'^{T})^T$, where the matrix

$S_4' \in D^{6\times 9}$ is defined by:

$$
S_4' = \begin{pmatrix}
-d_2 d_3 d_4 & 0 & 0 & 0 & -d_2 d_3 d_4 & d_3^2 d_4 - d_4^3 & -d_3^3 + d_3 d_4^2 & -2 d_2 d_3 d_4 & d_2 d_3^2 + d_2 d_4^2 \\[4pt]
-d_1 d_3 d_4 & 0 & 0 & d_3^2 d_4 - d_4^3 & -d_1 d_3 d_4 & 0 & 0 & -2 d_1 d_3 d_4 & d_1 d_3^2 + d_1 d_4^2 \\[4pt]
-d_2^2 d_4 + d_3^2 d_4 & 0 & -d_3^3 + d_3 d_4^2 & 0 & -d_2^2 d_4 + 2 d_3^2 d_4 - d_4^3 & 0 & -2 d_2 d_3^2 + 2 d_2 d_4^2 & -2 d_2^2 d_4 + d_3^2 d_4 + d_4^3 & 2 d_2^2 d_3 - 2 d_3 d_4^2 \\[4pt]
d_2^2 d_3 + d_3^3 & 0 & 0 & 0 & d_2^2 d_3 + 2 d_3^3 - d_3 d_4^2 & -2 d_2 d_3^2 + 2 d_2 d_4^2 & 0 & 2 d_2^2 d_3 + d_3^3 + d_3 d_4^2 & -2 d_2^2 d_4 - 2 d_3^2 d_4 \\[4pt]
d_2 d_3 d_4 & 0 & 0 & 0 & 2 d_2 d_3 d_4 & -d_2^2 d_4 + d_4^3 & -d_2^2 d_3 - d_3 d_4^2 & d_2 d_3 d_4 & d_2^3 - d_2 d_4^2 \\[4pt]
d_2^3 + d_2 d_3^2 & 0 & 0 & 0 & d_2^3 + 2 d_2 d_3^2 + d_2 d_4^2 & -2 d_2^2 d_3 + 2 d_3 d_4^2 & -2 d_2^2 d_4 - 2 d_3^2 d_4 & 2 d_2^3 + d_2 d_3^2 - d_2 d_4^2 & 0
\end{pmatrix}
$$

But, we have $S_4' = F_4 S_4$, where $F_4 \in D^{6\times 10}$ is the matrix:

$$\begin{pmatrix}
0 & 0 & 0 & d_3 & -d_4 & 0 & 0 & 0 & 0 & -d_1 \\
d_4 & 0 & 0 & 0 & 0 & 0 & 0 & 0 & -d_3 & d_2 \\
0 & 0 & -d_3 & d_2 & 0 & -d_4 & 0 & 0 & -d_1 & 0 \\
-d_1 & 0 & d_4 & 0 & d_2 & -2\,d_3 & -d_3 & 0 & 0 & 0 \\
0 & 0 & -d_2 & 0 & d_4 & 0 & 0 & d_1 & 0 & 0 \\
0 & d_1 & 0 & d_4 & d_3 & -d_2 & -d_2 & 0 & 0 & 0
\end{pmatrix}$$

which shows that $\operatorname{im}_D(\overline{S}_4) = \operatorname{im}_D(S_4)$. Similarly, we get that $\ker_D(S_4) = \operatorname{im}_D(T_4)$, where the matrix T_4 is defined by:

and $\ker_D(T_4) = \operatorname{im}_D(U_4)$, where the matrix U_4 is:

$$\begin{pmatrix} -2\,d_3 & 0 & 0 & -d_2 & d_4 & 0 & d_1 & 0 & -2\,d_3 \\ 2\,d_1 & -d_2 & -d_4 & 0 & 0 & 0 & d_3 & 0 & 0 \\ 0 & d_1 & 0 & d_3 & 0 & -d_4 & 0 & -2\,d_2 & 0 \\ 0 & 0 & d_1 & 0 & -d_3 & -d_2 & 0 & -2\,d_4 & 2\,d_4 \end{pmatrix}$$

Finally, we can check that U_4 has full row rank, i.e., $\ker_D(U_4) = 0$, which shows that the D-module $M = D^4/(D^9\,R_4) = D^4/(D^{10}\,W_4)$, associated with the system, admits the following finite free resolution:

$$0 \to D^4 \xrightarrow{U_4} D^9 \xrightarrow{T_4} D^{10} \xrightarrow{S_4} D^9 \xrightarrow{R_4} D^4 \to M \to 0$$

Finally, we note that a free resolution of M can be computed in 0.616 CPU seconds when $n = 4$.

We have the corresponding *most striking differential sequence*:

$$0 \to \hat{\Theta} \to 4 \xrightarrow{1} 9 \xrightarrow{2} 10 \xrightarrow{2} 9 \xrightarrow{1} 4 \to 0$$

163

1.4 Weyl tensor in dimension 5

Let us consider $n = 5$ and the euclidean metric $\omega_{ij} = \delta^i_j$ for $1 \leq i < j \leq 5$. If $D = \mathbb{Q}[d_1, d_2, d_3, d_4, d_5]$ denotes the commutative ring of PD operators in $d_i = \frac{\partial}{\partial x_i}$ for $i = 1, \ldots, 5$, then the system can be rewritten as $W_5\, \eta = 0$, where $\eta = (\xi^1 \quad \ldots \quad \xi^5)^T$ and:

$$
W_5 =
\begin{pmatrix}
\frac{8}{5}d_1 & -\frac{2}{5}d_2 & -\frac{2}{5}d_3 & -\frac{2}{5}d_4 & -\frac{2}{5}d_5 \\
d_2 & d_1 & 0 & 0 & 0 \\
d_3 & 0 & d_1 & 0 & 0 \\
d_4 & 0 & 0 & d_1 & 0 \\
d_5 & 0 & 0 & 0 & d_1 \\
-\frac{2}{5}d_1 & \frac{8}{5}d_2 & -\frac{2}{5}d_3 & -\frac{2}{5}d_4 & -\frac{2}{5}d_5 \\
0 & d_3 & d_2 & 0 & 0 \\
0 & d_4 & 0 & d_2 & 0 \\
0 & d_5 & 0 & 0 & d_2 \\
-\frac{2}{5}d_1 & -\frac{2}{5}d_2 & \frac{8}{5}d_3 & -\frac{2}{5}d_4 & -\frac{2}{5}d_5 \\
0 & 0 & d_4 & d_3 & 0 \\
0 & 0 & d_5 & 0 & d_3 \\
-2/5\, d_1 & -\frac{2}{5}d_2 & -\frac{2}{5}d_3 & \frac{8}{5}d_4 & -\frac{2}{5}d_5 \\
0 & 0 & 0 & d_5 & d_4 \\
-\frac{2}{5}d_1 & -\frac{2}{5}d_2 & -\frac{2}{5}d_3 & -\frac{2}{5}d_4 & \frac{8}{5}d_5
\end{pmatrix}
\in D^{15 \times 5}.
$$

If we note $\lambda_5 = \begin{pmatrix} 1 & 0 & 0 & 0 & 0 & 1 & 0 & 0 & 0 & 1 & 0 & 0 & 1 & 0 & 1 \end{pmatrix}$, then we have $\lambda_5\, W_5 = 0$, which shows that the D-module $\mathrm{im}_D(W_5)$, generated by the rows of W_5, can be generated by the first 14th rows of W_5, i.e., we have $\mathrm{im}_D(W_5) = \mathrm{im}_D(R_5)$, where R_5 is the following matrix:

$$R_5 = \begin{pmatrix} \frac{8}{5}d_1 & -\frac{2}{5}d_2 & -\frac{2}{5}d_3 & -\frac{2}{5}d_4 & -\frac{2}{5}d_5 \\ d_2 & d_1 & 0 & 0 & 0 \\ d_3 & 0 & d_1 & 0 & 0 \\ d_4 & 0 & 0 & d_1 & 0 \\ d_5 & 0 & 0 & 0 & d_1 \\ -\frac{2}{5}d_1 & \frac{8}{5}d_2 & -\frac{2}{5}d_3 & -\frac{2}{5}d_4 & -\frac{2}{5}d_5 \\ 0 & d_3 & d_2 & 0 & 0 \\ 0 & d_4 & 0 & d_2 & 0 \\ 0 & d_5 & 0 & 0 & d_2 \\ -\frac{2}{5}d_1 & -\frac{2}{5}d_2 & \frac{8}{5}d_3 & -\frac{2}{5}d_4 & -\frac{2}{5}d_5 \\ 0 & 0 & d_4 & d_3 & 0 \\ 0 & 0 & d_5 & 0 & d_3 \\ -2/5\,d_1 & -\frac{2}{5}d_2 & -\frac{2}{5}d_3 & \frac{8}{5}d_4 & -\frac{2}{5}d_5 \\ 0 & 0 & 0 & d_5 & d_4 \end{pmatrix} \in D^{14 \times 5}.$$

Using the OREMODULES package, we can show that the left kernel of R_5, i.e., the D-module $\ker_D(R_5)$, is generated by the rows of a matrix $S_5 = (S_A \quad S_B) \in D^{35 \times 14}$, where the matrices $S_A \in D^{35 \times 7}$ and $S_B \in D^{35 \times 7}$ are respectively defined by:

$$
\left(
\begin{array}{ccccccc}
d_4^2 - d_4 d_2 & 0 & 0 & 0 & 0 & -d_4 d_2 & d_3 d_4 \\
d_2 d_3 & 0 & 0 & 0 & 0 & d_2 d_3 & d_4^2 - d_5^2 \\
-d_4 d_1 & 0 & d_3 d_4 & -d_3^2 + d_5^2 & -d_5 d_4 & -d_4 d_1 & 0 \\
d_1 d_3 & 0 & d_4^2 - d_5^2 & -d_3 d_4 & d_3 d_5 & d_1 d_3 & 0 \\
d_3 d_4 & 0 & 0 & 0 & 0 & 2 d_3 d_4 & -d_4 d_2 \\
d_3^2 - d_4^2 & 0 & 0 & 0 & 0 & 2 d_3^2 - 2 d_4^2 & -2 d_2 d_3 \\
d_2^2 - 2 d_3^2 + d_4^2 & 0 & 0 & 0 & 0 & d_2^2 - 3 d_3^2 + 3 d_4^2 - d_5^2 & 2 d_2 d_3 \\
-d_4 d_1 & d_4 d_2 & 0 & -d_2^2 + d_5^2 & -d_5 d_4 & -2 d_4 d_1 & 0 \\
d_1 d_2 & -d_3^2 + 2 d_4^2 - d_5^2 & d_2 d_3 & -2 d_4 d_2 & d_2 d_5 & d_1 d_2 & d_1 d_3 \\
d_5 d_4 & 0 & 0 & -d_1 d_5 & -d_4 d_1 & 0 & 0 \\
d_3 d_5 & 0 & -d_1 d_5 & 0 & -d_1 d_3 & 0 & 0 \\
2 d_3 d_4 & 0 & -d_4 d_1 & -d_1 d_3 & 0 & d_3 d_4 & 0 \\
2 d_3^2 - 2 d_4^2 & 0 & -2 d_1 d_3 & 2 d_4 d_1 & 0 & d_3^2 - d_4^2 & 0 \\
d_2 d_5 & -d_1 d_5 & 0 & 0 & -d_1 d_2 & 0 & 0 \\
2 d_4 d_2 & -d_4 d_1 & 0 & -d_1 d_2 & 0 & d_4 d_2 & 0 \\
3 d_3^2 - 3 d_4^2 & -2 d_1 d_2 & 0 & 2 d_4 d_1 & 0 & d_1^2 - d_2^2 + 4 d_3^2 - 5 d_4^2 + d_5^2 & -2 d_2 d_3 \\
d_1^2 - 6 d_3^2 + 6 d_4^2 - d_5^2 & 2 d_1 d_2 & 2 d_1 d_3 & -6 d_4 d_1 & 2 d_1 d_5 & d_2^2 - 6 d_3^2 + 6 d_4^2 - d_5^2 & 2 d_2 d_3 \\
0 & d_5 d_4 & 0 & 0 & -d_4 d_2 & 0 & 0 \\
0 & d_3 d_5 & 0 & 0 & -d_2 d_3 & 0 & -d_1 d_5 \\
0 & d_3 d_4 & 0 & -d_2 d_3 & 0 & 0 & -d_4 d_1 \\
0 & d_3^2 - d_4^2 & -d_2 d_3 & d_4 d_2 & 0 & 0 & -d_1 d_3 \\
0 & d_2 d_5 & 0 & -d_5 d_4 & -d_2^2 + d_4^2 & -d_1 d_5 & 0 \\
0 & d_2 d_3 & -d_2^2 + d_4^2 & -d_3 d_4 & 0 & -d_1 d_3 & d_1 d_2 \\
0 & -d_1 d_3 & -d_1 d_2 & 0 & 0 & -d_2 d_3 & d_1^2 - 2 d_4^2 + d_5^2 \\
0 & 0 & d_5 d_4 & 0 & -d_3 d_4 & 0 & 0 \\
0 & 0 & d_3 d_5 & -d_5 d_4 & -d_3^2 + d_4^2 & 0 & 0 \\
0 & 0 & d_2 d_5 & 0 & -d_2 d_3 & 0 & -d_1 d_5 \\
0 & 0 & d_4 d_2 & -d_2 d_3 & 0 & 0 & -d_4 d_1 \\
0 & 0 & 0 & d_3 d_5 & -d_3 d_4 & 0 & 0 \\
0 & 0 & 0 & d_2 d_5 & -d_4 d_2 & 0 & 0 \\
0 & 0 & 0 & 0 & 0 & d_5 d_4 & 0 \\
0 & 0 & 0 & 0 & 0 & d_3 d_5 & -d_2 d_5 \\
0 & 0 & 0 & 0 & 0 & 0 & d_5 d_4 \\
0 & 0 & 0 & 0 & 0 & 0 & d_3 d_5 \\
0 & 0 & 0 & 0 & 0 & 0 & 0
\end{array}
\right) ,
$$

and:

$$\begin{pmatrix}
-d_3^2+d_5^2 & -d_5 d_4 & -2 d_4 d_2 & d_2 d_3 & 0 & -d_4 d_2 & -d_2 d_5 \\
-d_3 d_4 & d_3 d_5 & d_2 d_3 & -d_4 d_2 & d_2 d_5 & 2 d_2 d_3 & 0 \\
0 & 0 & -2 d_4 d_1 & d_1 d_3 & 0 & -d_4 d_1 & -d_1 d_5 \\
0 & 0 & d_1 d_3 & -d_4 d_1 & d_1 d_5 & 2 d_1 d_3 & 0 \\
-d_2 d_3 & 0 & d_3 d_4 & d_2^2-d_5^2 & d_5 d_4 & d_3 d_4 & d_3 d_5 \\
2 d_4 d_2 & 0 & d_2{}^2+d_3^2-d_4^2-d_5^2 & 0 & 2 d_3 d_5 & -d_2^2+d_3^2-d_4^2+d_5^2 & -2 d_5 d_4 \\
-4 d_4 d_2 & 2 d_2 d_5 & -2 d_3^2+2 d_5^2 & 2 d_3 d_4 & -4 d_3 d_5 & 3 d_2^2-3 d_3^2+d_4^2-d_5^2 & 2 d_5 d_4 \\
d_1 d_2 & 0 & -d_4 d_1 & 0 & 0 & -d_4 d_1 & -d_1 d_5 \\
-2 d_4 d_1 & d_1 d_5 & 0 & 0 & 0 & 3 d_1 d_2 & 0 \\
0 & 0 & -d_5 d_4 & d_3 d_5 & d_3 d_4 & 0 & d_1^2-d_3^2 \\
0 & 0 & 0 & d_5 d_4 & d_1^2-d_4^2 & -d_3 d_5 & d_3 d_4 \\
0 & 0 & d_3 d_4 & d_1^2-d_5^2 & d_5 d_4 & d_3 d_4 & d_3 d_5 \\
0 & 0 & d_1^2+d_3^2-d_4^2-d_5{}^2 & 0 & 2 d_3 d_5 & -d_1^2+d_3^2-d_4{}^2+d_5^2 & -2 d_5 d_4 \\
d_5 d_4 & d_1^2-d_4^2 & 0 & 0 & 0 & -d_2 d_5 & d_4 d_2 \\
d_1^2-d_5^2 & d_5 d_4 & d_4 d_2 & 0 & 0 & d_4 d_2 & d_2 d_5 \\
6 d_4 d_2 & -2 d_2 d_5 & 3 d_3^2-3 d_5^2 & -4 d_3 d_4 & 6 d_3 d_5 & -d_1^2-4 d_2^2+5 d_3{}^2-2 d_4^2+2 d_5^2 & -4 d_5 d_4 \\
-6 d_4 d_2 & 2 d_2 d_5 & -5 d_3^2+5 d_5^2 & 6 d_3 d_4 & -10 d_3 d_5 & 4 d_1{}^2+4 d_2^2-8 d_3^2+3 d_4^2-3 d_5{}^2 & 6 d_5 d_4 \\
-d_1 d_5 & 0 & 0 & 0 & 0 & 0 & d_1 d_2 \\
0 & 0 & 0 & 0 & d_1 d_2 & 0 & 0 \\
0 & 0 & 0 & d_1 d_2 & 0 & 0 & 0 \\
d_4 d_1 & 0 & d_1 d_2 & 0 & 0 & -d_1 d_2 & 0 \\
0 & d_1 d_2 & 0 & 0 & 0 & d_1 d_5 & -d_4 d_1 \\
0 & 0 & 0 & -d_4 d_1 & 0 & d_1 d_3 & 0 \\
2 d_3 d_4 & -d_3 d_5 & -d_2 d_3 & 2 d_4 d_2 & -d_2 d_5 & -3 d_2 d_3 & 0 \\
0 & 0 & 0 & -d_1 d_5 & 0 & 0 & d_1 d_3 \\
0 & 0 & -d_1 d_5 & 0 & d_1 d_3 & d_1 d_5 & -d_4 d_1 \\
0 & d_1 d_3 & 0 & 0 & 0 & 0 & 0 \\
d_1 d_3 & 0 & 0 & 0 & 0 & 0 & 0 \\
0 & 0 & 0 & -d_1 d_5 & d_4 d_1 & 0 & 0 \\
-d_1 d_5 & d_4 d_1 & 0 & 0 & 0 & 0 & 0 \\
-d_2 d_5 & -d_4 d_2 & -d_5 d_4 & d_3 d_5 & d_3 d_4 & 0 & d_2{}^2-d_3^2 \\
0 & -d_2 d_3 & 0 & d_5 d_4 & d_2^2-d_4^2 & -d_3 d_5 & d_3 d_4 \\
0 & -d_3 d_4 & 0 & -d_2 d_5 & 0 & 0 & d_2 d_3 \\
-d_5 d_4 & -d_3^2+d_5^2 & -d_2 d_5 & 0 & d_2 d_3 & d_2 d_5 & -d_4 d_2 \\
d_3 d_5 & -d_3 d_4 & 0 & -d_2 d_5 & d_4 d_2 & 0 & 0
\end{pmatrix}$$

Similarly, we can prove that we have $\ker_D(S_5) = \operatorname{im}_D(T_5)$, where the matrix $T_5 = (T_A \quad T_B \quad T_C) \in D^{35 \times 35}$, where T_A, T_B, T_C are successively defined by:

$$
\begin{bmatrix}
0 & 0 & 0 & 0 & 0 & -d_3 & 0 & 0 & 0 & 0 & d_1 & 0 & 0 & d_4 \\
0 & 0 & 0 & 0 & 0 & 0 & 0 & 0 & 0 & 0 & 0 & d_1 & 0 & -d_5 \\
0 & d_1 & 0 & 0 & 0 & d_5 & 0 & 0 & 0 & 0 & 0 & 0 & 0 & 0 \\
0 & 0 & -d_2 & d_4 & d_4 & 0 & 0 & 0 & 0 & 0 & 0 & -d_3 & 0 & 0 \\
0 & 0 & -d_2 & 2d_4 & d_4 & 0 & 0 & 0 & d_3 & 0 & 0 & 0 & -d_3 & 0 \\
d_5 & d_5 & 2d_4 & -d_2 & 0 & -d_3 & 0 & 0 & -d_4 & 0 & 0 & 0 & 0 & 0 \\
-d_4 & 0 & -d_5 & 0 & 0 & 0 & 0 & -d_3 & 0 & 0 & 0 & 0 & 0 & 0 \\
0 & 0 & 0 & d_4 & 2d_4 & 0 & 0 & 0 & 0 & -d_2 & 0 & 0 & -d_3 & 0 \\
0 & 0 & 0 & 0 & -d_3 & 0 & 0 & 0 & d_4 & 0 & -d_2 & 0 & d_4 & 0 \\
0 & 0 & 0 & 0 & 0 & 0 & 0 & 0 & 0 & 0 & -d_2 & -d_5 & 0 & 0 \\
0 & -d_2 & 0 & d_5 & d_5 & 0 & 0 & 0 & 0 & 0 & 0 & 0 & 0 & 0 \\
0 & 0 & d_1 & 0 & 0 & d_4 & 0 & 0 & 0 & 0 & 0 & 0 & -d_3 & 0 \\
0 & 0 & 0 & d_1 & 0 & 0 & 0 & 0 & 0 & 0 & 0 & 0 & 0 & 0 \\
0 & 0 & 0 & d_4 & -d_4 & 0 & 0 & 0 & 2d_3 & 0 & 0 & 0 & d_3 & 0 \\
0 & 0 & 0 & -d_4 & -2d_4 & 0 & 0 & 0 & d_3 & 0 & 0 & 0 & d_3 & 0 \\
0 & 0 & 0 & d_3 & d_3 & 0 & 0 & 0 & 0 & 0 & 0 & 0 & -d_4 & 0 \\
0 & 0 & 0 & 0 & 0 & 0 & 0 & 0 & 0 & 0 & 0 & 0 & d_1 & 0 \\
0 & 0 & 0 & 0 & d_1 & 0 & 0 & 0 & 0 & 0 & 0 & 0 & 0 & 0 \\
0 & 0 & -3d_3 & 0 & 0 & 2d_4 & d_5 & 0 & -2d_2 & 2d_3 & 2d_4 & d_5 & -d_2 & d_1 \\
d_5 & d_5 & d_4 & -d_2 & -d_2 & d_1 & 0 & 0 & 0 & 0 & 0 & 0 & 0 & 0
\end{bmatrix}
$$

$$
\begin{pmatrix}
0 & 0 & -d_5 & 0 & 0 & 0 & 0 & d_5 & 0 & 0 & 0 & 0 & 0 & 0 & 0 & 0 & 0 & 0 & 0 & 0 & 0 \\
0 & 0 & d_2 & 0 & 0 & 0 & 0 & 0 & d_4 & 0 & 0 & 0 & 0 & 0 & 0 & 0 & 0 & 0 & d_5 & 0 & -d_3 \\
0 & 0 & -d_4 & 2d_3 & 0 & 0 & d_1 & 2d_4 & -d_2 & 0 & 0 & 0 & 0 & 0 & 0 & 0 & 0 & 0 & 0 & 0 & 0 \\
0 & 0 & 0 & -d_4 & 0 & 0 & 0 & -d_3 & 0 & 0 & 0 & 0 & 0 & 0 & 0 & 0 & 0 & 0 & 0 & 0 & 0 \\
0 & 0 & 0 & d_2 & 0 & 0 & 0 & 0 & -d_3 & 0 & 0 & 0 & 0 & 0 & 0 & 0 & 0 & 0 & 0 & -d_5 & d_4 \\
0 & 0 & 0 & -d_3 & 0 & d_1 & 0 & -d_4 & 0 & 0 & 0 & 0 & 0 & 0 & 0 & 0 & 0 & 0 & 0 & 0 & 0 \\
0 & 0 & 0 & 0 & -d_5 & 0 & 0 & 0 & 0 & 0 & -d_4 & d_5 & 0 & 0 & 0 & 0 & 0 & 0 & 0 & 0 & 0 \\
0 & 0 & 0 & 0 & -d_5 & 0 & 0 & 0 & 0 & -d_3 & 0 & d_5 & 0 & 0 & 0 & 0 & 0 & 0 & 0 & 0 & 0 \\
0 & 0 & 0 & 0 & d_1 & 0 & 0 & d_3 & 0 & 0 & 0 & 0 & 0 & 0 & 0 & 0 & 0 & 0 & 0 & 0 & -d_2 \\
0 & 0 & 0 & 0 & 0 & -d_5 & 0 & 0 & 0 & d_4 & -d_3 & 0 & d_5 & 0 & 0 & 0 & 0 & 0 & 0 & 0 & 0 \\
0 & 0 & 0 & 0 & 0 & -d_5 & d_5 & 0 & 0 & d_4 & 0 & 0 & 0 & -d_2 & 0 & d_5 & 0 & 0 & 0 & 0 & 0 \\
0 & 0 & 0 & 0 & 0 & 0 & 0 & 0 & 0 & 0 & 0 & 0 & 0 & 0 & 0 & 0 & 0 & 0 & d_4 & -d_3 & 0 \\
0 & 0 & 0 & 0 & 0 & 0 & 0 & 0 & 0 & 0 & 0 & 0 & 0 & 0 & 0 & 0 & 0 & 0 & -d_3 & 0 & d_5 \\
0 & 0 & 0 & 0 & 0 & 0 & 0 & 0 & 0 & 0 & 0 & 0 & 0 & 0 & 0 & 0 & 0 & 0 & d_4 & -d_3 & 0 \\
0 & 0 & 0 & 0 & 0 & 0 & 0 & 0 & 0 & 0 & 0 & 0 & 0 & 0 & 0 & 0 & 0 & 0 & -d_3 & 0 & d_5 \\
0 & 0 & 0 & 0 & 0 & 0 & 0 & 0 & 0 & 0 & 0 & 0 & 0 & 0 & 0 & 0 & 0 & 0 & -d_2 & 0 & 0 \\
0 & 0 & 0 & 0 & 0 & 0 & 0 & 0 & 0 & 0 & 0 & 0 & 0 & 0 & 0 & 0 & 0 & 0 & 0 & -d_4 & d_5 \\
0 & 0 & 0 & 0 & 0 & 0 & 0 & 0 & 0 & 0 & 0 & 0 & 0 & 0 & 0 & 0 & 0 & 0 & 0 & -d_4 & d_5 \\
0 & 0 & 0 & 0 & 0 & 0 & 0 & 0 & 0 & 0 & 0 & 0 & 0 & 0 & 0 & 0 & 0 & 0 & 0 & -d_2 & 0 \\
0 & 0 & 0 & 0 & 0 & 0 & 0 & 0 & 0 & 0 & 0 & 0 & 0 & 0 & 0 & 0 & 0 & 0 & 0 & 0 & 0 \\
0 & 0 & 0 & 0 & 0 & 0 & 0 & 0 & 0 & 0 & 0 & 0 & 0 & 0 & 0 & 0 & 0 & 0 & 0 & 0 & 0
\end{pmatrix}
$$

$$
\begin{pmatrix}
0 & 0 & 0 & 0 & 0 & 0 & 0 & 0 & 0 & d_1 & 0 & 0 & 0 & -d_4 & d_3 \\
0 & 0 & 0 & 0 & 0 & 0 & 0 & 0 & 0 & 0 & 0 & 0 & -d_3 & 0 & d_3 \\
0 & 0 & 0 & -d_3 & 0 & 0 & 0 & 0 & 0 & 0 & 0 & 0 & 0 & -2d_5 & 0 \\
0 & 0 & 0 & 0 & 0 & 0 & 0 & 0 & 0 & 0 & 0 & 0 & 0 & 0 & 0 \\
0 & 0 & 0 & 0 & 0 & 0 & 0 & 0 & 0 & 0 & -d_5 & 0 & 0 & 0 & 0 \\
2d_4 & 0 & 0 & 0 & 0 & 0 & 0 & d_3 & 0 & 0 & 0 & 0 & 0 & 0 & 0 \\
0 & 0 & 0 & d_4 & 0 & 0 & 0 & d_1 & 0 & 0 & 0 & 0 & d_5 & 0 & -d_5 \\
0 & 0 & 0 & d_5 & 0 & 0 & d_1 & 0 & 0 & 0 & 0 & 0 & d_4 & 0 & 0 \\
0 & 0 & d_1 & -d_2 & 0 & 0 & 0 & 0 & 0 & 0 & 0 & d_5 & 0 & 0 & 0 \\
d_1 & 0 & 0 & d_3 & 0 & 0 & 0 & 0 & 0 & 0 & 0 & 0 & 0 & d_5 & 0 \\
0 & 0 & 0 & d_4 & 0 & 0 & 0 & 0 & 0 & 0 & 0 & 0 & d_5 & 0 & 0 \\
0 & 0 & 0 & d_5 & 0 & 0 & 0 & 0 & 0 & 0 & 0 & 0 & d_4 & 0 & -d_4 \\
0 & 0 & 0 & 0 & 0 & 0 & 0 & 0 & 0 & 0 & 0 & d_5 & 0 & 0 & 0 \\
-d_3 & 0 & 0 & 0 & 0 & 0 & 0 & 0 & 0 & 0 & 0 & 0 & 0 & 0 & 0 \\
0 & -d_4 & 0 & 0 & 0 & d_4 & 0 & 0 & -d_3 & d_2 & 0 & 0 & 0 & 0 & 0 \\
2d_4 & 0 & 0 & 0 & 0 & 0 & 0 & 0 & 0 & -d_5 & 0 & 0 & 0 & 0 & 0 \\
0 & -d_5 & -d_3 & 0 & 0 & 2d_5 & 0 & 0 & 0 & 0 & 0 & 0 & 0 & 0 & 0 \\
0 & 0 & d_4 & 0 & -d_5 & 0 & 0 & d_2 & d_5 & 0 & 0 & 0 & 0 & 0 & 0 \\
-d_3 & 0 & 0 & 0 & 0 & 0 & d_5 & -d_4 & 0 & 0 & 0 & 0 & 0 & 0 & 0 \\
-d_2 & 0 & d_3 & 0 & 0 & -d_5 & 0 & 0 & 0 & 0 & 0 & 0 & 0 & 0 & 0 \\
0 & 0 & 0 & 0 & 0 & 0 & 0 & 0 & d_1 & 0 & 0 & d_4 & 0 & 0 & -d_2 \\
0 & 0 & 0 & 0 & d_1 & 0 & 0 & 0 & 0 & 0 & d_3 & 0 & -d_2 & 0 & 0 \\
0 & 0 & 0 & 0 & 0 & 0 & 0 & 0 & -d_5 & 0 & 0 & 0 & 0 & 0 & 0 \\
0 & 0 & 0 & 0 & 0 & d_1 & 0 & 0 & 0 & 0 & -d_4 & d_3 & 0 & -d_2 & 0 \\
0 & d_1 & 0 & 0 & 0 & 0 & 0 & 0 & 0 & 0 & 0 & d_3 & 0 & -d_2 & 0 \\
d_5 & 0 & 0 & 0 & 0 & d_2 & 0 & 0 & 0 & 0 & 0 & 0 & 0 & 0 & 0 \\
0 & 0 & 0 & 0 & d_2 & 0 & 0 & -d_5 & 0 & 0 & 0 & 0 & 0 & 0 & 0 \\
d_5 & 0 & 0 & 0 & 0 & 0 & d_3 & 0 & 0 & -d_4 & 0 & 0 & 0 & d_1 & 0 \\
0 & 0 & 0 & 0 & 0 & 0 & d_4 & -d_5 & 0 & 0 & 0 & 0 & d_1 & 0 & 0 \\
0 & d_4 & 0 & 0 & d_3 & -d_4 & 0 & 0 & 0 & d_1 & 0 & 0 & 0 & 0 & 0 \\
0 & 0 & 0 & 0 & 0 & 0 & 0 & 0 & d_2 & 0 & 0 & 0 & 0 & 0 & 0 \\
0 & 0 & 0 & 0 & 0 & 0 & d_4 & -d_5 & 0 & 0 & 0 & 0 & 0 & 0 & d_1 \\
0 & d_3 & 0 & 0 & 0 & 0 & 0 & 0 & d_4 & 0 & 0 & d_1 & 0 & 0 & 0 \\
0 & -d_3 & d_5 & 0 & -d_4 & 0 & d_2 & 0 & 0 & 0 & 0 & 0 & 0 & 0 & 0 \\
0 & 0 & 0 & 0 & 0 & 0 & -d_4 & d_5 & 0 & d_3 & 0 & 0 & 0 & 0 & 0
\end{pmatrix}
$$

We have $\ker_D(T_5) = \operatorname{im}_D(U_5)$, where $U_5 = \begin{pmatrix} U_A & U_B & U_C \end{pmatrix} \in D^{14 \times 35}$ where U_A, U_B are defined by:

$$
U_A = \begin{pmatrix}
-d_2\,d_3 & 0 & 0 & 0 & -d_3 d_5 & 0 & -d_2\,d_5 & d_4\,d_2 & 0 & 0 & 0 & 0 & -d_5\,d_4 & 0 & -d_1\,d_3 \\
-d_3\,d_5 & d_3\,d_5 & 0 & d_2\,d_3 & -d_2\,d_3 & 0 & -d_2{}^2+d_4{}^2 & d_5\,d_4 & -d_4\,d_2 & -d_3\,d_4 & d_3{}^2-d_4{}^2 & -d_5\,d_4 & 0 & 0 & 0 \\
-d_2{}^2+d_5{}^2 & d_4{}^2-d_5{}^2 & -d_5\,d_4 & -d_2\,d_5 & 0 & 0 & 0 & 0 & 0 & -d_5\,d_4 & -d_3\,d_5 & d_3\,d_4 & 0 & 0 & d_1\,d_2 \\
-d_1\,d_3 & 0 & 0 & 0 & 0 & d_3\,d_5 & -d_1\,d_5 & d_4\,d_1 & 0 & 0 & 0 & 0 & 0 & d_5\,d_4 & 0 \\
d_1\,d_5 & -d_1\,d_5 & -d_4\,d_1 & 0 & -d_1\,d_2 & -d_2{}^2+d_4{}^2 & 0 & 0 & 0 & -2\,d_4\,d_1 & d_1\,d_3 & 0 & 0 & -d_3\,d_4 & 0 \\
-d_1\,d_2 & 0 & 0 & -d_1\,d_5 & d_1\,d_5 & d_2\,d_5 & 0 & 0 & 0 & 0 & 0 & 0 & 0 & 0 & d_1{}^2-d_4{}^2 \\
0 & d_3\,d_4 & -d_3\,d_5 & 0 & 0 & 0 & d_5 d_4 & -d_2{}^2+d_5{}^2 & -d_2\,d_5 & -2\,d_3\,d_5 & -d_5\,d_4 & d_3{}^2-d_5{}^2 & d_2\,d_5 & 0 & 0 \\
0 & d_1\,d_3 & 0 & 0 & 0 & -d_3\,d_5 & d_1\,d_5 & 0 & 0 & 0 & -d_1\,d_5 & 0 & 0 & 0 & 0 \\
0 & -d_4\,d_1 & 0 & 0 & 0 & d_5\,d_4 & 0 & 0 & 0 & -d_1\,d_5 & 0 & d_1\,d_3 & 0 & d_3\,d_5 & -d_4\,d_2 \\
0 & 0 & d_1\,d_3 & 0 & 0 & -d_3\,d_4 & d_4\,d_1 & -d_1\,d_5 & 0 & d_1\,d_3 & -d_4\,d_1 & d_1\,d_5 & -d_1\,d_2 & -d_2{}^2+d_3{}^2 & 0 \\
0 & 0 & 0 & d_1\,d_3 & 0 & d_2\,d_3 & -d_1\,d_2 & 0 & -d_4\,d_1 & 0 & 0 & 0 & d_4\,d_1 & d_4\,d_2 & 0 \\
0 & 0 & 0 & 0 & 0 & 0 & 0 & -d_1\,d_2 & -d_1\,d_5 & 0 & 0 & 0 & 0 & -d_2\,d_5 & -d_3\,d_4 \\
0 & 0 & 0 & 0 & 0 & 0 & 0 & 0 & 0 & 0 & -d_1\,d_5 & d_4\,d_1 & 0 & d_5\,d_4 & 0 \\
0 & 0 & 0 & 0 & 0 & 0 & 0 & 0 & 0 & 0 & 0 & 0 & 0 & 0 & -d_2\,d_3
\end{pmatrix},
$$

$$
U_B = \begin{pmatrix}
0 & 0 & -d_1\,d_5 & 0 & 0 & -d_3{}^2+d_5{}^2 & d_4{}^2-d_5{}^2 & 0 & d_3\,d_4 & 0 & 0 & 0 & 0 & 0 \\
d_1\,d_3 & 0 & d_1\,d_2 & d_4\,d_1 & 0 & d_2\,d_5 & -d_2\,d_5 & 0 & 0 & 0 & 0 & 2\,d_1\,d_5 & 0 & 0 \\
d_1\,d_5 & 0 & 0 & 0 & 0 & -d_2\,d_3 & 0 & 0 & 0 & d_4\,d_2 & -d_4\,d_1 & 0 & 0 & 0 \\
0 & 0 & 0 & 0 & 0 & 0 & 0 & 0 & 0 & 0 & 0 & 0 & -d_3\,d_4 & -d_4{}^2+d_5{}^2 \\
d_1{}^2-d_4{}^2 & -d_4\,d_2 & d_2\,d_3 & d_3\,d_4 & -2\,d_4\,d_2 & 0 & 0 & d_2\,d_3 & 0 & 0 & -d_5\,d_4 & 2\,d_3\,d_5 & d_5\,d_4 & -2\,d_3\,d_5 \\
0 & d_5\,d_4 & d_3\,d_5 & 0 & d_5\,d_4 & d_1\,d_3 & 0 & d_3\,d_5 & -d_4\,d_1 & d_4\,d_1 & 0 & 0 & -d_4\,d_2 & 0 \\
0 & 0 & 0 & -d_1\,d_5 & 0 & 0 & -d_4 d_2 & 0 & -d_2\,d_3 & d_2\,d_3 & d_1\,d_3 & 2\,d_4\,d_1 & 0 & 0 \\
0 & 0 & -d_2\,d_5 & -d_5\,d_4 & 0 & 0 & -d_1\,d_2 & -d_2\,d_5 & 0 & 0 & d_3\,d_4 & d_1{}^2-d_3{}^2+d_4{}^2-d_5{}^2 & 0 & -d_2{}^2+d_3{}^2 \\
-d_5\,d_4 & 0 & 0 & -d_3\,d_5 & -d_2\,d_5 & 0 & 0 & 0 & -d_1\,d_2 & 0 & d_1{}^2-d_5{}^2 & 2\,d_3\,d_4 & -d_2{}^2+d_5{}^2 & -2\,d_3\,d_4 \\
d_3\,d_4 & d_2\,d_3 & 0 & d_1{}^2-d_3{}^2 & d_2\,d_3 & 0 & 0 & -d_4\,d_2 & 0 & 0 & -d_3\,d_5 & 0 & d_3\,d_5 & 0 \\
0 & 0 & d_1{}^2-d_3{}^2 & 0 & d_3\,d_4 & -d_1\,d_5 & d_1\,d_5 & -d_3{}^2+d_4{}^2 & 0 & 0 & 0 & 0 & 0 & 2\,d_2\,d_5 \\
0 & d_3\,d_5 & 0 & 0 & 2\,d_3\,d_5 & 0 & d_4\,d_1 & d_5\,d_4 & 0 & d_1\,d_3 & 0 & 0 & d_2\,d_3 & 2\,d_4\,d_2 \\
0 & 0 & -d_2\,d_5 & -d_5\,d_4 & 0 & -d_1\,d_2 & 0 & -d_2\,d_5 & 0 & 0 & 0 & d_4{}^2-d_5{}^2 & 0 & -d_4{}^2+d_5{}^2 \\
-d_3\,d_5 & 0 & -d_2\,d_5 & -d_5\,d_4 & 0 & 0 & 0 & 0 & 0 & 0 & d_3\,d_4 & d_4{}^2-d_5{}^2 & 0 & 0
\end{pmatrix},
$$

and U_C is defined by:

$$
\begin{pmatrix}
0 & 0 & 0 & 0 & d_4 d_1 & 2 d_1 d_2 \\
0 & -2 d_1 d_5 & 0 & 0 & 0 & 2 d_1 d_5 \\
0 & 2 d_1 d_3 & 0 & 0 & 0 & 0 \\
0 & 0 & d_3{}^2 - d_5{}^2 & 0 & 0 & d_1{}^2 - d_4{}^2 \\
-d_2 d_5 & 0 & 0 & 0 & 0 & 0 \\
-d_4{}^2 + d_5{}^2 & 0 & 2 d_2 d_3 & -d_3 d_4 & -d_3 d_4 & 0 \\
0 & 0 & 0 & 0 & d_1 d_2 & 0 \\
d_2 d_3 & 0 & -d_3{}^2 + d_5{}^2 & 0 & d_4 d_2 & 0 \\
-d_4 d_2 & -2 d_3 d_4 & 2 d_3 d_4 & d_2 d_3 & d_2 d_3 & 0 \\
0 & -2 d_5 d_4 & 2 d_5 d_4 & d_2 d_5 & 0 & 2 d_5 d_4 \\
-d_3 d_5 & 0 & -2 d_2 d_5 & d_5 d_4 & 0 & 0 \\
-d_3 d_4 & 0 & 0 & -d_3{}^2 + d_5{}^2 & d_1{}^2 - d_3{}^2 & 2 d_4 d_2 \\
0 & d_1{}^2 - d_4{}^2 & -d_2{}^2 + d_4{}^2 & d_4 d_2 & d_4 d_2 & 0 \\
0 & -d_3{}^2 + d_5{}^2 & 0 & 0 & d_4 d_2 & d_2{}^2 - d_5{}^2
\end{pmatrix}
$$

Moreover, we have $\ker_D(U_5) = \operatorname{im}_D(V_5)$, where $V_5 \in D^{5 \times 14}$ is the full row rank matrix defined by:

$$
\begin{pmatrix}
-d_2 & d_5 & d_3 & 0 & 0 & 0 & -d_4 & 0 & 0 & 0 & 0 & 0 & 0 & 2 d_1 \\
d_1 & 0 & 0 & -2 d_2 & 0 & d_3 & 0 & 0 & 0 & 0 & d_5 & -d_4 & 0 & 0 \\
0 & d_1 & 0 & -2 d_5 & -d_3 & 0 & 0 & -2 d_5 & 0 & -d_4 & -d_2 & 0 & 2 d_5 & 0 \\
0 & 0 & d_1 & 0 & -d_5 & -d_2 & 0 & 0 & d_4 & 0 & 0 & 0 & -2 d_3 & 0 \\
0 & 0 & 0 & 0 & 0 & 0 & d_1 & -2 d_4 & -d_3 & d_5 & 0 & -d_2 & 0 & 2 d_4
\end{pmatrix} .
$$

Hence, the D-module $M = D^5/(D^{14} R_5) = D^5/(D^{15} W_5)$ admits the following finite free resolution:

$$0 \to D^5 \xrightarrow{V_5} D^{14} \xrightarrow{U_5} D^{35} \xrightarrow{T_5} D^{35} \xrightarrow{S_5} D^{14} \xrightarrow{R_5} D^5 \to M_5 \to 0$$

We have the corresponding *striking differential sequence*:

$$0 \to \hat{\Theta} \to 5 \xrightarrow{1} 14 \xrightarrow{2} 35 \xrightarrow{1} 35 \xrightarrow{2} 14 \xrightarrow{1} 5 \to 0$$

As a very general comment and a research problem, it should be interesting to understand the kind of "*symmetry* " existing between the beginning and the end of the differential sequences or differential resolutions obtained. Nevertheless, we point out again that we have been able, in the preceding Appendix, to obtain all these results without any computer or even indices;

REFERENCES FOR THE APPENDIX 2

[1] F. CHYZAK, A. QUADRAT and D. ROBERTZ: Effective algorithms for parametrizing linear control systems over Ore algebras, Appl. Algebra Engrg. Comm. Comput., 16, 319-376, 2005.

[2] F. CHYZAK, A. QUADRAT and D. ROBERTZ: OREMODULES: A symbolic package for the study of multidimensional linear systems, Springer, Lecture Notes in Control and Inform. Sci., 352, 233-264, 2007.
http://wwwb.math.rwth-aachen.de/OreModules

[3] A. QUADRAT: An Introduction to Constructive Algebraic Analysis and its Applications, Les cours du CIRM, Journées Nationales de Calcul Formel (2010), 1(2), 281-471, 2010.

[5] J. J. ROTMAN: An Introduction to Homological Algebra, Springer, 2009.

APPENDIX 3

ABSTRACT

The main purpose of this Appendix 3 is to revisit the mathematical foundations of General Relativity and Gauge Theory in the light of the formal theory of systems of partial differential equations ([13]) and Lie pseudogroups created by D.C. Spencer (1912-2001) in 1970 that has been already described in the main core of this book. For this, let us start with a very general comment.

When constructing inductively the Janet sequences for two *involutive* (care must be taken) systems $R'_q \subset R_q \subset J_q(E)$, the Janet sequence for R'_q *projects onto* the Janet sequence for R_q, that is we may define inductively epimorphisms $F'_r \to F_r$ for $r = 0, 1, ..., n$. Such a result can also be obtained from the general formula allowing to define the Janet bundles *in this case only* (care again) by chasing in the following commutative diagram:

$$
\begin{array}{ccccccccc}
 & & 0 & & & & & & \\
 & & \downarrow & & & & & & \\
0 \to & \wedge^r T^* \otimes R'_q + \delta(\wedge^{r-1}T^* \otimes S_{q+1}T^* \otimes E) & \to & \wedge^r T^* \otimes J_q(E) & \to & F'_r & \to 0 \\
 & \downarrow & & \| & & \downarrow & & \\
0 \to & \wedge^r T^* \otimes R_q + \delta(\wedge^{r-1}T^* \otimes S_{q+1}T^* \otimes E) & \to & \wedge^r T^* \otimes J_q(E) & \to & F_r & \to 0
\end{array}
$$

It will follow from the exactness of the short exact vertical sequences in the DIAGRAM II below that the Spencer sequence starting with $C'_0 = R'_q$ is *contained into* the Spencer sequence starting with $C_0 = R_q$. Also, the kernel of the canonical projection $F'_r \to F_r$ is isomorphic to the cokernel C_r/C'_r of the canonical injection $C'_r \subset C_r \subset C_r(E)$ and we may say that *Janet and Spencer play at see-saw*. This result is not evident at all as it highly depends on involution according to the example presented in Appendix 1.

Therefore, when $E = T$ and we consider finite type Lie equations determined by Lie groups of transformations, then $C'_r = \wedge^r T^* \otimes R'_q \subset \wedge^r T^* \otimes R_q = C_r \subset C_r(T)$. Accordingly, comparing the classical and conformal Killing systems, the Riemann bundle projects onto the Weyl bundle and we shall prove that the kernel of this projection, namely the Ricci bundle, *only depends on* the n nonlinear *elations* of the conformal group. In addition to that, we have also seen in Appendix 1 that the Riemann and Weyl tensors are just obstructions to the 2-acyclicity in the Spencer δ-cohomology of the first order symbols of the Killing and conformal Killing systems respectively. It will therefore be possible to compute the number of linearly independent components of these tensors and to compare them *without using indices or combinatorial methods*. These *purely mathematical results* contradict the formulation of Einstein equations but cannot be imagined within the classical tensorial framework because *explicit indices do not appear any longer*.

In addition, the *Vessiot structure equations*, first discovered in 1903 ([14]), are still not known by the mathematical community for reasons which are not scientific at all, in particular because of an *"affair"* involving the best french mathematicians at the beginning of the last century (H. Poincaré, E. Picard, G. Darboux, J. Drach, E. Cartan, E. Vessiot, P. Painlevé, E. Borel, ...). Such a story, known by everybody at that time but then deliberately forgotten, explains why Cartan never spoke about the work of Vessiot on Lie pseudogroupds and conversely. It has been impossible afterwards to touch the *"legend"* established around Cartan. The corresponding private documents have been given to me by M. Janet, who had also never been quoted by Cartan in his 1930 letters to Einstein, as a gift before he died in 1983, because I made his name and work known. I have now brought them to the library of Ecole Normale Supérieure in Paris where they can be consulted under request (See my Gordon and Breach books "Differential Galois Theory " (1983) and "Lie Pseudogroups and Mechanics " (1988) for more details and translated letters).

1) FIELD/MATTER COUPLINGS:

As the infinitesimal deformation tensor of elasticity theory is equal to half of the Lie derivative $\Omega = (\Omega_{ij} = \Omega_{ji}) = \mathcal{L}(\xi)\omega$ of the euclidean metric ω with respect to the displacement vector ξ, a general quadratic lagrangian may contain, apart from its standard purely elastic or electrical parts well known by engineers in finite element computations, a coupling part $c^{ijk}\Omega_{ij}E_k$ where $E = (E_k)$ is the electric field. The corresponding induction $D = (D^k)$ becomes:

$$D_0^k = \epsilon E^k \longrightarrow D^k = D_0^k + c^{ijk}\Omega_{ij}$$

and is therefore modified by an electric polarization $P^k = c^{ijk}\Omega_{ij}$, brought by the deformation of the medium. In all these formulas and in the forthcoming ones the indices are raised or lowered by means of the euclidean metric. If this medium is homogeneous, the components of the 3-tensor c are constants and the corresponding coupling, called *piezoelectricity*, is only existing if the medium is non-isoptropic (like a crystal), because an isotropic 3-tensor vanishes identically.

In the case of an homogeneous isotropic medium (like a transparent plastic), one must push the coupling part to become cubic by adding $\frac{1}{2}d^{ijkl}\Omega_{ij}E_kE_l$ with $d^{ijkl} = \alpha\omega^{ij}\omega^{kl} + \beta\omega^{ik}\omega^{jl} + \gamma\omega^{il}\omega^{jk}$ from Curie's law. The corresponding coupling, called *photoelasticity*, has been discovered by T.J. Seebeck in 1813 and D. Brewster in 1815. With $\delta = \beta + \gamma$, the new electric induction is:

$$D_0^k = \epsilon E^k \longrightarrow D^k = D_0^k + (\alpha\, tr(\Omega)\omega^{kr} + \delta\omega^{ik}\omega^{jr}\Omega_{ij})E_r$$

As Ω is a symmetric tensor, we may choose an orthogonal frame at each point of the medium in such a way that the deformation tensor becomes diagonal with $\Omega = (\Omega_1, \Omega_2, \Omega_3)$ where the third direction is orthogonal to the elastic plate. We get:

$$D^i = D_0^i + (\alpha\, tr(\Omega) + \delta\Omega_i)E^i$$

for $i = 1, 2$ without implicit summation and there is a change of the dielectric constant $\epsilon \longrightarrow \epsilon + \alpha\, tr(\Omega) + \delta\Omega_i$ along each proper direction in the medium, corresponding to a change $n \longrightarrow n_i$ of the refraction index. As there is no magnetic property of the medium and $\Omega \ll 1$, we obtain in first approximation:

177

$$\epsilon\mu_0 c^2 = n^2 \implies n_1^2 - n_2^2 \simeq 2n(n_1 - n_2) = \mu_0 c^2 \delta(\Omega_1 - \Omega_2) \implies n_1 - n_2 \sim \Omega_1 - \Omega_2$$

where μ_0 is the magnetic constant of vacuum, c is the speed of light in vacuum and n is the refraction index. The speed of light in the medium becomes c/n_i and therefore depends on the polarization of the beam. As the light is crossing the plate of thickness e put between two polarized filters at right angle, the entering monochromatic beam of light may be decomposed along the two proper directions into two separate beams recovering together after crossing with a time delay equal to:

$$e/(c/n_1) - e/(c/n_2) = (e/c)(n_1 - n_2)$$

providing interferences and we find back the Maxwell phenomenological law of 1850:

$$\sigma_1 - \sigma_2 \sim \Omega_1 - \Omega_2 = \frac{k\lambda}{eC}$$

where σ is the stress tensor, k is an integer, λ is the wave length of the light used and C is the photoelastic constant of the medium ivolved in the experience.

Such a result proves, *without any doubt for anybody doing this experiment*, that the deformation $\Omega = \mathcal{L}(\xi)\omega$ and the electromagnetic field $F = dA$, using standard notations in the space-time formulation of electromagnetism, are *on equal footing* in a lagrangian formalism. However, as $\Omega \in S_2 T^* = F_0$ is in the *Janet sequence* and $F \in \wedge^2 T^*$ *cannot* appear at this level, the main purpose of this Appendix is to prove that *another differential sequence must be used*, namely the *Spencer sequence*. The idea has been found *totally independently*, by the brothers E. and F. Cosserat in 1909 ([2]) and by H. Weyl in 1916 ([15]) by using the conformal group of space time, but the first ones were only dealing with the *translations* and *rotations* while the second was only dealing with the *dilatation* and the non-linear *elations* of this group, with no real progress during the last hundred years. Using the group inclusions *Poincaré* ⊂ *Weyl* ⊂ *Conformal*, we obtain:

THEOREM 1: (*Elasticity*) Refering to their book ([2], p 137 for $n = 3$, p 167 for $n = 4$)), the *Cosserat couple-stress equations* discovered in 1909, namely:

$$\partial_r \sigma^{i,r} = f^i \quad , \quad \partial_r \mu^{ij,r} + \sigma^{i,j} - \sigma^{j,i} = m^{ij}$$

are *exactly* described by the formal adjoint of the first Spencer operator $D_1 : R_1 \to T^* \otimes R_1$. Introducing $\phi^{r,ij} = -\phi^{r,ji}$ and $\psi^{rs,ij} = -\psi^{rs,ji} = -\psi^{sr,ij}$, they can be *parametrized* by the formal adjoint of the second Spencer operator $D_2 : T^* \otimes R_1 \to \wedge^2 T^* \otimes R_1$:

$$\sigma^{i,j} = \partial_r \phi^{i,jr} \quad , \quad \mu^{ij,r} = \partial_s \psi^{ij,rs} + \phi^{j,ir} - \phi^{i,jr}$$

Proof: When $n = 2$, lowering the indices by means of the constant metric ω, we just need to look for the factors of ξ_1, ξ_2 and $\xi_{1,2}$ in the integration by parts of the sum:

$$\sigma^{1,1}(\partial_1 \xi_1 - \xi_{1,1}) + \sigma^{1,2}(\partial_2 \xi_1 - \xi_{1,2}) + \sigma^{2,1}(\partial_1 \xi_2 - \xi_{2,1}) + \sigma^{2,2}(\partial_2 \xi_2 - \xi_{2,2}) + \mu^{12,r}(\partial_r \xi_{1,2} - \xi_{1,2r})$$

in order to find $(\partial_r \sigma^{i,r})\xi_i + (\partial_r \mu^{12,r} + \sigma^{1,2} - \sigma^{2,1})\xi_{1,2}$ up to sign because we have $\xi_{ij}^k = 0, \forall i, j, k = 1, 2$.

Finally, setting $\phi^{1,12} = \phi^1, \phi^{2,12} = \phi^2, \psi^{12,12} = \phi^3$, we obtain the parametrization:

$$\sigma^{1,1} = \partial_2 \phi^1, \sigma^{1,2} = -\partial_1 \phi^1, \sigma^{2,1} = -\partial_2 \phi^2, \sigma^{2,2} = \partial_1 \phi^2,$$
$$\mu^{12,1} = \partial_2 \phi^3 + \phi^1, \mu^{12,2} = -\partial_1 \phi^3 - \phi^2$$

by means of the formal adjoint of the second Spencer operator D_2, in a coherent way with the Airy parametrization obtained when $\phi^1 = \partial_2 \phi, \phi^2 = \partial_1 \phi, \phi^3 = -\phi$.

$$\text{Q.E.D.}$$

THEOREM 2: (*Electromagnetism*) When $n = 4$ (care), the *Maxwell field equations* only depend on the second Spencer operator $D_2 : T^* \otimes \hat{R}_2 \longrightarrow \wedge^2 T^* \otimes \hat{R}_2$ projecting onto $d : \wedge^2 T^* \longrightarrow \wedge^3 T^*$ and are parametrized by the first Spencer operator $D_1 : \hat{R}_2 \longrightarrow T^* \otimes \hat{R}_2$ projecting onto $d : T^* \longrightarrow \wedge^2 T^* (\exists \ potential)$. The *Maxwell induction equations* only depend on the formal adjoint of the first Spencer operator $ad(D_1) : \wedge^3 T^* \otimes \hat{R}_2^* \longrightarrow \wedge^4 T * \otimes \hat{R}_2^*$ inducing $ad(d) : \wedge^2 T^* \longrightarrow \wedge^3 T^*$ and are parametrized by the formal adjoint of the second Spencer operator $ad(D_2) : \wedge^2 T^* \otimes \hat{R}_2^* \longrightarrow \wedge^3 T^* \otimes \hat{R}_2^*$ inducing $ad(d) : T^* \longrightarrow \wedge^2 T^* (\exists \ pseudopotential)$.

Proof: As the Spencer operator is $D : J_3(T) \longrightarrow T^* \otimes J_2(T) : \xi_3 \longrightarrow j_1(\xi_2) - \xi_3$ in the present situation, using the fact that the jets of strict order 3 vanish for the conformal Killing equations while setting $\xi_{ri}^r = A_i$ for the jets of strict order 2, we obtain:

$$(\partial_i \xi_{rj}^r - \xi_{rij}^r) - (\partial_j \xi_{ri}^r - \xi_{rij}^r) = \partial_i \xi_{rj}^r - \partial_j \xi_{ri}^r = \partial_i A_j - \partial_j A_i = F_{ij}$$

as the image under the Spencer map $\delta : T^* \otimes T^* \longrightarrow \wedge^2 T^*$ of a 1-form with value in the symbol $\hat{g}_2$, in a coherent way with Weyl but one must point out that the above framework has not been known before 1975. In particular, the key DIAGRAM I below proves that the kernel $S_2 T^*$ of the previous map in the split short exact δ-sequence:

$$0 \longrightarrow S_2 T^* \stackrel{\delta}{\longrightarrow} T^* \otimes T^* \stackrel{\delta}{\longrightarrow} \wedge^2 T^* \longrightarrow 0$$

is the *Ricci tensor*, a result that could not have even been imagined by Weyl but points out its purely conformal origin, *specifically determined by the jets of strict order two.*

The dual procedure is exactly similar to the one used in the previous Theorem and will not be repeated. Acordingly, and though striking it may look like, *there is no conceptual difference between the Cosserat couple-stress equations and the Maxwell induction equations.*

Q.E.D.

It follows from the preceding results and comments that the *field* is a section of C_1 killed by D_2 and parametrized by D_1 while the *induction* is a section of $\wedge^n T^* \otimes C_1^*$ and the induction equations are described by $ad(D_1)$ and parametrized by $ad(D_2)$. The electromagnetic field has therefore *nothing to do* with the so-called Cartan curvature which is a section of C_2 and the lagrangian *must* be defined on C_1 but *not* on C_2, exactly like in the Cosserat elasticity. It follows that the *Vessiot structure equations* providing $\mathcal{D}_1$ in the *Janet sequence* have "nothing to do" with the *Cartan structure equations* providing D_2 in the *Spencer sequence* of the key DIAGRAM II.

As a summary, using the results of Appendix 1 while introducing the respective 2-cocycles of the Spencer δ-cohomology of g_1 and $\hat{g}_1$, we obtain the following commutative and exact diagram:

$$
\begin{array}{ccccccccc}
& & 0 & & 0 & & 0 & & \\
& & \downarrow & & \downarrow & & \downarrow & & \\
0 \to & & Z^2(g_1) & \to & \wedge^2 T^* \otimes g_1 & \overset{\delta}{\to} & \wedge^3 T^* \otimes T & \to 0 \\
& & \downarrow & & \downarrow & & \| & & \\
0 \to & & Z^2(\hat{g}_1) & \to & \wedge^2 T^* \otimes \hat{g}_1 & \overset{\delta}{\to} & \wedge^3 T^* \otimes T & \to 0 \\
& & \downarrow & & \downarrow & & \downarrow & & \\
0 \to & & \wedge^2 T^* & = & \wedge^2 T^* & \to & 0 & & \\
& & \downarrow & & \downarrow & & & & \\
& & 0 & & 0 & & & &
\end{array}
$$

because the only infinitesimal generator $\theta = x^i \partial_i$ of the dilatation is providing $\xi_r^r \neq 0$ and thus $dim(\hat{g}_1) - dim(g_1) = 1$. As $g_2 = 0$, using the isomorphism $0 \to T^* \otimes \hat{g}_2 \overset{\delta}{\to} B^2(\hat{g}_1) \to 0$ existing because $\hat{g}_3 = 0$, we obtain the following commutative and exact diagram relating the so-called *candidate* vector bundles $Riemann = H^2(g_1) = Z^2(g_1)$ and $Weyl = H^2(\hat{g}_1) = Z^2(\hat{g}_1)/B^2(\hat{g}_1)$ to the bundle $Ricci \simeq S_2 T^*$ in its right short split exact column:

DIAGRAM I

$$
\begin{array}{ccccccccc}
& & & & & & 0 & & \\
& & & & & & \downarrow & & \\
& & & & 0 & & Ricci & & \\
& & & & \downarrow & & \downarrow & & \\
& & 0 & \to & Z^2(g_1) & \to & Riemann & \to 0 & \\
& & \downarrow & & \downarrow & & \downarrow & & JANET \\
0 \to & T^* \otimes \hat{g}_2 & \overset{\delta}{\to} & & Z^2(\hat{g}_1) & \to & Weyl & \to 0 & \\
& \downarrow & & & \downarrow & & \downarrow & & \\
0 \to S_2 T^* \overset{\delta}{\to} & T^* \otimes T^* & \overset{\delta}{\to} & & \wedge^2 T^* & \to & 0 & & \\
& \downarrow & & & \downarrow & & & & \\
& 0 & & & 0 & & & & \\
& & SPENCER & & & & & &
\end{array}
$$

For the sake of completeness, we now recall in this new framework the Janet/Spencer commutative diagram with short exact columns, already met in Section 4:

DIAGRAM II

$$SPENCER \quad SEQUENCE$$

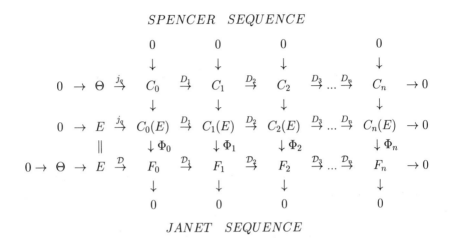

$$JANET \quad SEQUENCE$$

This diagram only depends on the left commutative square $\mathcal{D} = \Phi \circ j_q$ and the epimorhisms $\Phi_r : C_r(E) \to F_r$ for $0 \leq r \leq n$ are successively induced by the canonical projection $\Phi = \Phi_0 : C_0(E) = J_q(E) \to J_q(E)/R_q = F_0$ when $R_q \subseteq J_q(E)$ is an involutive system of order q on E.

Finally, we present a few explicit engineering or mathematical examples leading to three very different problems, *totally unrelated at first sight*:

EXAMPLE 1: Let a rigid bar be able to slide along an horizontal axis with reference position x and attach two pendula, one at each end, with lengths l_1 and l_2, having small angles θ_1 and θ_2 with respect to the vertical. If we project Newton law with gravity g on the perpendicular to each pendulum in order to eliminate the respective tensions while using the notation $d = d_t = \frac{d}{dt}$, we get the two equations:

$$d^2x + l_1 d^2\theta_1 + g\theta_1 = 0, \qquad d^2x + l_2 d^2\theta_2 + g\theta_2 = 0$$

As an *experimental fact* that can be checked by everybody, starting from an arbitrary movement of the pendula, one can stop them by moving the bar conveniently, if and only if $l_1 \neq l_2$.

More generally, we can bring the OD equations describing the behaviour of a mechanical or electrical system to the *Kalman form* $\dot{y} = Ay + bu$ with *input* $u = (u^1, ..., u^p)$ and *output* $y = (y^1, ..., y^m)$. We say that the system is *controllable* if, for any given $y(0), y(T), T < \infty$, one can find $u(t)$ such that a coherent *trajectory* $y(t)$ may be found. In 1963, R.E. Kalman discovered that the system is controllable if and only if $rk(B, AB, ..., A^{m-1}B) = m$. Surprisingly, such a *functional definition* admits therefore a *formal test* which is only valid for Kalman type systems, that is with first order systems with constant coefficients having a specific form and is thus far from being intrinsic.

PROBLEM 1: Is a system of OD or PD equations " CONTROLLABLE " (answer must be YES or NO) and how can we define controllability ?.

EXAMPLE 2: The Cauchy equations for the stress in continuum mechanics with dimension $n = 2$ are $\partial_1 \sigma^{11} + \partial_2 \sigma^{21} = 0, \partial_1 \sigma^{12} + \partial_2 \sigma^{22} = 0$ with $\sigma^{12} = \sigma^{21}$. Their parametrization $\sigma^{11} = \partial_{22}\phi, \sigma^{12} = \sigma^{21} = -\partial_{12}\phi, \sigma^{22} = \partial_{11}\phi$ has been discovered by Airy in 1862 and the *Airy function* ϕ may be used in order to construct dams ([6], Introduction). When $n = 3$, Beltrami found a parametrization with 6 potentials while Maxwell and Morera discovered different parametrizations with 3 potentials (exercise). Similarly, as we already saw, the first set of Maxwell equations for the electromagnetic field are parametrized by the 4-potential while the second set of Maxwell equations for the electromagnetic induction can also be parametrized by the so-called pseudopotential.

More generally, if a differential operator $\xi \xrightarrow{\mathcal{D}} \eta$ is given, a *direct problem* is to find (generating) *compatibility conditions* (CC) as an operator $\eta \xrightarrow{\mathcal{D}_1} \zeta$ such that $\mathcal{D}\xi = \eta \Rightarrow \mathcal{D}_1\eta = 0$. Conversely, if a differential operator $\eta \xrightarrow{\mathcal{D}_1} \zeta$ is given, the *inverse problem* will be to find $\xi \xrightarrow{\mathcal{D}} \eta$ such that $\mathcal{D}_1$ generates the CC of $\mathcal{D}$ and we shall say that $\mathcal{D}_1$ is *parametrized* by $\mathcal{D}$.

PROBLEM 2: Is an operator " PARAMETRIZABLE " (answer must be YES or NO) and how can we find a parametrization ?.

EXAMPLE 3: If $l_1 = l_2 = l$ in the first Example, setting $\theta = \theta_1 - \theta_2$, we get $d^2\theta + g\theta = 0$ by substraction, that is $(ld^2 + g)\theta = 0$, but no such *autonomous* element can be found if $l_1 \neq l_2$.

Let A be a *unitary ring*, that is $1, a, b \in A \Rightarrow a + b, ab \in A, 1a = a$ and even an *integral domain*, that is $ab = 0 \Rightarrow a = 0$ or $b = 0$. We say that $M = {}_A M$ is a *left module* over A if A acts on M wth $x, y \in M \Rightarrow ax, x + y \in M, \forall a \in A$ and we denote by $hom_A(M, N)$ the set of morphisms $f : M \to N$ such that $f(ax) = af(x)$. We define the *torsion submodule* $t(M) = \{x \in M \mid \exists 0 \neq a \in A, ax = 0\} \subseteq M$.

PROBLEM 3: Is a module M TORSION-FREE, that is $t(M) = 0$ (answer must be YES or NO) and how can we test such a property ?.

We have proved in ([5],[6]) and many papers that the above three problems are indeed identical and that *only the solution of the third will provide the solution of the two others* by means of new (difficult) mathematical methods mixing *homological algebra* and *differential geometry* in the framework of *"Algebraic Analysis"*.

EXAMPLE 4: (*Einstein equations revisited*) In particular, a parametrizing challenge for the Einstein equations has been proposed in 1970 by J. Wheeler for 1000 \$ while the author was studying at Princeton University and solved *negatively* in 1995 by the author who only received 1 \$ (See [7]). With more details, if $n = 4$, ω is the Minkowski metric and $\phi = GM/r$ is the gravitational potential, then $\phi/c^2 \ll 1$ and a perturbation Ω of ω may satisfy in vacuum the 10 second order *Einstein equations* for the 10 Ω:

$$\omega^{rs}(d_{ij}\Omega_{rs} + d_{rs}\Omega_{ij} - d_{ri}\Omega_{sj} - d_{sj}\Omega_{ri}) - \omega_{ij}(\omega^{rs}\omega^{uv}d_{rs}\Omega_{uv} - \omega^{ru}\omega^{sv}d_{rs}\Omega_{uv}) = 0$$

the double duality test can be described as follows, using the fact that, contrary to the Ricci operator, the Einstein operator is self-adjoint because it comes from a variational procedure, the sixth terms being exchanged between themselves under ad. For example, we have:

$$\lambda^{ij}(\omega^{rs}d_{ij}\Omega_{rs}) \overset{ad}{\leftrightarrow} (\omega^{rs}d_{ij}\lambda^{ij})\Omega_{rs} = (\omega_{ij}d_{rs}\lambda^{rs})\Omega^{ij}$$

184

and the adjoint of the first operator is the sixth. Accordingly, one has the following diagram where we have 20 generating *Riemann* CC instead of the 10 *Einstein* CC:

$$\text{Riemann} \quad 20 \quad \overset{\text{Bianchi}}{\longrightarrow} \quad 20$$

$$\nearrow \qquad \downarrow \qquad\qquad\qquad \downarrow$$

$$4 \quad \overset{\text{Killing}}{\longrightarrow} \quad 10 \quad \overset{\text{Einstein}}{\longrightarrow} \quad 10 \quad \overset{\text{div}}{\longrightarrow} \quad 4$$

$$4 \quad \overset{\text{Cauchy}}{\longleftarrow} \quad 10 \quad \overset{\text{Einstein}}{\longleftarrow} \quad 10$$

Such a result also puts a shadow on General Relativity by showing that the Cauchy CC on the left has *nothing to do* with the so-called *"div "* on the right which is known to be induced by the Bianchi identities. However, as the negative answer to the parametrization problem for Einstein equations essentially depends on *biduality* and many diagrams, it is not necessary to explain why it has never been acknowledged by physicists, though it is perfectly known by the control community for completely different reasons, namely because *a control system is controllable if and only if it is parametrizable* ([5], p 341)(See also [16], p 34-39).

EXAMPLE 5: We invite the reader to find out by himself the following *injective* 4th-order parametrization of the double pendulum equations whenever $l_1 \neq l_2$:

$$
\begin{aligned}
-l_1 l_2 d^4 \phi - g(l_1 + l_2) d^2 \phi - g^2 \phi &= x \\
l_2 d^4 \phi + g d^2 \phi &= \theta_1 \\
l_1 d^4 \phi + g d^2 \phi &= \theta_2
\end{aligned}
$$

leading to $g^2(l_1 - l_2)\phi = -(l_1 - l_2)x + (l_2)^2\theta_2 - (l_1)^2\theta_1$ with no other possibility .
(*Hints*: Using the double duality test, the ajoint operator of the control system/operator is:

$$
\begin{aligned}
x &\rightarrow d^2 \lambda^1 + d^2 \lambda^2 = \mu \\
\theta_1 &\rightarrow l_1 d^2 \lambda^1 + g\lambda^1 = \mu^1 \\
\theta_2 &\rightarrow l_2 d^2 \lambda^2 + g\lambda^2 = \mu^2
\end{aligned}
$$

with the unique generating CC:

$$\nu \equiv l_2 d^4 \mu^1 + l_1 d^4 \mu^2 - l_1 l_2 d^4 \mu - g(l_1 + l_2) d^2 \mu + g d^2 (\mu^1 + \mu^2) - g^2 \mu = 0$$

and the corresponding formal adjoint operator provides the parametrization.)

185

It is important to notice that the previous methods are also valid for systems or operators with variable coefficients.

EXAMPLE 6: Similarly, the Lie operator defining the infinitesimal contact transformations of Example 1.3 admits the injective parametrization:

$$-x^3\partial_3\phi + \phi = \xi^1, \quad -\partial_3\phi = \xi^2, \quad \partial_2\phi - x^3\partial_1\phi = \xi^3 \quad \Rightarrow \quad \xi^1 - x^3\xi^2 = \phi$$

.

Finally, the crucial reason for being able to deal with *duality* in this new framework is coming from delicate arguments of *homological algebra* saying that the *extension modules* $ext^i(M)$ for a (differential) module M do not depend on the resolution allowing to define M and conversely, the only mathematical quantities which do not depend on the resolution of a module are the extension modules (See [11], chapter 7, Axioms for details). As it can be shown that the Spencer sequence *for any Lie group of transformations considered as a Lie pseudogroup* is locally isomorphic to a tensor product of the *Poincaré sequence* by the Lie algebra of the group involved, this standard but difficult theorem proves that the parametrization property is existing in *any other* formally exact differential sequence used as a resolution.

EXAMPLE 7: The formally exact differential sequence $2 \overset{Killing}{\longrightarrow} 3 \overset{Riemann}{\longrightarrow} 1 \to 0$ is not a Janet sequence because the first order Killing operator is not involutive as we have proved that it admits $dim(H^2(g_1)) = n^2(n^2 - 1)/12$ *second order* CC, that is exactly 1 when $n = 2$. Taking the adjoint sequence, we obtain the formally exact sequence $0 \leftarrow 2 \overset{Cauchy}{\longleftarrow} 3 \overset{Airy}{\longleftarrow} 1$ (care) showing that the Airy parametrization is nothing else than the formal adjoint of the linearized Riemann tensor considered as a second order linear operator, a very surprising result indeed.

All these purely mathematical results lead therefore to revisit the mathematical foundations of General Relativity and Gauge Theory, *though striking it may look like for well established theories.*

MAIN REFERENCES FOR THE APPENDIX 3:

[1] J.E. BJORK: Analytic D-Modules and Applications, Kluwer, Springer, 1993.

[2] E. COSSERAT and F. COSSERAT: Théorie des Corps Déformables, Hermann, Paris, 1909.

[3] M. JANET: Les Systèmes d'Equations aux Dérivées Partielles, Journal de Math. Pures et Appliquées, 3, 1920, 65-151.

[4] M. KASHIWARA: Algebraic Study of Systems of Partial Differential Equations, Mémoires de la Société Mathématique de France 63, 1995, (Transl. from Japanese of his 1970 Master's Thesis).

[5] J.-F. POMMARET: Partial Differential Equations and Group Theory, New Perspectives for Applications, Mathematics and its Applications 293, Kluwer, Springer, 1994.
http://dx.doi.org/10.1007/978-94-017-2539-2

[6] J.-F. POMMARET: Partial Differential Control Theory, Kluwer, Springer, 2001, 957 pp.

[7] J.-F. POMMARET: The Mathematical Foundations of General Relativity Revisited, Journal of Modern Physics, 4, 2013, 223-239.
http://dx.doi.org/10.4236/jmp.2013.48A022
http://arxiv.org/abs/1306.2818

[8] J.-F. POMMARET: The Mathematical Foundations of Gauge Theory Revisited, Journal of Modern Physics, 2014, 5, 157-170.
http://dx.doi.org/10.4236/jmp.2014.55026
http://arxiv.org/abs/1310.4686

[9] J.-F. POMMARET: From Thermodynamics to Gauge Theory: the Virial Theorem Revisited, in " Gauge Theories and Differential geometry ", NOVA Science Publishers, 2015, Chapter 1, 1-44.
http://arxiv.org/abs/1504.04118

[10] A. QUADRAT, D. ROBERTZ: A Constructive Study of the Module Structure of Rings of Partial Differential Operators, Acta Applicandae Mathematicae, 133, 2014, 187-234.
http://hal-supelec.archives-ouvertes.fr/hal-00925533

[11] J.J. ROTMAN: An Introduction to Homological Algebra, Pure and Applied Mathematics, Academic Press, 1979.

[12] J.-P. SCHNEIDERS: An Introduction to D-Modules, Bull. Soc. Roy. Sci. Liège, 63, 1994, 223-295.

[13] D.C. SPENCER: Overdetermined Systems of Partial Differential Equations, Bull. Amer. Math. Soc., 75, 1965, 1-114.

[14] E. VESSIOT: Sur la théorie des groupes infinis, Ann. Ec. Norm. Sup., 20, 1903, 411-451 (Can be obtained from   http://www.numdam.org).

[15] H. WEYL: Space, Time, Matter, Springer, Berlin, 1918, 1958, Dover, 1952, Blanchard, Paris, 1958.

[16] E. ZERZ: Topics in Multidimensional Linear Systems Theory, Lecture Notes in Control and Information Sciences (LNCIS) 256, 2000, Springer.

編
辑
手
记

世界著名数学家彼得·拉克斯(Peter Lax)曾指出:

数学和物理的关系尤其牢固.其原因在于,数学的课题毕竟是一些问题,而许多数学问题是在物理中产生的,并且不止于此,许多数学理论正是为了处理深刻的物理问题而发展出来的.

本书正是这样一部反映为了处理深刻的物理问题而发展出来的数学理论的专著.

本书的中文书名或可译为《代数结构和几何结构的形变理论》.

本书的作者为珍 — 弗朗索瓦·蓬马雷特(Jean-Francois Pommaret)教授.他毕业于法国"精英大学"和巴黎第六大学.他的研究方向为偏微分方程组的形式理论与李伪群,并将研究成果应用于工程学和理论物理学.他写过5本专著,发表过150多篇论文.

正如本书作者所介绍的 S. 李(S. 李) 在 1890 年发现了李伪群,将其命名为偏微分方程组的变换解群. 在之后的 50 年,只有 E. 嘉当(E. Cartan) 和 E. 韦西奥(E. Vessiot) 研究过这些群,但是"韦西奥结构方程"直到今天仍是未知的. 1920 年,关于偏微分方程组的"形式理论"已经被 M. 雅内(M. Janet) 所倡导. 物理学家 E. 伊诺努(E. Inonu) 和 E. P. 魏格纳(E. P. Wigner) 在 1953 年通过考虑在速度的洛伦兹复合中作为参数的光速,引入了"李代数形变"的概念. 这个想法导致了"代数结构的形变理论"和计算机代数的第一个应用的出现. 几年之后,"几何结构的形变理论"被 D. C. 斯潘塞(D. C. Spencer) 和他的同事引入,他们使用了偏微分方程的形式理论,该理论是他们为了研究"李伪群"而发展起来的. 这两个形变理论之间的联系是一个猜想,还没得到证实. 本书首次通过使用新的数学方法来解决这个猜想. 该选题将会引起数学和物理学专业的学生和研究者的兴趣.

为了使读者能够快速地了解本书的基本内容,本书的版权编辑特为我们翻译了本书的目录,如下:

本书的主角是备受专业数学家推崇的李.

　　李（李，Marius Sophus，1842—1899），挪威人.1842 年 12 月 17 日生于诺弗沃尔杰德.1865 年毕业于克里斯蒂安尼亚大学（现奥斯陆大学）.为了继续深造,他于 1870 年来到巴黎,在那里结识了达布（Darboux）等数学家.还与克莱因（Klein）一起进行过数学研究.1877 年至 1886 年间,他在克里斯蒂安尼亚大学数学系工作.1886 年起任莱比锡大学教授.他还是巴黎科学院院士以及其他许多科学院和学术团体的成员.1899 年 2 月 18 日逝世.

　　李对数学的最著名的贡献是建立了连续变换群的理论,开创了代数学的新分支——李群与李代数.他的贡献还涉及微分方程、微分几何等领域.

　　1869 年至 1870 年间,李与克莱因在共同研究球面几何学问题时,得到了连续群的概念.后来,克莱因用这个概念对几何学

191

进行分类,提出了著名的埃尔兰根纲领.而李本人则着手研究连续变换群,即所谓局部李变换群的概念.他是从研究微分方程入手的,他观察到,用过去的方法可积分出来的常微分方程,其大多数在某些类型的连续变换群下是不变的.因此,他想到用连续变换群能够阐明微分方程的解,并将它们进行分类.1870 年,他向巴黎科学院提出了关于阐述变换理论的长篇报告,为连续变换群的经典理论奠定了基础.1874 年,李引进了他的变换群的一般理论.1880 年左右,李又发展了连续群论.1883 年,在另一篇关于连续群的论文中,他引进了无限连续群的概念,这篇论文刊登在一个没有名气的挪威杂志上.这种群是借助一组微分方程来定义的,所得的变换不是只依赖于有限多个连续的参变量,而是依赖于任意的函数.这样一来,当时已为人所知的四种主要类型的群,有两种以李的名字而命名,即有限连续李群和由微分方程定义的无限连续李群.另外,当初李与克莱因定义一个变换群只具有封闭性,其他性质或由变换的性质推出,或作为变换自明的性质.而李在自己的工作中认识到,每个元素的逆元素的存在性应作为一条公设放在群的定义之中.

李代数产生于李对连续变换群结构的研究.李于 1876 年引进了无穷小变换,即将点移动一个无穷小距离的变换,并确定了变换的李群.李从研究 r 个参数的有限单连续群的结构开始,接着发现了李代数的 4 种主要类型.嗣后,李代数又有更深刻的发展.

李一方面利用几何学的概念、分析学尤其是微分方程论的方法,发展自己的李群与李代数的理论,另一方面又将自己的理论运用于微分方程以及微分几何等领域中.但是,他的这些工作的意义在当时没有受到足够的重视.到 20 世纪中期,经过嘉当、外尔(Weyl)等人努力完成李群论之后,人们也弄清了作为李群的拓扑群的特性.这样李的贡献对数学史的巨大影响才被人们所公认.所以,现代数学中有许多以他的名字命名的概念和定理.除李群、李代数之外,在微分几何中有张量场的李导数、微分形式的李导数;在球面几何中有李切触圆变换、李线球变换;在拓扑学中有李变换群、李局部变换群、李基本定理等. 李的主要著作有《变换群理论》(1888 ~1893)、《连续群》(1893)、《微分方程》(1891).

1897 年,喀山大学数学物理学会因李的群论在几何论证方面的应用,授予他罗巴切夫斯基国际奖.

为了使初学者对李群有所了解,我们拟引用三篇经典文献,如下:
文献一:《从同伦论的观点看李群》①.

我打算通过对近年来在有限 H—空间方面的一些研究做一综述,从回顾李群的一些重要事实以及同伦论的某些基本事实开始.

那么,什么是李群呢? 李群首先是一个群 G,其次,它是解析流形(因此它的转换函数(transition function)可展为幂级数),再次,若取映射 $\theta:G\times G\rightarrow G$ 为 $\theta(x,y)=x\cdot y^{-1}$,则要求 θ 是解析映射. 由于具有流形的乘积结构的 $G\times G$ 是解析流形,因此要求 θ 为解析映射是有意义的.

李群的例子有:圆周 S^1,R^n,行列式非 0 的 $n\times n$ 阶矩阵组成的一般线性群 $GL(n,R)$,酉群 $U(n)$ 和正交群 $O(n)$. 因此李群的例子非常多.

李群方面最早获得的事实之一,是它们可以完全分类. 这个分类可以粗略地叙述如下:为简单起见,设我们讨论的李群单连通. 取李群 G,并作对应的李代数 g. g 是李群在 G 上左不变向量场的全体,即给定 G 在单位元处的一个切向量,它可唯一扩充为 G 上的一个左不变向量场. 左不变向量场在一个反交换的"括号"(bracket) 运算下封闭.

从群 G 过渡到李代数 g 的重要性在于:两个李群如果同构,那么它们的李代数也同构. 反过来,若两个李代数同构,则对应

①　原题:*Lie Groups from a Homotopy Point of View*,译自:*Lecture Notes in Mathematics*,1989,1370:1-18.

的李群"局部同构",即在单位元的邻域里,存在着"局部"同构,它将一个李群的这种邻域变为另一个李群的单位元的邻域. 在单连通的情形,这个局部同构可扩充为第一个李群到第二个李群的同构. 因为在李群为单连通时,李群与李代数之间存在着一一对应的关系.

因此,单连通李群的分类归结为李代数的分类. 后者主要由嘉当和基灵(Killing)完成. 他们对"单"李代数做了完全分类,"过渡"回李群便得到对应的"单"李群. 我们知道有 4 族单李群 A_n, B_n, C_n 和 D_n,以及 5 个例外单李群,称为 G_2, F_4, E_6, E_7 和 E_8.

这样便完成了单李群的完全分类. 而它们的上同调的计算,历史上的做法是通过分类定理逐个进行. 整个计算的框架可总结为图 1.

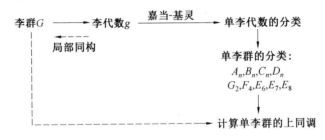

图 1

今天我要讲的是:是否有其他计算李群的上同调的方法,这个方法不依赖于嘉当和基灵的有关李代数的复杂的分类定理. 为此我要介绍李群的一些拓扑性质.

第一批性质之一是岩泽健吉(Iwasawa Kenkichi)[1] 注意到的,他证明了每个李群可分解为一个紧李群与一个欧氏空间的乘积. 这样每个李群有紧李群的伦型,因此它有有限复形的伦型. 以后,"有限"用来表示任意一个有有限复形伦型的空间.

---

① S. Helgason, *Differential geometry and symmetric spaces*, Academic Press, 1962.

第二个事实在历史上曾经是很有趣的,称为希尔伯特第五问题.希尔伯特第五问题用下面的方式刻画李群.

定理 1(蒙哥马利(Montgomery),格利森(Gleason),齐平(Zippin)) 每个局部欧氏的拓扑群是李群.

注意到拓扑群是这样一个拓扑空间:它是一个群且其乘法与逆映射连续.在希尔伯特第五问题中,李群的解析结构①变了,但群结构保持不变,即有一个包含关系:

解析流形 $\subset C^\infty$ 流形 $\subset C^n$ 流形 $\subset$ 拓扑流形.

希尔伯特第五问题是说:如果我们将解析结构减弱为拓扑结构,群流形仍为李群.

对这些问题的研究最近已集中在同伦论方面,因此我们介绍一下同伦的概念.若有两个拓扑空间 X 和 Y 及两个连续映射 $f,g:X \to Y$,则我们说"f 和 g 同伦"(记为 $f \simeq g$)是指存在映射

$$F:X \times I \to Y$$

使得

$$F(x,0) = f(x), \; F(x,1) = g(x)$$

形象地说,这意味着给定点 x,可用一道路联结点 $f(x)$ 和 $g(x)$(图 2),并且这个道路作为 x 的函数连续.

$f(x)$ $\qquad\qquad\qquad\qquad$ $g(x)$

图 2

在初等拓扑的课程里我们就知道同伦的映射在上同调上诱导出相同的同态.类似地,我们说空间 X 和 Y 有相同的"伦型"(记为 $X \simeq Y$)是指存在映射

$$f:X \to Y \text{ 和 } g:Y \to X$$

使得

$$gf \simeq \mathrm{id}_X, \; fg \simeq \mathrm{id}_Y$$

此时,$H^*(X)$ 同构于 $H^*(Y)$.因此,如果要计算李群的上同调,

---

① 原文为 topology,疑误.

那么只需要考虑与这个李群有同样伦型的空间.

现在回顾希尔伯特第五问题,流形结构改变了,但群结构保持不变.为什么我们不让群结构也变化呢? 这里的意思是我们可引入"同伦单位元"的概念,这就是说有一个拓扑空间 X 及某个二元运算

$$X \times X \xrightarrow{\mu} X , \ \mu(x_1, x_2) = x_1 x_2$$

以及一个特殊的点 $e \in X$,我们希望 e 在 X 中起类似于单位元的作用.

一个同伦单位元 e 应该具有性质:映射

$$\mu(e, \ ) : X \rightarrow X$$
$$\mu(\ , e) : X \rightarrow X$$

同伦于单位映射 1_X. 直观地说,对每个 $x \in X$,存在如图 3 所示的道路,这些道路应该随 x 而连续变化.

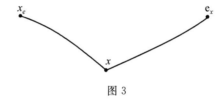

图 3

带有配对(pairing)μ 以及同伦单位的空间 X 称为"H-空间",用以表示对首先给出这个概念的海因茨·霍普夫(Heinz Hopf)① 的敬意.

注意群总是结合的,因此,还有另一个要扩充的概念,即"同伦结合"的概念.给定 H-空间(X, e, μ) 及点 $x, y, z \in X$,我们有两种不同的方式将它们配对

$$(xy)z = \mu(\mu(x, y), z)$$
$$x(yz) = \mu(x, \mu(y, z))$$

若有一条道路联结 $(xy)z$ 与 $x(yz)$(图 4),并且它为 x, y, z 三个变量的函数连续,则称 (X, e, μ)"同伦结合".

① R. Kane, The homology of Hopf spaces.

图 4

我们还可继续考虑 4 个点 x, y, z, ω 的情况,此时我们得到图 5.

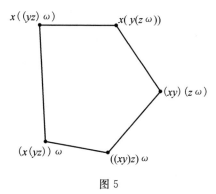

图 5

五边形的边对应于 (X, e, μ) 的同伦结合性. 如果给定的 H — 空间同伦结合,那么在把这个五边形缩为一个点时,将有一个深一层的障碍. 粗略地讲,若这个收缩存在,并且为 x, y, z, ω 的函数连续,则 (X, e, μ) 称为"a_4 — 空间". 事实上,有这样的 H — 空间:它不是同伦结合的,或虽同伦结合但不是 a_4 — 空间.

菅原正博(Sugawara Masahiro)和斯特谢夫(Stasheff)研究了这些概念,实际上,H — 空间可以细分. 若记 H — 空间为 a_2 — 空间,同伦结合 H — 空间为 a_3 — 空间,斯特谢夫对每个正整数 $n \geqslant 2$ 定义了 a_n — 空间.

我们定义"a_∞ — 空间"为这样一个空间:它对每个正整数 $n \geqslant 2$ 是 a_n — 空间. 下列定理属于斯特谢夫:

定理 2 a_∞ — 空间有回路(loop)空间 ΩB 的伦型.

米尔诺(Milnor)的下列定理则描绘了拓扑群的伦型:

定理 3 拓扑群有回路空间 ΩB 的伦型.

因此有空间的细分:

李群 $\subset$ 有限拓扑群 $\subset$ 有限回路空间 $\subset$ 有限 a_n — 空间 $\subset$ 有限 H — 空间.

一个合理的问题是：有限 H — 空间的上同调有些什么性质？

给定有限 H — 空间 (X,e,μ)，应用上同调到映射

$$\mu:X \times X \to X$$

我们得到同态

$$\Delta = \mu^* : H^*(X) \to H^*(X \times X)$$

如果系数是域，Kunneth 定理告诉我们

$$H^*(X \times X) \cong H^*(X) \otimes H^*(X)$$

此外 $H^*(X)$ 对称积是环，故我们得同态

$$\Delta:H^*(X) \to H^*(X) \otimes H^*(X)$$
$$v:H^*(X) \otimes H^*(X) \to H^*(X)$$

Δ 称为"倒代数结构"(coalgebra structure). 当 Δ 是代数同态时，称 $H^*(X)$ 是霍普夫代数.

下面的定理是霍普夫代数定理.

定理 4 （1）（霍普夫）：作为代数

$$H^*(X;\mathbf{Q}) \cong \Lambda(x_1,\cdots,x_l)$$

这里 $\mathrm{degree}(x_i)$ 为奇数.

（2）（博雷尔（Borel））：作为代数

$$H^*(X;\mathbf{Z}_2) \cong \otimes \Lambda(x_i) \otimes \mathbf{Z}_2[y_j]/y_j^{2^{f_j}}$$

由此立即知道某些空间不能有李群结构. 例如，偶维球不是李群或有限 H — 空间，因为它的有理系数上同调不合乎要求.

第二个工具是博雷尔结构定理的加强定理. 博雷尔结构定理只与 $H^*(X;\mathbf{Z}_2)$ 的环结构有关，而没有考虑可能的倒代数结构. 若设 $H_*(X;\mathbf{Z}_2)$ 是结合环，我们就可对倒代数结构得到若干结果.

定义

$$\overline{\Delta}:IH^*(X;\mathbf{Z}_2) \to IH^*(X;\mathbf{Z}_2) \otimes IH^*(X;\mathbf{Z}_2)$$

为

$$\overline{\Delta}x = \Delta x - |\otimes x - x \otimes|$$

称为"约化倒乘积"(reduced coproduct). 设

$$\xi:H^*(X;\mathbf{Z}_2) \to H^*(X;\mathbf{Z}_2)$$

是平方映射. 令

$$R = \{x \in IH^*(X;\mathbf{Z}_2) \mid \bar{\Delta}x \in \xi H^*(X;\mathbf{Z}_2) \otimes$$
$$H^*(X;\mathbf{Z}_2)\}$$

注意 R 本身是一个倒代数. 这里我们用到 $H_*(X;\mathbf{Z}_2)$ 结合这一事实.

令 $S(R)$ 为倒代数 R 生成的自由、交换霍普夫代数. 定义 I 是由形为

$$\xi x - x \otimes x \quad (x \in R)$$

的项所生成的理想, 则 $I \subset S(R)$ 是霍普夫理想, 且有如下定理:

定理 5(Lin)　　作为霍普夫代数

$$H^*(X;\mathbf{Z}_2) \cong S(R)/I$$

这个定理说明倒代数结构由子模 R 决定, 而博雷尔分解中的所有生成元可在 R 中选取. 进一步可证明有一个正合列

$$0 \to \xi H^*(X;\mathbf{Z}_2) \to R \to QH^*(X;\mathbf{Z}_2) \to 0$$

这里

$$QH^*(X;\mathbf{Z}_2) = IH^*(X;\mathbf{Z}_2)/IH^*(X;\mathbf{Z}_2)^2$$

是"不可分元"所构成的模.

第三个工具是斯廷罗德代数 $a(2)$. 存在着函数 $H^*(\ ;\mathbf{Z}_2)$ 间的一系列自然变换

$$Sq^i : H^*(\ ;\mathbf{Z}_2) \to H^*(\ ;\mathbf{Z}_2), Sq^i \in a(2)$$

给定连续映射 $f: X \to Y$, 图 6 交换.

$$
\begin{array}{ccc}
H^l(X;\mathbf{Z}_2) & \xrightarrow{Sq^i} & H^{l+i}(X;\mathbf{Z}_2) \\
\uparrow{f^*} & & \uparrow{f^*} \\
H^l(Y;\mathbf{Z}_2) & \xrightarrow{Sq^i} & H^{l+i}(Y;\mathbf{Z}_2)
\end{array}
$$

图 6

下面我要列出本文所需的另一些事实. 嘉当公式

$$Sq^i(x \bigcup y) = \sum_{l=0}^{i} Sq^l x \bigcup Sq^{i-l} y$$

若 $\mathrm{degree}(x) = k$, 则 $Sq^k x = x^2$.

如果

$$QH^*(X;\mathbf{Z}_2) = IH^*(X;\mathbf{Z}_2)/IH^*(X;\mathbf{Z}_2)^2$$

那么注意 Sq^i 诱导出自然变换

$$Sq^i : QH^*(\quad;\mathbf{Z}_2) \rightarrow QH^{*+i}(\quad;\mathbf{Z}_2)$$

我们要问：

问题 1 当 X 是有限 H 一 空间时，Sq^i 如何作用在 $QH^*(X;\mathbf{Z}_2)$ 上？

我们有如下一些结果，设 $H_*(X;\mathbf{Z}_2)$ 是结合环，则任给整数 n，找它的二进展开式中第一个不出现的 2 的幂次

$$n = 1 + 2 + \cdots + 2^{r-1} + 2^{r+1} + \cdots + 2^s$$
$$= 2^r + 2^{r+1}k - 1 \quad (\text{对某个 } k \geqslant 0)$$

定理 6（Lin） 对 $k > 0$，有：

(1) $QH^{2^r+2^{r+1}k-1}(X;\mathbf{Z}_2) = Sq^{2^r k} QH^{2^r+2^r k-1}(X;\mathbf{Z}_2)$；

(2) $Sq^{2^r} QH^{2^r+2^{r+1}k-1}(X;\mathbf{Z}_2) = 0$；

(3) $\sigma^* QH^{2^r+2^{r+1}k-1}(X;\mathbf{Z}_2) \subset Sq^{2^r} H^*(\Omega X;\mathbf{Z}_2)$，这里 $\sigma^*:$ $H^*(X;\mathbf{Z}_2) \rightarrow H^*(\Omega X;\mathbf{Z}_2)$ 是 同 纬 映 象 同 态（suspension homomorphism）；

(4) 若 $t = 2^{i_1} + 2^{i_2} + \cdots + 2^{i_l}$，这里 $i_1 < i_2 < \cdots < i_l$，则

$$Sq^t QH^*(X;\mathbf{Z}_2) = Sq^{2^{i_1}} Sq^{2^{i_2}} \cdots Sq^{2^{i_l}} QH^*(X;\mathbf{Z}_2)$$

下面给一个例子说明如何应用定理 5 和定理 6.

E_8 的模 2 上同调为

$$H^*(E_8;\mathbf{Z}_2) = \mathbf{Z}_2[x_3,x_5,x_9,x_{15}]/(x_3^{16},x_5^8,x_9^4,x_{15}^4) \otimes$$
$$\Lambda(x_{17},x_{23},x_{27},x_{29})$$

由定理 5，所有 x_i 可在 R 中选取，故有

$$\overline{\Delta} x_i \in \xi H^*(E_8;\mathbf{Z}_2) \otimes R$$

从定理 6 中的 (1) 和 (4)，得

$$r = 1: \quad x_5 = Sq^2 x_3$$
$$x_9 = Sq^4 x_5$$
$$x_{17} = Sq^8 x_9$$
$$x_{29} = Sq^2 Sq^4 Sq^8 x_{15}$$
$$r = 2: \quad x_{27} = Sq^4 Sq^8 x_{15}$$
$$r = 3: \quad x_{23} = Sq^8 x_{15}$$

用定理 6 中的(3)做一简单计算可证

$$x_{17} = Sq^2 x_{15}$$

这就完全刻画了 $a(2)$ 在 $H^*(E_8; \mathbf{Z}_2)$ 的生成元上的作用.
可证明 x_{15} 不是本原元(primitive).

定理 6 有许多应用,其中(1)有下列推论:

推论 1(Lin) 第一个非零的 mod 2 上同调群的度数是 $2^i -$
1.事实上,它只能是 1,3,7 或 15.

由于第一个非零的 mod 2 上同调与第一个非零的同伦群出
现在同一度数上,因此我们有:

推论 2(Lin) 有限 H — 空间的第一个非零的同伦群出现
在度数 1,3,7 或 15 上,若 $H^*(H; \mathbf{Z})$ 没有 2 挠,则第一个非零的
同伦群出现在度数 1,3 或 7 上.

推论 2 是亚当斯(Adams)的霍普夫不变量为 1 的定理的推
广,这个定理说的是:在球面里,只有 S^1, S^3 或 S^7 是 H — 空间.
托玛斯(Thomas)在 $H^*(X; \mathbf{Z}_2)$ 为本原元生成时曾获得过这种
类型的定理.

当 $r = 0$ 时,定理 6 表明

$$QH^{2k}(X; \mathbf{Z}_2) = Sq^k QH^k(X; \mathbf{Z}_2)$$

注意到当 $\deg x = k$ 时,$Sq^k x = x^2$,故 $QH^{\text{even}}(X; \mathbf{Z}_2) = 0$. 可以证
明这等价于:

推论 3(Lin) $H_*(\Omega X; \mathbf{Z})$ 没有 2 挠.

当 X 是李群时,博特(R. Bott)用几何方法证明了推论 3. 他
的证明方法涉及莫尔斯理论,因而不适用于有限 H — 空间.

另一个简单应用是:

推论 4(Lin) 当 $n \neq 2^i - 1$ 时,mod 2 胡雷维奇映射

$$h_n \otimes \mathbf{Z}_2 : \pi_n(X) \otimes \mathbf{Z}_2 \rightarrow H_n(X; \mathbf{Z}_2)$$

为 0.

由于某些自旋群(spin group)的 mod 2 胡雷维奇映射在度
数 $2^i - 1$ 时非平凡,因此这个结果是最佳的.

推论 5(Lin) 例外李群的 mod 2 上同调可由它的有理上
同调决定.事实上,作为斯廷罗德代数上的代数,例外李群的

mod 2 上同调由其有理上同调决定.

推论 5 是用 Bockstein 谱序列来证明的. 历史上,计算例外李群的 mod 2 上同调是非常困难的. 关于 E_8 的结果是荒木不二洋(Araki Huzihiro)和 Shikata 于 1961 年宣布的,但计算的细节用到李群的性质直到 1984 年才发表.

下面转到亚当斯和 Wilkerson 最近提出的问题. 这个问题最广泛的提法应该像下面这样.

问题 2 给定有限回路空间 ΩB,ΩB 的 mod 2 上同调同构于某个李群的 mod 2 上同调吗?

为使 $H^*(\Omega B;\mathbf{Z}_2)$ 成为有限维向量空间,对 $H^*(B;\mathbf{Z}_2)$ 有些限制. 近年来,有两个霍普夫代数被认为可能是有限回路空间的 mod 2 上同调,但不是任何李群的 mod 2 上同调. 这两个霍普夫代数是

$$A_1 = \mathbf{Z}_2[x_7]/x_7^4 \bigotimes \Lambda(x_{11},x_{13})$$
$$A_2 = \mathbf{Z}_2[x_{15}]/x_{15}^4 \bigotimes \Lambda(x_{23},x_{27},x_{29})$$

由定理 6,下列生成元必须由斯廷罗德运算联系

$$x_{11} = sq^4 x_7, \quad x_{13} = sq^2 x_{11}$$
$$x_{23} = sq^8 x_{15}, \quad x_{27} = sq^4 x_{23}, \quad x_{29} = sq^2 x_{27}$$

有几个试图用 K 理论证明 A_1 不是 $H-$空间的 mod 2 上同调的尝试,但都没能成功. U. 祖特尔(U. Suter)证明了(未发表)A_2 不是有限 $H-$空间的 mod 2 上同调. 最近我们得到:

定理 7(Lin) A_1 不是 $H-$空间的 mod 2 上同调;

(Lin-Williams)(祖特尔) A_2 不是 $H-$空间的 mod 2 上同调.

下面我把注意力转到另一个问题上,即:

问题 3 给定有限回路空间 ΩB,ΩB 的有理上同调同构于某个李群的有理上同调吗?

由霍普夫定理知:有霍普夫代数同构

$$H^*(\Omega B;\mathbf{Q}) \cong \Lambda(y_{n_1-1},\cdots,y_{n_r-1})$$

其中 $n_1 \leqslant n_2 \leqslant \cdots \leqslant n_r$,$n_r$ 为偶.

由博雷尔的一个定理知道,B 的有理上同调是生成元

$x_{n_1}, \cdots, x_{n_r}$ 上的多项式代数

$$H^*(B;\mathbf{Q}) = \mathbf{Q}[x_{n_1}, \cdots, x_{n_r}]$$

我们称数组 $[n_1, \cdots, n_r]$ 为 ΩB 的"型"(type),称 r 为 ΩB 的"秩"(rank),因此问题 3 归结为验证有限回路空间的型与李群的型相同.

关于有理上同调与 $\bmod p$ 上同调的关系我们有:

定理 8(博雷尔,布劳德(Browder))   若对每个 i,有 $p \nmid n_i$,则

$$H^*(B;\mathbf{Z}_p) = \mathbf{Z}_p[x_{n_1}, \cdots, x_{n_r}]$$

近年来获得的两个重要结果使我们有可能回答第 3 个问题,第一个结果是 Clark 和 Ewing 给出的,本质上他们构造出了一张能够实现的型的表,第二个是亚当斯和 Wilkerson 给出的.

定理 9   若对每个 i,有 $p \nmid n_i$,则 $[n_1, \cdots, n_r]$ 必定是 Clark-Ewing 表中的型的并.

J. Aguade 注意到 Clark-Ewing 表中绝大多数型只在这样的素数 p 时出现,其中 $p \equiv 1 \pmod{m}$,而这里的 m 满足:存在 i,使得 $m \mid n_i$.用迪利克雷定理找素数 p 使得下列条件成立:对任何 $m \mid d, p \not\equiv 1 \pmod{m}$,其中 d 为 $n_1, \cdots, n_r$ 的最大公因子.从而可删掉表中很多型,并可得下列定理:

定理 10(Aguade)   若 $H^*(\Omega B;\mathbf{Z})$ 无挠,型 $[n_1, \cdots, n_r]$ 满足 $n_1 < \cdots < n_r$,则 $H^*(\Omega B;\mathbf{Q})$ 同构于某个李群的有理上同调.

定理 11(Lin)   设 ΩB 的型为 $[n_1, \cdots, n_r]$,则:

(1) $[n_1, \cdots, n_r]$ 是单李群的型以及数组 $[4,24]$,$[12,16]$ 的并;

(2) $[n_1, \cdots, n_r]$ 是单李群的型及数组 $[4,16]$,$[4,24]$,$[4,48]$ 的并.

定理 11 告诉我们:从有理的观点看,有限回路空间和李群几乎相同,在某种意义上,有四个例外单回路空间的型 $[4,24]$,$[12,16]$,$[4,16]$ 和 $[4,48]$.

还有一个用同伦论来刻画单李群的方法.

我们可以定义具有性质 $\pi_3(\Omega B) = \mathbf{Z}$ 的有限回路空间 ΩB 为

"单回路空间".注意在李群的场合,这个定义可作为单李群的定义.从定理 11 中的(2)可得:

定理 12(Lin) 每个单回路空间有与单李群相同的有理上同调.

现在我们回到本节开头提到的概念,并总结一下用来计算李群的上同调的基本策略.

原来的方法是先从李群过渡到它们的李代数,然后应用嘉当 — 基灵分类获得所有单李代数,再回过来便获得对应的单李群,最后逐个计算所有单李群的上同调.图 1 描述了这个过程.

希尔伯特第五个问题说的是,我们可以改变流形上的解析结构,但仍得到李群(图 7).

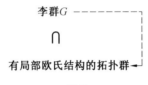

图 7

从有限回路空间出发,可用亚当斯,Wilkerson,Clark 和 Ewing 的结果来计算其有理上同调.因此我们有图 8.

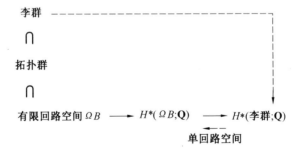

图 8

最后从有限 H — 空间 X 出发,我们研究 mod 2 上同调 $H^*(X;\mathbf{Z}_2)$(它是斯廷罗德代数上的霍普夫代数)以获取有关它的不可分元、倒代数结构以及连通度的性质.因此我们有图 9.

将这些图放在一起,就得到图 10.

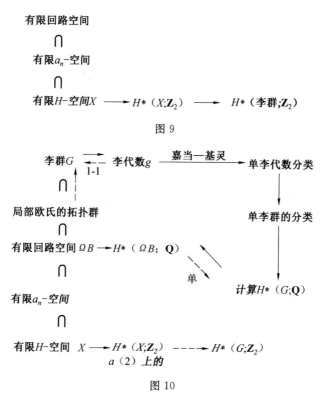

有限回路空间

∩

有限a_n-空间

∩

有限H-空间X ⟶ $H*(X;\mathbf{Z}_2)$ ⟶ $H*($李群$;\mathbf{Z}_2)$

图 9

李群G ⟵ 李代数g ⟶（嘉当—基灵）⟶ 单李代数分类

$1\text{-}1$

↓

单李群的分类

局部欧氏的拓扑群

∩

有限回路空间ΩB ⟶ $H*(\Omega B;\mathbf{Q})$

单

有限a_n-空间

计算$H*(G;\mathbf{Q})$

∩

有限H-空间 X ⟶ $H*(X;\mathbf{Z}_2)$ ----⟶ $H*(G;\mathbf{Z}_2)$

$a(2)$上的

图 10

我希望我已说清楚了以下这一观点：尽管李群有不可更改的解析结构和群结构，但还有一种从同伦论观点来研究它的方法，它不涉及不可更改的群结构，可仍然能得到李群的许多拓扑结果.

文献二：《紧李群的基本几何结构》①.

本文献对紧李群的一些基本结构定理给出了一个几何的证明，目的是要从一个新的角度来考察这些结果，对其中一些在文

---

① 原题：*The elementary geometric structure of compact* 李 *groups*. 译自：*Bull. London Math. Soc.*, 1998, 30：337-364.

献中难以找到证明的结果给出证明,并说明我们的方法从同伦的角度处理 p — 紧群时也同样有效.

1 引论

一个紧李群 G 是带有光滑乘法映射 $G \times G \to G$ 的紧可微流形,其中乘法映射给出 G 的群结构(逆映射必定是光滑的).例如,G 是由单位长度复数组成的乘法群,或者是由 n 维欧氏空间中的旋转所组成的群 SO_n.紧李群在拓扑、代数、分析中处处可见,本文献的目的是从在几何中与同伦论有关的视角来研究它们的基本结构,这时同调群、欧拉示性数、莱夫谢茨数起重要的作用.我们将尽可能地避免无穷小构造以及使用李代数.我们采用这种不同寻常的途径有几点理由.首先,论证相对来说不那么复杂,而且可能使读者对这些结构定理有新的认识.其次,我们能证明一些在以往的文献中往往被省略的结果.比如,我们给出了一个计算连通群的中心的公式(定理 9):计算出了一个紧李群的极大环面的一个子群在该紧李群中的中心化子的连通分支所成的群(定理 10),以及连通李群可分裂为乘积的判别准则(定理 11).最后,我们这种研究紧李群的方法来源于 p — 紧群(紧李群的同伦类似)的结构理论的研究,在那里,我们还得借助于额外的一些同伦论手段,因为在研究 p — 紧群时,没有像李代数这样的分析对象.

第二节引进环面,它其实恰为连通交换紧李群.第三节是关于莱夫谢茨数以及欧拉示性数的介绍.第四节对紧李群 G 证明极大环面 T 的存在性和本质上的唯一性.第五节考虑 T(或 G)的外尔群,并证明,若 G 连通,则分类空间 BG 的有理系数上同调同构于 W 在 $H^*(BT; \mathbf{Q})$ 上的自然作用的不变量的环,从而证明 W 是考克斯特群(定理 6).第六节简要地讨论了极大秩子群.第七节和第八节收集了 G 的中心,T 的子群在 G 中的中心化子以及这些中心化子的连通分支的一些性质.第九节用一些涉及 T 的一般条件来讨论连通李群 G 分裂为乘积的问题.第十节利用第九节的结果来证明任何中心平凡且连通的李群 G 可写成单因子的乘积.最后一节包含了第五节用到的一个代数定理的一个简短证明(属于 J. 莫尔(J. Moore)).

记号及术语. 我们假设所考虑的流形都是光滑的. 一个紧流形,如果没有边界,就称为闭的. 紧李群 G 在流形 M 上的一个(光滑)作用是通常意义上的一个群作用,并且作用映射 $G \times M \to M$ 光滑. 紧李群之间的同态(或同构)是可微的群同态(或群同构).

若 G 是李群,则 T_eG 代表 G 在单位元 e 处的切空间. 若 ϕ: $G \to H$ 是李群间的同态,则微分 $\mathrm{d}\phi$ 给出一个线性映射 $(\mathrm{d}\phi)_e$: $T_eG \to T_eH$. 偶尔,我们也把实向量空间 V 当作(非紧)李群,其乘法是向量的加法;此时单位元 e 是 V 的零元,并且有自然的同构 $T_eV \cong V$. 若 K 是 G 的子群,则 $\mathrm{C}_G(K)$ 和 $\mathrm{N}_G(K)$ 分别代表 K 在 G 中的中心化子和正规化子.

几何背景. 当然,我们必须假设流形理论和微分几何中的一些结果是已知的. 这些结果中的绝大部分会在用到它们的地方提到. 我们会多次用到紧李群的任何闭子群是紧李子群这样一个事实,即 G 的任何闭子群也是 G 的一个紧光滑子流形. 我们还要用到以下事实:若 G 是紧李群,H 是 G 的紧李子群,则陪集空间 G/H 是光滑流形,并且投影 $G \to G/H$ 是结构群为 H 的光滑主纤维丛;若 H 是 G 的正规子群,则 G/H 是李群,而 $G \to G/H$ 是同态.

与其他工作的关系. 本文的所有结果都是已知的,尽管其中某些相对于其他而言并不太为人所知. 到第四节为止,我们的论证大体上类似于亚当斯的论证,而后者本身大都取自经典文献. 自那以后,我们的方法就与通常的大不相同了. 第五节中用来建立外尔群 W 与 $H^*(BG; \mathbf{Q})$ 之间关系的方法似乎是新的. 定理 10 对中心化子做了极其细致的分析,而第九节和第十节中处理乘积分解的几何方法则是我们以前未见到的.

2 环面

定义 1 环面是同构于

$$\mathbf{R}^n/\mathbf{Z}^n \cong (\mathbf{R}/\mathbf{Z})^n$$

的紧李群,其中 $n \geqslant 0$. 数 n 称为环面的秩.

显然,环面的秩等于它作为流形的维数.下面的定理解释了我们为何对环面感兴趣.

定理 1 任何连通紧交换李群同构于环面.

定理 2 设 V 是 n 维向量空间,G 是李群,而且 $\varphi: V \to T_e G$ 是向量空间之间的映射.若 $n = 1$ 或 G 是交换群,则存在唯一的同态 $\varphi: V \to G$ 使得 $\varphi = (\mathrm{d}\phi)_e$.

定理 1 的证明 令 G 是连通紧交换李群,$V = T_e G$,且 $\varphi: V \to V$ 是单位映射.由定理 2,存在同态 $\phi: V \to G$ 使得 $\varphi = (\mathrm{d}\phi)_e$.核 $K = \phi^{-1}(e)$ 是 V 的一个闭子群.由于 $(\mathrm{d}\phi)_e$ 是 V 的单位映射,隐函数定理表明 ϕ 给出 V 中单位元的一个邻域 U 与 G 中单位元的一个邻域 $\phi(U)$ 之间的一个微分同胚.特别地,$K \bigcap U = \{e\}$,因此 K 是 V 的离散子群.由于 G 连通,集合 $\phi(U)$ 生成 G,因此 $\phi: V \to G$ 是满射.从而易知 ϕ 是复选映射,因而 K 同构于 $\pi_1 G$.群 $K \subset V$ 有限生成(因为 G 是子群),交换且无挠,因而 $K \cong \mathbf{Z}^n$,对某个 n,利用 K 在 V 中的离散性质,直接可以证明 $n = \dim V$,并且可选 V 的一组基,使得 K 是 V 中由某些基元生成的子群.

命题 1 设 T 和 T' 是环面,则自然映射
$$\mathrm{Hom}(T, T') \to \mathrm{Hom}(\pi_1 T, \pi_1 T')$$
$$\cong \mathrm{Hom}(H_1(T; \mathbf{Z}), H_1(T'; \mathbf{Z}))$$
$$\cong \mathrm{Hom}(H^1(T'; \mathbf{Z}), H^1(T; \mathbf{Z}))$$
是同构.特别地,同态 $T \to T'$ 构成的空间是离散的.

证明 本命题是定理 1 的推论,后者实际上是说,若 T 是环面,其基本群为 π,则在 T 与 $(\mathbf{R} \otimes \pi)/\pi$ 之间有一个自然的同构.

命题 2 设 T 是环面,并且 $f: T \to T$ 是单射,则 f 是 T 的自同构.

证明 取 $T = \mathbf{R}^n/\mathbf{Z}^n$.由命题 1 的证明知,$f$ 可提升为向量空间之间的同态 $\tilde{f}: \mathbf{R}^n \to \mathbf{R}^n$.易知 $\tilde{f}$ 是单射,因此,根据线性代数的知识,它也是满射,由此 f 是满射.

命题3 设 $1 \to T_1 \to G \to T_2 \to 1$ 是紧李群的正合列,其中 T_1,T_2 是环面,则 G 也同构于环面.

证明 序列 $T_1 \to G \to T_2$ 是空间的纤维化序列,由它的同伦长正合列知 G 是连通的.要证明 G 是环面,只要证明 G 是交换群(定理1).由于 T_1 是 G 的正规子群,G 在自身上的共轭作用诱导出一个(连续)同态 $G \to \mathrm{Aut}(T_1)$.但是 $\mathrm{Aut}(T_1)$ 是离散群(命题1)且 G 连通,因而上述同态为平凡同态,且 T_1 包含在 G 的中心里.这样 $\forall g \in G$,构造映射

$$C_g : T_2 \to T_1, C_g(x) = g\widetilde{x}g^{-1}\widetilde{x}^{-1}$$

其中 $\widetilde{x}$ 是 $x \in T_2$ 在 G 中的一个原象.易知 $C_g(x)$ 不依赖于 $\widetilde{x}$ 的选择,因此对应 $g \to C_g$ 定义了从 G 到由 T_2 到 T_1 的同态所组成的空间的一个连续映射.如前所述,G 连通,而 T_2 到 T_1 的同态的空间是离散的,因此同态 C_g 不依赖于 g.对单位元 e 计算一下 C_e 得知,$\forall g \in G,C_g$ 是平凡同态.由此立即知道 G 是交换群.

命题4 设 G 是正维数紧李群,则 G 包含一个同构于环面的非平凡李子群.

证明 由于 G 的维数大于 0,T_eG 非平凡.令 $\varphi: \mathbf{R} \to T_eG$ 为一个非平凡同态.由定理2,存在同态 $\phi: \mathbf{R} \to G$ 使得 $\varphi = (\mathrm{d}\phi)_e$.令 K 是 $\phi(\mathbf{R})$ 在 G 中的闭包,易知 K 是 G 的一个非平凡子群,并且 K 为交换,而且连通(因为 $\mathbf{R}$ 具有这些性质).如前所述,K 也是 G 的李子群,再由定理1,K 是环面.

引理1 若 K 是环面,则存在 T 中的元素 x 使得循环子群 $\langle x \rangle$ 是 T 的稠密子群.

3 莱夫谢茨数以及欧拉示性数

本文献中的许多论证都依赖于莱夫谢茨数和欧拉示性数的初等性质.本节就是要介绍这些性质.注意一个(带边和不带边)光滑紧流形可三角剖分为一个有限单纯复形.

定义2 设 X 是有限 CW − 复形.X 的欧拉示性数记为 $\chi(X)$,定义为和 $\sum_i (-1)^i \mathrm{rk}_{\mathbf{Q}} H^i(X;\mathbf{Q})$.若 $f: X \to X$ 是映射,

则 f 的莱夫谢茨数记为 $\Lambda(f)$，定义为和 $\sum_i(-1)^i \operatorname{tr}_\mathbf{Q} H^i(f;\mathbf{Q})$.

其中 $\operatorname{rk}_\mathbf{Q} H^i(X;\mathbf{Q})$ 表示有理域上的向量空间 $H^i(X;\mathbf{Q})$ 的维数，而 $\operatorname{tr}_\mathbf{Q} H^i(f;\mathbf{Q})$ 表示 f 在这个向量空间上诱导出的自同态的迹. 显然 X 的欧拉示性数等于 X 的恒同映射的莱夫谢茨数，并且 f 的莱夫谢茨数只依赖于 f 的同伦类. 下面是经典的莱夫谢茨不动点定理.

命题 5 设 X 是有限 CW－复形，映射 $f:X\to X$ 没有不动点，则 $\Lambda(f)=0$.

以下是命题 5 的一个简单推论.

推论 设 G 是紧李群，则 $\chi(G)\neq 0\Leftrightarrow G$ 是离散群 $\Leftrightarrow G$ 是有限群.

证明 若 G 是有限群，则 $\chi(G)=\#(G)\neq 0$. 若 G 不是有限群，则 G 的单位分支非平凡，那么左乘以单位分支中任何非平凡元，就得到 G 的一个同伦于恒同映射且没有不动点的映射. 由命题 5，$\chi(G)=0$.

莱夫谢茨数还有下面的性质：

命题 6 设有限 CW－复形 X 可写成并 $A\bigcup_B C$，其中 A 和 C 是 X 的子复形，而子复形 B 是 A 和 C 的交. 令 $f:X\to X$ 满足 $f(A)\subset A,f(C)\subset C$（因而 $f(B)\subset B$），则

$$\Lambda(f)=\Lambda(f\mid_A)+\Lambda(f\mid_C)-\Lambda(f\mid_B)$$

证明 若 $g=(g_U,g_V,g_W)$ 是 $\mathbf{Q}$ 上的有限维向量空间的正合列 $0\to U\to V\to W\to 0$ 的自同态，则

$$\operatorname{tr}(g_V)=\operatorname{tr}(g_U)+\operatorname{tr}(g_W)$$

这是迹对短正合列的可加性，要证明这一点，只要把 U 的一组基扩充为 V 的一组基，然后相对于这组基把 g 表示为一个矩阵即可. 现在 f 诱导出迈尔－菲托里斯序列的自同态

$$\cdots\to H^i(X;\mathbf{Q})\xrightarrow{j^*} H^i(A;\mathbf{Q})\bigoplus$$

$$H^i(C;\mathbf{Q})\xrightarrow{k^*} H^i(B;\mathbf{Q})\longrightarrow H^{i+1}(X;\mathbf{Q})\to\cdots$$

令 U^* 为 j^* 的象，V^* 为 k^* 的象，W^* 为 j^* 的核. 由迹的可加性

以及莱夫谢茨数的定义立即可得下列公式. 如果 f_{U^i} 是 f 在 U^i 上诱导出的自同态, 命

$$\Lambda(f_{U^*}) = \sum (-1)^i \mathrm{tr}(f_{U^i})$$

那么

$$\Lambda(f\mid_A) + \Lambda(f\mid_C) = \Lambda(f_{U^*}) + \Lambda(f_{V^*})$$
$$\Lambda(f\mid_B) = \Lambda(f_{V^*}) - \Lambda(f_{W^*})$$
$$\Lambda(f) = \Lambda(f_{W^*}) + \Lambda(f_{U^*})$$

将第一个方程减去第二个方程, 再与第三个方程比较, 即得命题 6.

注意 对恒同映射应用命题 6 得到熟知的欧拉示性数的可加性. 由此可知, 若 $F \to E \to B$ 是一个适当的纤维丛, 比如紧流形上纤维也是紧流形的光滑纤维丛, 则 $\chi(E) = \chi(F)\chi(B)$. 要证明这一点, 只要按 B 的骨架 B_i 的逆象构造 E 的一个滤子, 然后用归纳法及可加性证明 $\chi(E_i) = \chi(B_i)\chi(F)$.

命题 7 设 M 是紧流形, 紧李群 G 光滑地作用在 M 上, f: $M \to M$ 是由 G 中某个元素 g 的作用给出的微分同胚, 令 N 是 f 的不动点集, 则 $\Lambda(f) = \chi(N)$.

证明 令 $H \subset G$ 是 g 生成的循环子群在 G 中的闭包, 则 H 是 G 的李子群, 因此 $N = M^H$ 是 M 的子流形. 令 A 为 N 的一个闭的管状邻域, 使得 $f(A) = A$, 再令 C 是 A 的内部在 M 中的补集, $B = A \cap C$. 空间 B 同胚于 N 在 M 中的法球丛的全空间. 由命题 6

$$\Lambda(f) = \Lambda(f\mid_A) + \Lambda(f\mid_C) - \Lambda(f\mid_B)$$

但是, 根据命题 5, $\Lambda(f\mid_C)$ 和 $\Lambda(f\mid_B)$ 都等于 0. 因为置入 $N \to A$ 是同伦等价, 并且它与 f 的作用可换, 而 f 在 N 上的作用是平凡作用. 因此

$$\Lambda(f\mid_A) = \Lambda(f\mid_N) = \chi(N)$$

注意 可以证明一个紧流形 M 的一个微分同胚 f 是由 M 上某个紧李群作用给出的, 当且仅当 f 相对于 M 上某个黎曼度量是等距的. 如果对微分同胚不做某种限制, 命题 7 的结论一般

不成立. 例如, 不难构造 S^2 的这样一个微分同胚, 它同胚于恒同映射, 而且只有一个不动点.

命题 8 设 X 是有限 CW—复形, 若一个环面连续作用在 X 上且没有不动点, 则 $\chi(X) = 0$.

证明 取 T 中的一个元素 α 使得 α 生成的子群 $\langle\alpha\rangle$ 在 T 中稠密 (引理 1). X 中 α 的不动点也是 $\langle\alpha\rangle$ 的不动点, 从而 α 在 X 上没有不动点. 由于 T 连通, 由 α 的作用给出的 X 的自映射同伦于恒同映射, 因此这个自映射的莱夫谢茨数等于 $\chi(X)$. 由命题 5, $\chi(X) = 0$.

命题 9 设 M 是光滑紧流形, 环面 T 光滑作用在 M 上, 其不动点集为 M^T, 则 $\chi(M^T) = \chi(M)$.

证明 其证明与命题 7 的证明本质上是一样的, M^T 也是 M 的子流形, 也可以找到 M^T 的在 T 作用下不变的闭管状邻域. 根据命题 8, A 的边界和 A 的内部在 M 中补集的欧拉示性数都等于 0. 由欧拉示性数的可加性加上 $\chi(A) = \chi(M^T)$ 这一事实即可得本命题.

4 极大环面

G 的同构于环面的闭子群 T 若满足 $\chi(G/T) \neq 0$, 则称 T 为紧李群 G 的极大环面. 本节的主要目的是要证明下面的定理.

定理 3 任何紧李群都有一个极大环面.

以下的论证并不是新的, 但我们所选择的这一特殊证明思路是为了将其推广到 p—紧群. 在证明定理 3 之前, 我们指出极大环面实际上在适当意义下是极大的, 而且它在共轭等价下是唯一的.

引理 2 设 T 是紧李群 G 的极大环面, 而 T' 是 G 的任何另一个环面子群, 则存在 $g \in G$, 使得
$$T' \subset gTg^{-1}$$

证明 T' 可以通过左平移作用在 G/T 上. 根据命题 8, 不动点集 $(G/T)^{T'}$ 非空. 若 $g \in G$ 是 $(G/T)^{T'}$ 的某个点的原象, 则

$T' \subset gTg^{-1}$.

定理 4　若 T 和 T' 是紧李群 G 的极大环面，则 T 和 T' 共轭，即存在 $g \in G$ 使得 $gTg^{-1} = T'$.

证明　由引理 2，存在 $g, h \in G$，使得 $T' \subset gTg^{-1}$，而 $T \subset hT'h^{-1}$. 根据命题 2，这两个包含都是等式.

注意　紧李群 G 的秩定义为 T 的秩（定义 1），其中 T 是 G 的任一个极大环面.

定理 3 的证明依赖于极大环面的另一个刻画.

定义 3　紧李群 G 的一个交换子群 A 称为在 G 中是自中心化的（self-centralizing），若 $A = C_G(A)$. A 称为几乎自中心化的（almost self-centralizing），若 A 在 $C_G(A)$ 中的指标是有限数.

命题 10　设 G 是一个紧李群，T 是 G 的环面子群，则 T 是 G 的极大环面，当且仅当 T 是几乎自中心化的.

注意　我们以后（命题 13）将证明：若 G 连通，则极大环面实际上是自中心化的.

命题 10 的证明　令 C 为中心化子 $C_G(T)$，N 为正规化子 $N_G(T)$. C 和 N 都是 G 的闭子群，从而是 G 的李子群. 考虑不动点集 $(G/T)^T$，其中 T 以左平移作用在 G/T 上. 这个不动点集等于 S/T，其中 S 是 G 中满足 $gTg^{-1} \subset T$ 的元素 g 的集合. 由命题 2，$S = N$ 且 $(G/T)^T = N/T$. 根据命题 9，$\chi(N/T) = \chi(G/T)$. 此外我们有一个纤维化

$$C/T \to N/T \to N/C$$

在这个纤维化中，底空间 N/C 同胚于离散空间 $\mathrm{Aut}(T)$（命题 1）的一个紧子空间，从而是有限集，由此易知

$$\chi(G/T) = \chi(N/T) = \chi(G/T) \cdot \#(N/C) \tag{1}$$

若 T 是几乎自中心化的，则 C/T 是有限群，公式（1）表明 $\chi(G/T) \neq 0$. 若 T 不是几乎自中心化的，则紧李群 C/T 的单位分支非平凡，$\chi(C/T) = 0$（命题 5 的推论），由公式（1）知 $\chi(G/T) = 0$.

定理 3 的证明　证明方法是对 G 的维数作归纳法. 若 G 的

维数是 0,换句话说,G 是有限群,则平凡子群是 G 的极大环面.

然后,设维数小于 G 的维数的任何紧李群都有极大环面.令 A 是 G 的一个同构于环面(命题4)的非平凡闭子群,令 C 是其中心化子 $C_G(A)$.由于 C 是 G 的闭子群,从而也是 G 的李子群.易知 A 是 C 的闭的正规子群,因而

$$\dim(C/A) = \dim C - \dim A < \dim C \leqslant \dim G$$

根据归纳假设,可以找到 C/A 的极大环面 $\overline{T}$.令 T 为 $\overline{T}$ 在 C 中的逆象,我们有短正合列

$$1 \to A \to T \to \overline{T} \to 1$$

由命题3知 T 是 G 的环面子群.由于

$$\chi(C/T) = \chi((C/A)/\overline{T})$$

不为 0,T 是 C 的极大环面,从而在 C 中是几乎自中心化的(命题10).由于 T 包含 A,T 在 G 中的中心化子等于 T 在 $C = C_G(A)$ 中的中心化子,因此 T 在 G 中是几乎自中心化的.根据命题10,T 是 G 的极大环面.

5 外尔群以及 BG 的上同调

本节我们要对每个紧李群定义一个特殊的有限群,称为外尔群.如果 G 连通,那么我们可以通过外尔群在极大环面上的作用计算 BG 的上同调,并从这个计算导出外尔群的一个很重要的代数性质:它在 $\mathrm{Aut}(\mathbf{Q} \otimes \pi_1 T)$ 中的象由反射生成.

定义 4 令 G 是紧李群,T 是它的极大环面,T 在 G 中的外尔群是商群 $N_G(T)/T$.

注意 外尔群 W 显然是紧李群.事实上,它是有限群,因为 T 是几乎自中心化的(命题10),而 $\mathrm{Aut}(T)$ 是离散群(命题1).T 的唯一性(定理4)意味着,在同构的意义下,W 只依赖于 G,因此,为方便起见,取定 T,并记 W 为 G 的外尔群.因为 T 是交换群,$N_G(T)$ 在 T 上的共轭作用诱导出 W 在 T 上的作用.

我们先来计算外尔群的阶数,从而说明它包含 G 的连通分支的个数作为因子.

命题 11 设 G 是紧李群,T 是极大环面,而 W 是其外尔群,

则 $\sharp(W) = \chi(G/T)$.

证明 如命题 10 的证明那样,W 可等同于不动点集 $(G/T)^T$. 利用命题 9 计算欧拉示性数即可.

若 G 是紧李群,令 G_e 表示 G 的单位分支.

命题 12 设 G 是紧李群,其极大环面为 T,而外尔群为 W. 令 N 表示 $N_G(T)$,$N' = N \bigcap G_e$,$W' = N' \bigcap T$ 为 N' 在 W 中的象. 则自然映射

$$N/N' = W/W' \to G/G_e = \pi_0 G$$

是一个同构.

注意 上述群 N' 是 T 在 G_e 中的正规化子.

命题 12 的证明 由上述构造知,映射是单射,故只需证明它是满射. 空间 G/T 微分同胚于 $\sharp(\pi_0 G)$ 个 G_e/T 的不交并. 由于

$$0 \ne \chi(G/T) = \sharp(\pi_0 G)\chi(G_e/T)$$

G/T 的每个连通分支 X 的欧拉示性数非 0,T 在 G/T 上的左平移作用限制为 T 在 X 上的作用. 由命题 8,这个作用至少有一个不动点,每个这样的不动点给出 N/T 中的一个元,这个元投影到 $\pi_0 G$ 中由 X 代表的连通分支.

以下我们要假设 G 是连通的.

命题 13 令 G 是连通紧李群,其极大环面为 T,而外尔群为 W. 则 T 在 G 中是自中心化的,等价地,W 在 T 上的共轭作用是忠实的.

证明 我们必须证明,若 $a \in G$ 与 T 交换,则 $a \in T$. 令 $A = \langle a, T \rangle$,考虑不动点集

$$(G/T)^A = ((G/T)^{\langle a \rangle})^T$$

这里,A 在 G/T 上的作用仍然是左平移作用. 由于 G 连通,a 在 G/T 上的作用同伦于单位映射,由命题 7

$$\chi((G/T)^{\langle a \rangle}) = \chi(G/T) \ne 0$$

再由命题 8(应用于 $M = (G/T)^{\langle a \rangle}$),可知 $(G/T)^A$ 非空. 若 $g \in G$ 投影到 $(G/T)^A$ 的一个点,则 $gAg^{-1} \subset T$. 由命题 2,必有 $gTg^{-1} =$

T,因此 $gAg^{-1} \subset T$ 导出 $A = T$,因而 $a \in T$.

注意 命题 13 的证明表明,若 G 连通,$a \in G$,S 是 G 中与 a 交换的环面,则 G 的子群 $\langle S, a \rangle$ 共轭于极大环面 T 的某子群. 取 $S = \{e\}$,可知 G 的每个元共轭于 T 的一个元.

下面我们来考虑计算 $H^*(BG; \mathbf{Q})$ 的问题.

命题 14 若 G 是连通紧李群,则有理系数上同调环 $H^*(BG; \mathbf{Q})$ 同构于有限个偶数维生成元的多项式代数.

注意 $H^*(BG; \mathbf{Q})$ 中多项式生成元的个数称为 G 的有理秩,我们将在下面证明(定理 5)它等于 G 的秩.

命题 14 的证明依赖于下面的标准计算.

命题 15 设 G 是连通拓扑群,则下面的几个条件等价:

(1) $H^*(G; \mathbf{Q})$ 是 $\mathbf{Q}$ 上的有限生成外代数;

(2) $H^*(BG; \mathbf{Q})$ 是 $\mathbf{Q}$ 上的有限生成多项式代数.

若 (1) 和 (2) 成立,且 $\{n_i\}$ 是 $H^*(BG; \mathbf{Q})$ 的任一组多项式生成元的次数的集合,则 $\{n_i - 1\}$ 是 $H^*(G; \mathbf{Q})$ 的任一组外代数生成元的次数的集合.

证明 这个结果可以从艾伦伯格 — 莫尔谱序列和 Rothenberg-Steenrod 谱序列导出,或者通过一个精细的论证也可从 Serre 谱序列推得. 另一种方法是注意空间 G 的有理同伦型是有限个艾伦伯格—麦克莱恩空间的乘积 $\prod_i K(\mathbf{Q}, 2n_i - 1)$. 特别地,$\mathbf{Q} \otimes \pi_i G$ 仅当 i 是奇数时才非 0,因此 $\mathbf{Q} \otimes \pi_i(BG)$ 仅当 i 为偶数时非 0,这样利用波斯尼科夫塔作归纳法容易证明,任何只在偶数维有理同伦群不为 0 的单连通空间有理等价于一个乘积 $\prod_i K(\mathbf{Q}, 2n_i)$,从而得到结论 (2). 反过来,若已知 (2),易知 BG 有理等价于一个乘积 $\prod_i K(\mathbf{Q}, 2n_i)$,因此 $G \sim \Omega BG$ 有理等价于 $\prod_i K(\mathbf{Q}, 2n_i - 1)$.

命题 14 的证明 若已知命题 15,只需注意霍普夫曾知道的事实,连通霍普夫代数 $H^*(G; \mathbf{Q})$,由于纯代数的理由,必定是有限个奇数维生成元上的外代数.

在以下两个引理中，G 是连通紧李群，而 T 是其极大环面. 存在置入映射 $\iota: T \to G$ 以及分类空间之间的诱导映射 $B\iota: BT \to BG$.

引理 3　有理上同调映射
$$(B\iota)^*: H^*(BG; \mathbf{Q}) \to H^*(BT; \mathbf{Q})$$
是一个单射.

证明　有一个纤维化
$$G/T \to BT \xrightarrow{B\iota} BG$$
其纤维有有限复形的伦型. 此时，贝克尔(Becker)和 Gottlieb 的转移(transfer)构造给出映射
$$\tau_*: H^*(BT; \mathbf{Q}) \to H^*(BG; \mathbf{Q})$$
复合 $\tau_* \cdot (B\iota)^*$ 作为 $H^*(BT; \mathbf{Q})$ 的自映射等于乘以 $\chi(G/T)$. 由于 $\chi(G/T) \neq 0$，$(B\iota)^*$ 为单射.

引理 4　$H^*(BT; \mathbf{Q})$ 作为环 $H^*(BG; \mathbf{Q})$ 上的模是有限生成的.

注意　若将 T 换为 G 的任何闭子群，这里的证明仍然有效.

引理 4 的证明　令 R 代表环 $H^*(BG; \mathbf{Q})$. 纤维丛 $G/T \to BT \to BG$ 的有理系数上同调塞尔谱序列是 R — 模的谱序列. 由于 $H^*(G/T; \mathbf{Q})$ 是有限维的，这个谱序列的 E_2 项是有限生成 R — 模. 根据归纳法以及 R 是诺特环可知每个 E_r 项，$r \geqslant 2$ 也是有限生成 R — 模. 由于这个谱序列只在某个水平带状区域里非 0，当 r 充分大时，$E_r = E_\infty$，因此 E_∞ 是有限生成 R — 模. 但是 E_∞ 是 $H^*(BT; \mathbf{Q})$ 的由 R — 模给出的有限滤子所对应的分次对象 (第 k 个滤子商为 $E_\infty^{*,k}$)，从而 $H^*(BT; \mathbf{Q})$ 是有限生成 R — 模.

命题 16　设 S 是 $\mathbf{Q}$ 上 s 个不定元的连通非负分次多项式代数，而 S 的一个分次子代数 R 同构于 r 个不定元的多项式代数. 再设 S 是有限生成 R — 模，则 $r = s$，且 S 是自由 R — 模.

我们需要有关纤维化的一个一般引理.

引理 5　设 $F \to E \to B$ 是纤维化，其中 B 单连通，且所有三

217

个空间的有理系数上同调群在每个次数都是有限维的. 设 $H^*(E;\mathbf{Q})$ 是自由 $H^*(B;\mathbf{Q})$ 模. 则自然映射

$$Q \otimes_{H^*(B;\mathbf{Q})} H^*(E;\mathbf{Q}) \rightarrow H^*(F;\mathbf{Q})$$

是一个同构.

证明 最简单的证明方法是用艾伦伯格—莫尔谱序列,也可以利用塞尔谱序列的乘法结构并对滤子做细致的考察证明所给纤维化的塞尔谱序列退化(否则 $H^*(E;\mathbf{Q})$ 不可能是自由 $H^*(B;\mathbf{Q})$ 模). 然后可验证所给映射为同构.

若 G 是紧李群, T 为其极大环面, 而 W 为其外尔群, 则由前面的构造知, T 的由 W 在 T 上作用所给出的自同构可扩充为 G 的内自同构. 由于一个群的内自同构平凡地作用在分类空间的上同调上, 限制映射

$$H^*(BG;\mathbf{Q}) \rightarrow H^*(BT;\mathbf{Q})$$

可提升为一个映射

$$H^*(BG;\mathbf{Q}) \rightarrow H^*(BT;\mathbf{Q})^W$$

定理 5 设 G 是连通紧李群, T 为其极大环面, 而 W 为其外尔群, 则:

(1) W 在 $H^*(BT;\mathbf{Q})$ 上的共轭作用是忠实的;

(2) T 的秩(定义 1)等于 G 的有理秩;

(3) 自然映射 $H^*(BG;\mathbf{Q}) \rightarrow H^*(BT;\mathbf{Q})^W$ 是一个同构.

证明 由命题 13, 共轭映射 $W \rightarrow \mathrm{Aut}(T)$ 是单射, 因此(由命题 1) W 忠实地作用在 $H^*(BT;\mathbf{Q})$ 上.

令 $R = H^*(BG;\mathbf{Q})$ 和 $S = H^*(BT;\mathbf{Q})$, 则 R 是 r 个生成元的多项式环, 其中 r 是 G 的有理秩, S 是 s 个生成元的多项式环, 其中 s 是 T 的秩. 映射 $R \rightarrow S^W \rightarrow S$ 是单射(引理 3), 并且 S 是有限生成 R—模(引理 4). 由命题 16, S 是, 比如说, t 个生成元的自由 R—模. 由引理 5, 存在同构 $H^*(G/T;\mathbf{Q}) \cong \mathbf{Q} \otimes_R S$. 由此易知 $H^*(G/T;\mathbf{Q})$ 只在偶数维不为 0, 并且 $t = \chi(G/T)$, 再由命题 11 知, $t = \#(W)$. 由构造, 很明显, t 是 S 的分次分式域 $F(S)$ 对 R 的分次分式域 $F(R)$ 的扩张度. 由于 $F(R) \subset F(S)^W$, $F(S)$ 对 $F(R)$ 的度数等于 $\#(W)$, 而且 W 忠实地作用在 $F(S)$ 上, 伽罗

瓦理论表明 $F(R)=F(S)^W$,因而,$S^W\subset F(R)$,扩张 $R\subset S^W$;换句话说,$\forall x\in S^W$,环 $R[x]$ 是有限生成 R—模(看清这一点的一个办法是注意 R 是诺特环,而且 $R[x]$ 包含在有限生成 R—模 S 中).由于像 R 这样的多项式代数在分式域 $F(R)$ 中是整闭的,由此可得同构 $R\cong S^W$.

现在我们能够研究外尔群的基本代数性质.

定义 5 设 V 是 $\mathbf{Q}$ 上的有限维向量空间,$\mathrm{Aut}(V)$ 中的元 α 称为反射,若它在 V 的某个余维 1 子空间上为恒同映射.$\mathrm{Aut}(V)$ 的子群 W 称为反射子群,若 W 是有限群并且 W 由它所包含的反射所生成.

在定义 5 的情形,$\mathrm{Aut}(V)$ 中的有限阶元 α 是一个反射,当且仅当存在 V 的一组基使得 α 相对于这组基的矩阵是对角矩阵 $\mathrm{diag}\{-1,1,\cdots,1\}$.特别地,$\alpha$ 的阶数为 2.

定理 6 设 G 是连通紧李群,其极大环面为 T,外尔群为 W,$V=\mathbf{Q}\otimes\pi_1 T$.则 W 在 T 上的共轭作用使得 W 成为 $\mathrm{Aut}(V)$ 的反射子群.

这是一个经典代数定理的推论,我们用到的这部分属于考克斯特(Coxeter),其逆命题是谢瓦莱(Chevalley)证明的.

定理 7 设 V 是 $\mathbf{Q}$ 上的有限维向量空间,W 是 $\mathrm{Aut}(V)$ 的一个有限子群.令 $V^\#$ 是 V 的对偶空间,$\mathrm{Sym}(V^\#)$ 为 $V^\#$ 上的对称代数,而 $\mathrm{Sym}(V^\#)^W$ 为 W 在 $\mathrm{Sym}(V^\#)$ 上的自然作用的不动点集,则 $\mathrm{Sym}(V^\#)^W$ 同构于 $\mathbf{Q}$ 上的多项式代数,当且仅当 W 由反射生成.

定理 6 的证明 由命题 13 及命题 1 知,映射 $W\to\mathrm{Aut}(V)$ 是单射.环 $\mathrm{Sym}(V^\#)$ 自然同构于 $H^*(BT;\mathbf{Q})$.由定理 5 及命题 14,$\mathrm{Sym}(V^\#)^W\cong H^*(BT;\mathbf{Q})$ 是多项式代数.由定理 7,群 W(当作 $\mathrm{Aut}(V)$ 的子群)由反射生成.

6 极大秩子群

这部分我们来研究紧李群 G 的一类叫作有极大秩的特殊子群.以后会知道它们是这样一些子群:G 的包含在极大环面中的

子群在 G 中的中心化子.

定义 6 紧李群 G 的闭子群 H 称为有极大秩, 若 H 包含 G 的一个极大环面.

由命题 2 不难知道 G 的闭子群 H 有极大秩, 当且仅当 H 的秩等于 G 的秩; 此时, H 的任何极大环面也是 G 的一个极大环面. 若 H 在 G 中有极大秩, 则 H 的外尔群可等同于 G 的外尔群的一个子群.

定理 8 设 G 是连通紧李群, H 是其极大秩闭子群, 则 $H = G$, 当且仅当 H 的外尔群等于 G 的外尔群.

我们给出的证明基于一个简单的维数计算. 设 V 是 $\mathbf{Q}$ 上秩为 r 的向量空间, W 是 $\mathrm{Aut}(V)$ 的反射子群 (定义 5). 令 $V^{\#}$ 为 V 的对偶, 将对称代数 $\mathrm{Sym}(V^{\#})$ 当作分次代数, 其中 $V^{\#}$ 的元的次数为 2. 不动点子代数 $\mathrm{Sym}(V^{\#})^{W}$ 同构于 r 个次数为, 比如说, $d_1, \cdots, d_r$ (定理 7) 的生成元的分次多项式代数 (不计置换的话, 序列 $d_1, \cdots, d_r$ 不依赖于 $\mathrm{Sym}(V^{\#})^{W}$ 的多项式生成元的选取). 定义整数 $\delta(W)$ (称为关联于 W 的维数) 为

$$\delta(W) = \sum_{i=1}^{r} (d_i - 1) = \left(\sum_{i=1}^{r} d_i\right) - r$$

引理 6 令 G 是连通紧李群, T 为其极大环面, W 为外尔群, 其中 W 是 $\mathrm{Aut}(\mathbf{Q} \otimes \pi_1 T)$ 的反射子群 (定理 6), 则 $\delta(W) = \dim(G)$.

证明 这是定理 5(3) 以及命题 15 的推论. 由于 G 是可定向 (实际上可平行化) 紧流形

$$\dim(G) = \max_{i} \{i : H^i(G; \mathbf{Q}) \neq 0\}$$

定理 8 的证明 若 $W(H) = W(G)$, 则由引理 6, $\dim(H) = \dim(G)$. 由于 H 和 G 是闭流形, 而且 H 是 G 的子流形, 因此 H 必须等于 G.

注意 设 W_i 是 $\mathrm{Aut}(V_i)$ 的反射子群, $i = 1, 2$, 则 $W_1 \times W_2$ 是 $\mathrm{Aut}(V_1 \times V_2)$ 的反射子群, 且有同构

$$\mathrm{Sym}(V_1^{\#} \times V_2^{\#})^{W_1 \times W_2} \cong \mathrm{Sym}(V_1^{\#})^{W_1} \otimes \mathrm{Sym}(V_2^{\#})^{W_2}$$

由此得知

$$\delta(W_1 \times W_2) = \delta(W_1) + \delta(W_2)$$

7 中心以及共轭形式

这部分我们要证明下列两个命题:

命题 17 若 G 是连通紧李群,则 G 的中心是 G 中所有极大环面的交.

命题 18 设 G 是连通紧李群,且 C 是 G 的中心,则 G/C 的中心平凡.

注意 在命题 18 中,G/C 称为是 G 的共轭形. 这个名称来自于下面的事实:G/C 是 G 在共轭表示中的象,共轭表示是 G 在 T_eG 上由 G 在自身上的共轭作用诱导出的作用给出的. 我们说连通紧李群是共轭形,如果它的中心平凡.

上述两命题依赖于一个本身就很有趣的结果.

命题 19 设 G 是连通紧李群,S 是 G 中的环面,则中心化子 $C_G(S)$ 连通.

证 令 $a \in C_G(S)$,我们要证明 a 属于 $C_G(S)$ 的单位分支. 令 T 是 G 的极大环面,令 A 为 G 的子群 $\langle a, S \rangle$. 利用命题 13 的证明方法可证明 A 共轭于 T 的一个子群,或等价地,存在 T 的某个包含 S 和 a 的共轭 T'. 由于 $T' \subset C_G(S)$ 连通,a 必定属于 $C_G(S)$ 的单位分支.

命题 17 的证明 设 x 属于 G 的中心,T 为 G 的极大环面. 由命题 13 的注,T 的某个共轭 T' 包含 x. 由于 x 属于 G 的中心,T 自身也包含 x,故 $x \in \bigcap_T T$.

反过来,设 x 不属于 G 的中心,则存在 $y \in G$ 使得 y 与 x 不可换,存在某个包含 y 的极大环面 T. 显然 $x \notin T$,因此 $x \notin \bigcap_T T$.

命题 18 的证明 令 T 是 G 的一个极大环面,由命题 17,T 包含 C. 由定理 1,T/C 是环面,那么微分同胚

$$(G/C)/(T/C) \cong G/T$$

证明 $\chi((G/C)/(T/C))$ 不为 0,因而 T/C 是 G/C 中的极大环面. 设 x 是 G/C 的非单位元,而 $\widetilde{x} \in G$ 投射到 x. 由于 $\widetilde{x} \notin C$,因而

G 中有不包含 $\widetilde{x}$ 的极大环面 T'(命题 17). 这样 T'/C 是 G/C 中不包含 x 的极大环面, 于是 x 不属于 G/C 的中心.

为未来的需要, 我们也给出以下结果:

引理 7 若 G 是连通紧李群, 则 G 的任何有限正规子群包含在 G 的中心里. 特别地, 若 G 是共轭形, 则 G 没有非平凡的有限正规子群.

证明 令 A 是 G 的有限正规子群. 由于 G 连通, 而 A 是离散群, G 在 A 上的共轭作用必为平凡作用, 因此 A 落在 G 的中心里.

8 关于中心的计算

在某种意义上, 前文给出了连通紧李群的中心的一个公式: 它是 G 的极大环面的交. 这个公式对于应用极不方便, 因此本节中我们用一个极大环面以及它的正规化子来表示中心. 为此我们对 G 的一个极大环面的任意子群 A 的中心化子 $C_G(A)$ 做一般的分析. 特别地, 我们确定了 $C_G(A)$ 的连通分支群.

在陈述主要定理之前, 必须先设定一些概念.

定义 7 设 G 是连通紧李群, 其极大环面和外尔群分别为 T 和 W. 若 $s \in W$ 是反射 (相对于 W 在 $\mathbf{Q} \otimes \pi_1 T$ 上的共轭作用), 则:

(1) 不动点子群 $F(s)$ 是 s 在 T 上作用的不动点集;

(2) 奇异环面 $H(s)$ 是 $F(s)$ 的单位分支;

(3) 奇异陪集 $K(S)$ 是 T 中形如 x^2 的元素所构成的子集, 其中 x 是 $N_G(T)$ 中投影到 $s \in W$ 的元;

(4) 奇异集合 $\sigma(s) = H(s) \bigcup K(s)$.

定理 9 设 G 是连通紧李群, 其极大环面和外尔群分别为 T 和 W. 则 G 的中心等于 $\bigcap_s \sigma(s)$, 其中 s 跑遍 W 中的反射.

注意 应用定理 9 来确定中心与确定 G 的中心的连通分支群有些关系; 由定理 9 很容易知道中心的单位分支等于不动点集 T^W 的单位分支, 或等价地, 等于 $N_G(T)$ 的中心的单位分支.

定理 9 可从下面更一般的计算得到:

定理 10 设 G 是连通紧李群,其极大环面和外尔群分别为 T 和 W,且 A 是 T 的子群.令 G' 是 $C_G(A)$ 的单位分支.则 G' 在 G 中有极大秩(定义 6),且 G' 的外尔群 W' 是 W 中由满足 $A \subset \sigma(s)$ 的反射 s 生成的子群.群 W' 不包含 W 的其他反射.

在定理 10 的情形,描述 $C_G(A)$ 的外尔群本身比描述它的单位分支的外尔群要容易得多.

命题 20 设 G 是连通紧李群,其极大环面和外尔群分别为 T 和 W,而 A 是 T 的子群,则 $C_G(A)$ 是 G 中的极大秩子群,且 $C_G(A)$ 的外尔群是 W 的子群.它由 W 的那些限制在 $A \subset T$ 上为恒同映射的元素组成.这里 W 共轭地作用在 T 上.

证明 显然,$C_G(A)$ 包含 G 的一个极大环面,即 T 自身.要计算 $C_G(A)$ 的外尔群只需注意到 T 在 $C_G(A)$ 中的正规化子是 A 在 $N_G(T)$ 中的中心化子.

注意 令 G, T, W, A 如定理 10 那样.结合命题 20 及定理 10 就给出了一个计算有限群 $\pi_0 C_G(A)$ 的公式,即(见命题 12)$\pi_0 C_G(A)$ 同构于商群 $W(A)/W_e(A)$,其中 $W(A)$ 是 W 中那些由共轭作用确定的 T 的自映射限制在 A 上为恒同的元素所构成的群,而 $W_e(A)$ 是 $W(A)$ 的由满足 $A \subset \sigma(s)$ 的反射 $s \in W$ 生成的子群.

我们现在来讨论 T 的在定义 7 中定义的子集.首先注意 $H(s)$ 是 $F(s)$ 的一个子群,$H(s)$ 在 $F(s)$ 中的指标为 1 或 2.这是因为,由于 s 是反射,商群 $T/H(s)$ 是圆周 $C = \mathbf{R}/\mathbf{Z}$,$s$ 在其上的作用为取逆,因此有正合列

$$0 \to H(s) \to F(s) \to C^{(s)} \to \cdots$$

其中群 $C^{(s)}$ 为 $\mathbf{Z}/2$.

其次,由于 $N_G(T)$ 中代表 s 的元素与它自己的平方交换,$K(s)$ 是 $F(s)$ 的子集.

最后,如名字所提示的,奇异陪集是 $H(s)$ 在 T 中的陪集.要证明这一点,设 x 和 y 是 $N_G(T)$ 中投影到 $s \in W$ 的两个元.则存在 $a \in T$ 使得 $x = ya$,因而 $x^2 = y^2 v(a)$,其中 $v(a) = a\sigma(a)$.$v:$ $T \to T$ 的象是 T 的连通子群,由于 s 在这个象上的限制为恒同映

射,因而它包含在 $H(s)$ 中.于是 x^2 和 y^2 在 $H(s)$ 的同一陪集中.反之,若 x 投影到 s,且 $a \in H(s)$,则(因为 $H(s)$ 是可除群)存在 $b \in H(s)$ 使得 $v(b) = b^2 = a$.因而 $y = xb$ 也投影到 s,从而 $y^2 = (xb)^2 = x^2 v(b) = x^2 a$.

以上这些事实说明奇异集 $\sigma(s) = H(s) \bigcup K(s)$,或者等于 $H(s)$,或者等于 $F(s)$,特别地,$\sigma(s)$ 是 T 的子群.下面列出的是这些群之间的三种可能的关系,并给出实现这些可能性的紧李群的例子.在每一种情形下,外尔群皆为 $\mathbf{Z}/2$,因此 s 只有一种选择:

(1) $H(s) \subsetneqq \sigma(s) = F(s)$: $SU(2)$;

(2) $H(s) = \sigma(s) \subsetneqq F(s)$: $SO(3)$;

(3) $H(s) = \sigma(s) = F(s)$: $U(2)$.

以下三个引理为我们提供了所需的有关 T 的上述子群的关键性质.在这三个引理中,G 是连通紧李群,其极大环面和外尔群分别为 T 和 W,$s \in W$ 是反射.

引理 8 设 K 是这些群 $H(s)$,$\sigma(s)$ 和 $F(s)$ 中的任一个,则 $C_G(K)$ 在 G 中有极大秩,且 $C_G(K)$ 的外尔群是 W 的子群 $\langle s \rangle$.

证明 $C_G(K)$ 的外尔群 W' 是 W 的某个子群,它包含 $\langle s \rangle$ (命题 20),通过共轭它忠实地作用在 T 上(命题 13),且在 $H(s)$ 上的限制为平凡作用.令 r 为 T 的秩,令 M 为 $\mathbf{Z}$ 上的那些 $r \times r$ 阶可逆矩阵所构成的群,这些矩阵的前 $r-1$ 列为单位矩阵.由命题 1,W' 同构于 M 的一个子群.易知取行列式诱导出 M 的任何有限子群与 $\{\pm 1\}$ 的一个子群之间同构,因此 W' 不比 $\langle s \rangle$ 大.

引理 9 群 $C_G(\sigma(s))$ 连通.

证明 令 $G' = C_G(H(s))$,由于命题 19,这个群连通,且外尔群为 $\langle s \rangle$ (引理 8).若 $\sigma(s) = H(s)$,则 $C_G(\sigma(s)) = G'$ 连通.设 $\sigma(s) \neq H(s)$,取 $\tilde{s} \in N_{G'}(T)$ 使得 $\tilde{s}$ 投影到 $s \in W$.又令 $x = \tilde{s}^2 \in K(s) = \sigma(s) \backslash H(s)$.则 $C_{G'}(x) = C_G(\sigma(s))$,且 $C_G(x)$ 的外尔群为 $\langle s \rangle$ (命题 20).为了证明 $C_G(\sigma(s))$ 连通,只需证明 $\tilde{s}$ 属于 $C_{G'}(x)$ 的单位分支(命题 12).这是因为 $\tilde{s}$ 包含在 G' 的某个环面 T' 内,

那么 T' 是 G' 的包含 $\tilde{s}$ 及 $x=\tilde{s}^2$ 的连通交换子群.

引理 10 若 $F(s)\neq\sigma(s)$,则 $C_G(F(s))$ 不连通.

证明 若 $F(s)\neq\sigma(s)$,则 $K(s)=H(s)$,因此可以选 $N_G(T)$ 中的元 $\tilde{s}$ 使得 $\tilde{s}$ 投影到 $s\in W$ 且 $x=\tilde{s}^2=e$. 由于 $H(S)$ 是可除群,因此作为交换群的入射 $Z-$ 模,于是有交换群同构

$$T\cong H(s)\times T/H(s)\cong H(s)\times \mathbf{R}/\mathbf{Z}$$

因为 $\sharp(F(s)/H(s))=2$,立即得知 $F(s)$ 包含 T 中所有的 2 阶元. 特别地,$C_G(F(s))\bigcap T$ 中所有 2 阶元都在中心里. 然而,$\tilde{s}\in C_G(F(s))$,$\tilde{s}\notin T$,$\tilde{s}$ 的阶为 2. 显然,$\tilde{s}$ 在 $C_G(F(s))$ 中不共轭于 T 的任何元,这样由命题 13 的注及命题 20 得知 $\tilde{s}$ 不属于 $C_G(F(s))$ 的单位分支.

定理 10 的证明 由定理 6,W' 是 W 的由反射生成的子群,剩下的是要确定 W 中的哪些反射 s 的确属于 W'.

设 $A\subset\sigma(s)$,由引理 9,在 $N_G(T)$ 中有 s 的代表元 $\tilde{s}$,它属于连通群 $C_G(\sigma(s))$,从而属于 G' 的单位分支,于是 $s\in W'$.

设 $A\not\subset\sigma(s)$,若 $A\not\subset F(s)$,则 s 甚至不属于 $C_G(A)$ 的外尔群(命题 20),因此 $s\notin W'$. 若 $A\subset F(s)$,则 $H(s)=\sigma(s)\subsetneqq F(s)$ 且 $\langle A,H(s)\rangle=F(s)$. 若 $s\in W'$,我们要证明这导致矛盾. 若 $s\in W'$,则存在元 $\tilde{s}\in N_{G'}(T)\subset N_G(T)$ 使得 $\tilde{s}$ 投影到 $s\in W$. 由于 $\tilde{s}$ 属于 $H(s)$ 的中心化子,$\tilde{s}$ 属于(命题 19)$C_G(F(s))$ 的连通子群 $C_{G'}(H(s))$. 根据引理 10,这不可能.

定理 9 的证明 令 $C=\bigcap_s\sigma(s)$,Z 为 G 的中心,由命题 17,$Z\subset T$. 设 A 是 T 的一个子群,若 $A\not\subset C$,则 $C_G(A)$ 的单位分支的外尔群真包含在 W 中(定理 10),因此 $A\not\subset Z$. 另外,由定理 10,$C_G(C)$ 的单位分支的外尔群包含 W 中所有的反射,由定理 6,这个外尔群等于 W,因此,由定理 8,$C_G(C)=G$ 且 $C\subset Z$. 由此推得 $C=Z$.

注意 令 $\Sigma=\bigcup_s\sigma(s)$,其中 s 跑遍 W 中所有反射,这是 T 的奇异子集. 读者如果愿意的话,可自己证明 Σ 是 T 中那些至少包

含在 G 的两个极大环面中的元的集合;等价地
$$T \setminus \Sigma = \{x \in T \mid C_G(x) \text{ 的单位分支为 } T\}$$

9 乘积分解

令 G 是紧李群,其极大环面和外尔群分别为 T 和 W. W 自然地作用在有限生成自由交换群 $L = \pi_1 T$ 上;从现在起,我们称这个 $W-$ 模为 G 的对偶权格(这个名称来自表示论,在那里模 $\mathrm{Hom}(L, \mathbf{Z})$ 被称为 G 的权格). 若 G 可以写成乘积形式 $\prod_i G_i$,其中每个 G_i 是连通紧李群,其外尔群和对偶权格分别为 W_i 和 L_i,则 W 同构于乘积群 $\prod_i W_i$,L 同构于乘积模 $\prod_i L_i$. 换句话说,群 G 的乘积分解反映为 $W-$ 模 L 的乘积分解. 我们称 L 的这种乘积分解被 G 的乘积分解所实现.

定理 11 设 G 是连通紧李群,其极大环面、外尔群以及对偶权格分别为 T, W 以及 L,则 L 作为 $W-$ 模的任何乘积分解可被 G 的乘积分解所实现.

第一步是在一个特殊情形,此时要找的某一个因子是环面,证明一个类似于定理 11 的结果.

命题 21 令 G 是连通紧李群,其外尔群及对偶权格分别为 W 和 L. 设 $L \cong L_1 \times L_2$ 是 L 作为 $W-$ 模的乘积分解,且 W 在 L_2 上的作用平凡. 则 L 的这个乘积分解可被 G 的唯一乘积分解 $G \cong G_1 \times G_2$ 所实现. 在这个分解中因子 G_2 是一个环面.

注意 命题 21 中的"唯一性"说的是 G_1 和 G_2 作为 G 的子群唯一确定.

引理 11 设 G 是连通紧李群,其极大环面、外尔群和对偶权格分别为 T, W 和 L. 则置入 $T \to G$ 诱导出映射
$$H_0(W; L) = H_0(W; \pi_1 T) \to \pi_1 G$$
这个映射是满射,且其核为有限群.

证明 设 $p: \widetilde{G} \to G$ 是 G 的连通有限复迭空间. 给定单位元 $e \in G$ 上的一个点 $\widetilde{e} \in \widetilde{G}$,在 $\widetilde{G}$ 上存在唯一的李群结构,使得 $\widetilde{e}$ 是

单位元,且 p 是同态(这可由复迭空间的初等性质推出). p 的核 K 是 $\widetilde{G}$ 的一个正规子群,因此 K 包含在 $\widetilde{G}$ 的中心里(引理 7),从而包含在 $\widetilde{G}$ 的任一极大环面里(命题 17). 商群 $\widetilde{T}/K$ 是 G 的极大环面(参见命题 18 的证明),因此,差一个共轭,我们可假设 $\widetilde{T}/K=T$. 则 $p^{-1}(T)=\widetilde{T}$,因此,特别地,p 限制为 T 的一个连通复迭映射. 然而 $\pi_1 G=\pi_2 BG$ 是有限生成交换群,因此若商群 $\pi_1 G/\mathrm{Im}(\pi_1 T)$ 非平凡,则从 $\pi_1 G$ 到一个有限群有一个非平凡满同态,它限制在 $\pi_1 T$ 上为平凡同态. 这样一个同态将导致 G 的一个连通有限复迭空间,使得它限制到 T 上为平凡(乘积)复迭. 由于不存在 G 的这种复迭空间,所以自然映射 $u:L=\pi_1 T\to\pi_1 G$ 必定是满射.

由胡雷维奇定理和同调纬悬定理,u 可等同于 $BT\to BG$ 诱导出的映射 $H_2(BT;\mathbf{Z})\to H_2(BG;\mathbf{Z})$,$N_G(T)$ 的元素共轭作用在 T 上,T 的这些自同构可扩充为 G 的内自同构;由于这些内自同构在 $H_*(BG;\mathbf{Z})$ 上的作用是平凡的,u 可扩充为一个满射

$$\bar{u}:H_0(W;L)\cong H_0(W;H_2(BT;\mathbf{Z}))\to$$
$$H_2(BG;\mathbf{Z})\cong\pi_1 G$$

由于对偶映射 $H^2(BG;\mathbf{Q})\to H^2(BT;\mathbf{Q})^W$ 为满射(定理 5),因此 $\bar{u}$ 的核为有限群.

现在我们来看看如何理解连通紧李群到 $\mathbf{R}/\mathbf{Z}$ 的同态.

命题 22 设 G 是连通紧李群,S^1 是圆周 $\mathbf{R}/\mathbf{Z}$,则自然映射

$$\alpha:\mathrm{Hom}(G,S^1)\to\mathrm{Hom}(\pi_1 G,\pi_1 S^1)=$$
$$\mathrm{Hom}(\pi_1 G,\mathbf{Z})=H^1(G;\mathbf{Z})$$

是一个双射.

注意 命题 22 在某种意义上是命题 1 的推广. 我们必须在这里证明这一点,但对 $p-$ 紧群的类似结论其证明却很简单:$p-$ 紧群之间的同态是其分类空间之间的映射,若目标空间是 $p-$ 完备圆周,后者等于上同调类.

命题 22 的证明依赖于德拉姆上同调理论. 称光滑流形 M 上

227

的闭 1 一形式 ω 是整的,若对 M 中的任何分段光滑回路 γ,有
$$\int_\gamma \omega \in \mathbf{Z}.$$

引理 12 设 G 是连通紧李群,令 S^1 为圆周群(单位元为 1),而 $\mathrm{d}\theta$ 为 S^1 上通常的左不变 1 一形,则对应 $f \mapsto f^*(\mathrm{d}\theta)$ 给出了双射

$$\{ f : G \to S^1 \mid f \text{ 光滑}, f(e)=1 \} \to$$
$$\{ G \text{ 上的闭整 } 1 \text{ 一形式} \}$$

满足 $f(e)=1$ 的光滑映射 $f : G \to S^1$ 是同态 $\Leftrightarrow f^*(\mathrm{d}\theta)$ 在 G 中元素的左平移作用下不变.

证明 给定 G 上整闭 1 一形式 ω,定义
$$\alpha : G \to S^1 = \mathbf{R}/\mathbf{Z}$$

为 $\alpha(g) = \int_\gamma \omega$,其中 γ 是 G 上从单位元到 g 的任一分段光滑道路.

映射 α 是定义好的,因为对 G 中的任何回路 γ,有 $\int_\gamma \omega \in \mathbf{Z}$. 那么 $\alpha(e)=1$ 且 $\alpha^*(\mathrm{d}\theta)=\omega$. 若 $\beta : G \to S^1$ 是任一满足 $\beta^*(\mathrm{d}\theta)=\omega$ 的光滑函数,则 $(\alpha/\beta)^*(\mathrm{d}\theta)=0$,从而 α/β 是常值函数;若 $\beta(e)=1$,则 $\beta=\alpha$.

若 $f : G \to S^1$ 是同态,则 $f^*(\mathrm{d}\theta)$ 是 G 上的左不变闭整 1 一形式,因为 $\mathrm{d}\theta$ 是 S^1 上的左不变闭整 1 一形式. 反之,设 ω 是 G 上左不变闭整 1 一形式. 令 g 和 g' 为 G 的两个元素,而 γ, γ' 是两条分别从 g, g' 到单位元的道路,定义 $\gamma + g\gamma'$ 为将 γ 接上 $g \cdot \gamma'$ 而得到的道路,它是从 gg' 到单位元的一条道路.则(α 如前)

$$\alpha(gg') = \int_{\gamma + g\gamma'} \omega$$
$$= \int_\gamma \omega + \int_{g\gamma'} \omega$$
$$= \alpha(g) + \alpha(g')$$

最后一个等式成立是因为 ω 是左不变的.特别地,α 是同态.

注意 作为练习,请验证两个映射 $f_1, f_2 : G \to S^1$ 同伦 $\Leftrightarrow$ 1 一形式 $f_1^*(\mathrm{d}\theta)$ 和 $f_2^*(\mathrm{d}\theta)$ 代表相同的上同调类.

引理 13 设连通紧李群 G 光滑作用于紧闭流形 M 上,ω 是

M 上的闭形式. 则存在闭形式 ω', 使得 ω 和 ω' 代表相同的上同调类, 而且在 G 左平移作用下不变.

证明概要 在流形 M 上有一个 G 不变黎曼度量. 由霍奇理论, ω 代表的上同调类有一个这样的代表元 ω', 使得 ω' 相对于从这个度量导出的拉普拉斯—贝尔特拉米算子而言是调和的. 由于 G 连通, $\forall g \in G$, 平移 L_g 在 M 的上同调上的作用平凡, 因此 ω' 与 $L_g^*(\omega')$ 代表相同的上同调类. 因为 L_g 是等距的, $L_g^*(\omega')$ 是调和的, 由 ω' 的唯一性, 得 $L_g^*(\omega') = \omega'$.

利用 G 上的 Haar 测度, 也可以直接从形式 $L_g^*(\omega)$, $g \in G$, 通过平均法 (即对哈尔测度的积分) 构造适当的 ω'.

命题 22 的证明 我们首先证明 α 是单射. 令 f_1 和 f_2 是两个不同的同态 $G \to S^1$, 由引理 12, 1—形式 $\omega_i = f_i^*(\mathrm{d}\theta)$, $i = 1, 2$ 是 G 上不同的闭左不变 1—形式. 差 $\omega = \omega_1 - \omega_2$ 就是 G 上的非 0 闭左不变 1—形式. 形式 ω 的上同调类非 0, 这是因为 G 上的任一光滑函数 f 在某点 $x \in G$ 取得极大值, 但 $(\omega)_x \neq 0 = (\mathrm{d}f)_x$. 故 ω_1 和 ω_2 的同调类不同, 于是 f_1 和 f_2 不同伦.

以下设 $f: G \to S^1$ 是一个映射, 我们必须证明 f 同伦于一个同态. 我们可以假设 f 光滑, 并令 $\omega = f^*(\mathrm{d}\theta)$. 由引理 12, 只要 ω 与某个左不形变式代表相同的上同调类. 这是引理 13 的特殊情形.

命题 21 的证明 令 T 是 G 的极大环面, 且
$$T_i = (\mathbf{R} \otimes L_i)/L_i \quad (i = 1, 2)$$
则 $T_G \cong T_1 \times T_2$ (参见命题 1 的证明). 由假设, W 在 T_2 上的共轭作用平凡 (命题 1). 由引理 11, 置入 $T \to G$ 诱导出满同态
$$H_0(W; L) \cong H_0(W; L_1 \times L_2) \cong H_0(W; L_1) \times L_2 \to \pi_1 G$$
且其核为有限群. 特别地, 该核包含在 $H_0(W; L_1)$ 中. 令 $c \in H^1(G; L_2) = \mathrm{Hom}(\pi_1 G, L_2)$ 是满足以下条件的唯一的上同调类: 它限制在 $H_0(W; L_1)$ 上为 0, 而限制在 L_2 上为恒同映射. 取 L_2 作为 $\mathbf{Z}$ 上的一组基, 使得可把 c 表示为 $H^1(G; \mathbf{Z})$ 的一组元素 $c_1, \cdots, c_{r_2}$, 或者甚至表示为 G 上的一组整闭 1—形式 $\omega_1, \cdots, \omega_{r_2}$. 由命题 22, 可假设每个形式 ω_i 左不变. 对这些形式积分 (引理

12) 给出同态 $G \to T_2$, 这些同态限制在 T_2 上为恒同映射(命题 1). 另外, 由定理 9 的注, 群 T_2 包含在 G 的中心里, 因此存在商同态 $G \to G/T_2$. 诱导映射 $T/T_2 \to G/T_2$ 是 G/T_2 的极大环面(参见命题 18 的证明). 乘积映射 $f: G \to T_2 \times G/T_2$ 是单射, 这是因为 $\mathrm{Ker}(f)$ 不能包含 T 的任何元素.

因为 f 的定义域以及取值的域维数相同, f 是一个同构. (也可证明 $(\mathrm{d} f)_e$ 可逆, 导出 f 在拓扑上是复迭映射. 再由于 f 单, 从而 f 是微分同胚.) 这给出 G 的所要求的分裂.

要验证唯一性, 设 $G = G/T_2 \times T_2 \cong G_1 \times G_2$ 是诱导出 L 的给定分解的一个乘积分解. 首先注意 $G_2 = T_2$. (任一适当的因子 G_2 必定包含 T_2, 因而等于 T_2(定理 8), 这是因为, 假设 W 平凡作用于 L_2 上, G_2 的外尔群平凡.) 选定因子 G_1 等于是选定了投影
$$G \cong (G/T_2) \times T_2 \to G/T_2$$
的一个截面, 并且令 $G = s(G/T_2)$. 要求乘积分解 $G \cong G_1 \times T_2$ 与 L 的乘积分解相容相当于要求 s 限制在 $T_1 \subset G/T_2$ 就是给定的置入 $T_1 \subset G$. 然而, 给定这样一个截面等于是给定一个同态 $\pi_1(G/T_2) \to \pi_1 T_2$(比较命题 22), 由于任何这样的同态由它在 $\pi_1 T_1$ 上的限制决定(引理 11), 这就推出了唯一性, 因为这时唯一符合要求的同态是平凡同态.

要从命题 21 推出定理 11, 需要了解环面的正规化子中的共轭以及群本身中的共轭之间的关系.

引理 14　设 G 是连通紧李群, 其极大环面为 T. 令 N 为正规化子 $\mathrm{N}_G(T)$, A 和 B 是 T 的子群. 若 A 和 B 在 G 中共轭, 则它们在 N 中也共轭. 特别地, T 的两个元素若在 G 中共轭, 那它们在 N 中也共轭.

证明　设 $g \in G$ 使得 $gAg^{-1} \subset B$. 环面 T 和 gTg^{-1} 都包含 B, 因此都包含在 $\mathrm{C}_G(B)$ 的单位分支里. 因为 T 和 gTg^{-1} 都是 G 中的极大环面, 它们也是 $\mathrm{C}_G(B)$ 中的极大环面. 由定理 4, 存在 $h \in \mathrm{C}_G(B)$ 使得 $h(gTg^{-1})h^{-1} = T$. 这里 $hg \in N$.

引理 15　令 G 是连通紧李群, 其极大环面为 T, 外尔群为 W, 对偶权格为 L. 设

$$\mathbf{Q}\otimes L\cong M_1\oplus\cdots\oplus M_n$$

是 $\mathbf{Q}\otimes L$ 作为 W－模的一个直和分解．令 W_i 是 W 中平凡地作用在$M_1\oplus\cdots\oplus\hat{M}_i\oplus\cdots\oplus M_n$ 上的元素所生成的子群．则：

(1) 当 $i\neq j$ 时，W_i 和 W_j 交换；

(2) 乘积映射 $W_1\times\cdots\times W_n\to W$ 是一个群同构．

证明　由定义，显然，若 $x\in W_i$，$y\in W_j$，$i\neq j$，则交换子$[x,y]\in W_i\cap W_j$，因而$[x,y]$平凡地作用在 L 上．由于W 在L 上的作用是忠实的(定理 5)，元素$[x,y]$是平凡的．简单地计算一下秩可知，若 $s\in W$ 是反射(定义 5)，则对某个 i，有 $s\in W_i$．(令 s_i 为 s 在 M_i 上的限制．由于 s 是反射，$s-1$ 的秩为 1，另外，它又等于 $\sum_i \mathrm{rk}(s_i-1)$．显然除了某一个 i 值之外，其他所有 s_i 都是恒同映射．)由于W 由反射生成(定理6)，(2) 中的乘积映射是满射，显然它还是单射．

定理 11 的证明　由归纳法，只需考虑两个因子的分解，$L\cong L_1\times L_2$．问题是要构造群分裂 $G\cong G_1\times G_2$．自然同构 $T\cong\mathbf{R}/\mathbf{Z}\otimes L$(见命题 1 的证明)给出分裂 $T\cong T_1\times T_2$，其中 $T_i\cong\mathbf{R}/\mathbf{Z}\otimes L_i$．令 H' 是 T_2 在G 中的中心化子，K' 是 T_1 的中心化子．记商群 H'/T_2 和 K'/T_1 分别为 H 和K．这些群连通(命题 19)，它们可组织在交换图 1 中．

$$\begin{array}{ccccccccc}
T_2 & \xrightarrow{k_2} & T & \xrightarrow{=} & T & \xleftarrow{=} & T & \xleftarrow{k_1} & T_1\\
{\scriptstyle =}\downarrow & & \downarrow{\scriptstyle i_{H'}} & & \downarrow{\scriptstyle i} & & \downarrow{\scriptstyle i_{K'}} & & \downarrow{\scriptstyle =}\\
T_2 & \xrightarrow{k_2'} & H' & \longrightarrow & G & \longleftarrow & K' & \xleftarrow{k_1'} & T_1\\
 & & \downarrow{\scriptstyle q_{H'}} & & & & \downarrow{\scriptstyle q_{K'}} & & \\
 & & H=H'/T_2 & & & & K=K'/T_1 & &
\end{array}$$

图 1

群 T 是 H' 和 K' 的极大环面．复合 $T_1\to T\to H'\to H$ 是极大环面的置入，对应的复合 $T_2\to K$ 也是一样(参见命题 18 的证明)，为简单起见，记 T_1 为 T_H，T_2 为 T_K．

格 $L = \pi_1 T$ 作为 H' 的外尔群 $W_{H'}$ 上的模可分解为乘积 $L_1 \times L_2$. 由于 H' 定义为中心化子,$W_{H'}$ 平凡地作用在因子 $L_2 = \pi_1 T_2$ 上. 由命题 21,存在唯一的同态 $H \to H'$,使得它是 $q_{H'}$ 的右逆,而限制在 T_H 上与已有同态 $T_H \to H'$ 相同;对 $g_{K'}$ 有一个类似的截面. 这些截面给出乘积分解 $H' \cong H \times T_K$ 与 $K' \cong K \times T_H$. 将这些截面与置入 $H' \to G$ 和 $K' \to G$ 复合得到同态 $H \to G$ 和 $K \to G$,这样使得我们可把 H 和 K 当作 G 的子群. H 的外尔群 W_H 同构于 W_G 的限制在 T_K 上为恒同的元素的子群;类似地,W_K 同构于 W_G 的限制在 T_H 上为恒同的元素的子群. 这两个子群互相可换,且乘积映射 $W_H \times W_K \to W_G$ 是群同构(引理 15).

由引理 14,T_H 的任一不等于单位元的元素在 G 中不能共轭于 T_k 的一个元素,因此由命题 13 的注,H 的任一不等于单位元的元素在 G 中不能共轭于 K 中的一个元素. 对此的另一种表达方式是说 H 在 G/K 上的左作用是自由的,这意味着双陪集空间 $H\backslash G/K$ 是流形,且存在主纤维丛

$$H \times K \to G \to H\backslash G/K$$

计算维数(引理 6 的注,引理 6) 得知 $\dim(H \times K) = \dim(G)$,因此 $\dim(H\backslash G/K) = 0$,即 $H\backslash G/K$ 是一个点,且乘法映射 $H \times K \to G$ 是微分同胚. 由此易知自然映射

$$H \cong (H \times T_K)/T_K = H'/T_K \to G/K \cong H$$

也是微分同胚. 很明显这个微分同胚关于 T_K 在 H'/T_K 及 G/K 上的左作用等变. 由于 T_K 在 H'/T_K 上的左作用平凡,从而它在 G/K 上的作用也平凡. 那么 K 在 G/K 上的左作用的核是 K 的包含 T_K 的一个正规子群,从而必定等于 K 自身. 换句话说,K 是 G 的正规子群;这样立即得知复合同态 $H \to G \to G/K$ 是同构. 对称地,H 在 G 中正规,由此易知乘积映射 $G \to G/H \times G/K \cong H \times K$ 是同构. 容易验证 G 的这个乘积分解诱导出 L 原来的分裂.

10　分裂为单因子的乘积

一个紧李群称为单群,如果它没有非平凡的闭正规子群. 这部分我们要证明下面的定理.

定理 12　中心平凡的任何连通紧李群同构于单紧李群的乘积.

利用类似的方法可证明若干相关的结果. 比如, 称一个连通紧李群为几乎单的, 若它的中心是有限群. 则对任何连通紧李群, 它的某一个有限复迭同构于乘积 $T \times G'$, 其中 T 是环面, 而 G' 是几乎单群的乘积. 特别地, 任何单连通紧李群同构于几乎单群的乘积.

为了从定理 11 推出定理 12, 我们需要以下两个辅助结果.

命题 23　令 G 是连通紧李群, 其外尔群和对偶权格分别为 W 和 L. 则 G 是单群, 当且仅当 G 的中心平凡, 且 $\mathbf{Q} \otimes L$ 是不可约 $\mathbf{Q}[W]$-模.

命题 24　令 G 是连通紧李群, 其外尔群和对偶权格分别为 W 和 L, 并设 G 的中心平凡. 则 $\mathbf{Q} \otimes L$ 可唯一地分裂为不可约 $\mathbf{Q}[W]$-子模的 (内) 直和

$$\mathbf{Q} \otimes L \cong M_1 + \cdots + M_n$$

令 $L_i = L \cap M_i$, 则 $\mathbf{Q} \otimes L \cong M_i$, 且加法映射

$$\alpha : L_1 \times L_2 \times \cdots \times L_n \to L$$

是 W-模同构.

定理 12 的证明　假设已证明了命题 23 和命题 24. 令 G 是连通紧李群, 其外尔群和对偶权格分别为 W 和 L. 由命题 24, L 分裂为 W-模的内直和 $L_1 + L_2 + \cdots + L_n$, 其中 W 在 $\mathbf{Q} \otimes L_i$ 上的作用是不可约的. 由定理 11, 这个分裂可实现为 G 的一个乘积分解 $\prod_i G_i$. 特别地, W 分裂为乘积 $\prod_i W_i$, 其中 W_i 是 G_i 的外尔群, 而 L_i 是 G_i 的对偶权格. W 在 L_i 上的作用可用投影 $W \to W_i$ 分解, 即 W_i 在 $\mathbf{Q} \otimes L_i$ 上的作用不可约, 再由命题 23, 得 G_i 是单群.

命题 23 的证明　由定理 11 和命题 24 易知任意单群 G 具有所述性质. 反之设 G 具有所述性质, 而 K 是 G 的闭正规子群. 若 K 是离散群, 则由紧性, K 是有限群, 再由引理 7, K 包含在 G 的中心里, 由命题的条件, K 平凡. 然后假设 K 的单位分支非平

凡,因此有一个非平凡的极大环面 T_K(定理8).可假设 $T_K \subset T$,其中 T 是 G 的极大环面(引理2).显然 T_K 是 $T \bigcap K$ 的单位分支,否则这个单位分支就是 K 中真包含 T_K 的连通交换子群,这不可能.令 $L_K = \pi_1 T_K$,因为 K 在 G 中正规,W 在 T 上的共轭作用将 T_K 映到自身,因此 $\mathbf{Q} \bigotimes L_K$ 是 $\mathbf{Q} \bigotimes L$ 的(非平凡)$W-$不变子空间.从 $\mathbf{Q} \bigotimes L$ 的不可约性推出

$$\mathbf{Q} \bigotimes L_K = \mathbf{Q} \bigotimes L$$

令 T' 为商群 T/T_K,使得有短正合列 $0 \to L_K \to L \to \pi_1 T' \to 0$.对它张量 $\mathbf{Q}$ 得 $\mathbf{Q} \bigotimes \pi_1 T' = 0$,因此由于 T' 是环面,T' 是平凡群.换句话说,$T_K = T, T \subset K$,由于 K 在 G 中正规且 G 中的每个元共轭于 T 中的一个元(命题13的注),$K = G$.

命题 24 的证明   由初等表示论知,$\mathbf{Q} \bigotimes L$ 可写成不可约 $\mathbf{Q}[W]-$子模 $\{M_i\}_{i=1}^n$ 的直和.从短正合列

$$0 \to L \to \mathbf{R} \bigotimes L \to T \to 0$$

得正合列

$$0 \to L^W \to (\mathbf{R} \bigotimes L)^W \to T^W \to \cdots$$

由于 G 的中心平凡,T^W 的单位分支平凡.这说明 $(\mathbf{R} \bigotimes L)^W = 0$,因此 $(\mathbf{Q} \bigotimes L)^W$ 平凡,因此 W 的平凡表示不可能出现在 $\mathbf{Q} \bigotimes L$ 的和项 $\{M_i\}$ 中.特别地,对 1 和 n 之间的每个 k,存在 W 中的反射 s_k,使得 s_k 非平凡地作用在 M_k 上(定理6).由反射的定义(定义5),M_k 在 $\{M_i\}$ 中在同构类的意义下出现且只出现一次.再利用初等表示论,因为 $\mathbf{Q} \bigotimes L$ 的不可约 $\mathbf{Q}[W]-$子模因子在 $\mathbf{Q} \bigotimes L$ 中只出现一次,$\mathbf{Q} \bigotimes L$ 分解为不可约子模的直和是唯一的.

下面考虑命题的断言中所构造的映射 α.由定义,α 是 $W-$模的映射.由于 α 可嵌入到下面的交换图 2 里,其中所有其他的映射都是单射,显然 α 也是单射.令 $D = \operatorname{coker}(\alpha)$,若 $x \in \mathbf{Q} \bigotimes L$,则 x 的某个倍数落在 L 中,这就是说每个商群 M_i/L_i 是挠群,因此 D 是挠群.由同调代数知有短正合列

$$0 \to \operatorname{Tor}(\mathbf{Q}/\mathbf{Z}, D) \to$$

$$\bigoplus_i (\mathbf{Q}/\mathbf{Z} \bigotimes L_i) \xrightarrow{\mathbf{Q}/\mathbf{Z} \bigotimes \alpha} \mathbf{Q}/\mathbf{Z} \bigotimes L \to 0$$

其中 $\operatorname{Tor}(\mathbf{Q}/\mathbf{Z}, D)$ 同构于 D,因此要证明 $D = 0$,只要证明

$\mathbf{Q}/\mathbf{Z}\otimes D$ 是单射. 由 L_i 的定义, 置入 $L_i \to L$ 的余核无挠, 这意味着映射 $\mathbf{Q}/\mathbf{Z}\otimes L_i \to \mathbf{Q}/\mathbf{Z}\otimes L$ 是单射, 这样 $\mathbf{Q}/\mathbf{Z}\otimes L_i$ 可看作 $\mathbf{Q}/\mathbf{Z}\otimes L$ 的子群.

$$
\begin{array}{ccc}
L_1 \times \cdots \times L_n & \longrightarrow & M_1 \times \cdots \times M_n \\
\alpha \downarrow & & \cong \downarrow \\
L & \longrightarrow & \mathbf{Q}\otimes L
\end{array}
$$

图 2

令 $W_i, i = 1, \cdots, n$, 像引理 15 那样, 使得 W 的每个反射落在唯一的 W_i 中. 取 $\boldsymbol{x} \in \ker(\mathbf{Q}/\mathbf{Z}\otimes\alpha)$, 使得 $\boldsymbol{x} = (x_1, \cdots, x_n)$, 其中 $x_i \in \mathbf{Q}/\mathbf{Z}\otimes L$ 且 $\sum x_i = 0$. 假如我们能证明 $\boldsymbol{x}$ 的每一分量 x_k 全为 0, 那么我们就完成证明了. 我们把 $\mathbf{Q}/\mathbf{Z}\otimes L$ 与 $\mathbf{R}/\mathbf{Z}\otimes L \cong T$ 中的挠元素的子群等同. 考虑 $\boldsymbol{x}$ 的某个分量 x_k, 并选反射 $s \in W$ 使得 $s \in W_i$. 若 $i \neq k$, 则 x_k 属于 $F(s)$ 的可除子群 $\mathbf{Q}/\mathbf{Z}\otimes L_k$, 因此 $x_k \in H(s)$. 若 $i = k$, 则 $x_k = -\sum\limits_{j \neq k} x_j$ 属于 $\sum\limits_{j \neq k} \mathbf{Q}/\mathbf{Z}\otimes L_j$, 因此 x_k 也属于 $F(s)$ 的一个可除子群, 从而也有 $x_k \in H(s)$. 结论是 $x_k \in \bigcap\limits_s H(s)$, 其中相交取遍 W 的所有反射 s. 由于这个交集 $\bigcap\limits_s H(s)$ 包含在 G 的中心里 (定理 9), 故它平凡, 从而 x_k 为 0.

11 交换代数中的一个定理

设 k 是一个域, S 是 k 上 s 个不定元的连通非负分次多项式代数, R 是 S 的分次 k-子代数, R 同构于 k 上 r 个不定元的多项式代数. 再设 S 是有限生成 R-模. 我们要证明 $r = s$ 且 S 实际上是自由 R-模; 从这可推出命题 16. 下面的证明是 J. 莫尔给出的, 并由 S. Halperin 解释给我们. 同样的结果对非分次的情形也成立, 但证明更复杂, 其证明要用到奥斯兰德 (Auslander) 和布赫斯包姆 (Buchsbaum) 关于深度 (depth) 和投影维数之间关系的定理.

令 $T = k[x_1, \cdots, x_t]$ 是 k 上连通非负分次多项式代数 (例如, $T = S$, $T = R$ 或 $T = R\otimes_k S$). 我们规定 k 通过自然映射 $T \to$

$T_0 = k$ 看作 T - 模. 令 $K(i)$ 是 $k[x_i]$ 上的链复形, 使得零维是 $k[x_i]$, 而一维也是 $k[x_i]$ (但现在的分次是原来的分次减掉 $|x_i|$); 微分是 $K(i)_1 \to K(i)_0$ 乘以 x_i. 易知 $K(i)$ 是 k 在 $k[x_i]$ 上的自由分解. 因此 $K = K(1) \bigotimes_k \cdots \bigotimes_k K(t)$ 是 k 在 T 上的自由分解. 我们有以下标准的引理.

引理 16 令 M 是 T 上的模, 则:

(1) $\mathrm{Tor}_0^T(k, M) = M/(x_1 M + \cdots + x_t M)$;

(2) $\mathrm{Tor}_t^T(k, M) = \{y \in M \mid x_1 y = \cdots = x_t y = 0\}$;

(3) $\mathrm{Tor}_i^T(k, M) = 0$, $i > t$.

证明 通过分解 K 直接验证即可.

引理 17 令 M 是 T 上的分次模, 且设 M 在负分次上皆为 0, 则:

(1) M 为 $0 \Leftrightarrow \mathrm{Tor}_0^T(k, M) = 0$;

(2) M 自由 $\Leftrightarrow \mathrm{Tor}_1^T(k, M) = 0$.

证明 若 M 不为 0, 则令 $m \in M$ 是最小次数的非零元. 从引理 16 立即知道 m 代表 $\mathrm{Tor}_0^T(k, M)$ 中一个非零元.

设 $\mathrm{Tor}_1^T(k, M) = 0$, 选取

$$\mathrm{Tor}_0^T(k, M) = M/(x_1 M + \cdots + x_t M)$$

的一组分次基 $\{\overline{y_\alpha}\}$, 并选取 $\{y_\alpha\} \in M$ 使得在映射 $M \to M/(x_1 M + \cdots + x_t M)$ 下 $\{y_\alpha\}$ 为 $\{\overline{y_\alpha}\}$. 令 F 是由 $\{y_\alpha\}$ 生成的自由 T - 模, 并考虑显然的映射 $f: F \to M$. Tor 的左正合性给出正合列

$$\mathrm{Tor}_0^T(k, T) \xrightarrow{\mathrm{Tor}_0^T(k, f)} \mathrm{Tor}_0^T(k, T) \longrightarrow$$
$$\mathrm{Tor}_0^T(k, \mathrm{coker}(f)) \longrightarrow 0$$

由构造, $\mathrm{Tor}_0^T(k, f)$ 是一个同构, 因而

$$\mathrm{Tor}_0^T(k, \mathrm{coker}(f)) = 0$$

从而, 如上面的证明, $\mathrm{coker}(f) = 0$, 于是 f 为满射. 从正合列

$$\mathrm{Tor}_1^T(k, M) \longrightarrow \mathrm{Tor}_0^T(k, \ker(f)) \longrightarrow$$
$$\mathrm{Tor}_0^T(k, F) \xlongequal{\cong} \mathrm{Tor}_0^T(k, M)$$

我们得到 $\mathrm{Tor}_0^T(k, \ker(f)) = 0$, 从而 $\ker(f) = 0$, 进而 f 为同构.

引理 18 设 M 是 $R \otimes_k S = R \otimes S$ 上的模，则有两个同调型的第一象限的谱序列

$$E_{i,j}^2 = \mathrm{Tor}_i^R(k, \mathrm{Tor}_j^S(k, M)) \Rightarrow \mathrm{Tor}_{i+j}^{R \otimes S}(k, M)$$

$$E_{i,j}^2 = \mathrm{Tor}_j^S(k, \mathrm{Tor}_i^R(k, M)) \Rightarrow \mathrm{Tor}_{i+j}^{R \otimes S}(k, M)$$

证明 这些是关联于同构

$$k \otimes_S (k \otimes_R M) \cong k \otimes_{R \otimes S} M \cong k \otimes_R (k \otimes_S M)$$

的复合函子谱序列. 要记住的是在我们的情形，所有的环都是交换的，因此区分左模和右模并不重要.

命题 16 的证明 令 $\{n_k\}$ 是 S 的多项式生成元的次数，$\{m_k\}$ 是 R 的生成元的次数. 那么庞加莱级数 $P_S(t) = \sum_i \dim_k(S_i) t^i$ 是 $\prod_{k \leqslant s} (1 - t^{m_k})^{-1}$，而

$$P_R(t) = \prod_{k \leqslant r} (1 - t^{n_k})^{-1}$$

令 N 是 S 作为 R—模的生成元的个数. 显然对 $0 < t < 1$，有不等式

$$P_R(t) \leqslant P_S(t) \leqslant N P_R(t)$$

由于当 $t = 1$ 时函数 $P_S(t)$ 有 s 阶的极点，而函数 $P_R(t)$ 有 r 阶的极点，对上面的不等式做一简单的估计得出 $r = s$.

下面来考虑引理 18 中的两个谱序列，我们要考虑 $M = S$ 的特殊情形，R 和 S 由乘法作用在 M 上. 第一个谱序列退化（因为 $M = S$ 是自由 S—模），且（参见引理 16(3)）有

$$\mathrm{Tor}_i^{R \otimes S}(k, S) = 0 \quad (i > r) \tag{1}$$

令 q 为使得 $\mathrm{Tor}_q^R(k, S) \neq 0$ 的最大整数. 由引理 17，我们只需证明 $q = 0$. 引理 18 的第二个谱序列只在一个长方形（$0 \leqslant i \leqslant s$，$0 \leqslant j \leqslant q$）里不为 0，在这个长方形的右上角的 E^2 项是 $E_{s,q}^2 = \mathrm{Tor}_s^S(k, \mathrm{Tor}_q^R(k, S))$. 由于位置的原因，$E_{s,q}^2$ 一直保持不变到 $E_{s,q}^\infty$，对我们来说只需证明 $E_{s,q}^2$ 不为 0；因为由此得知 $\mathrm{Tor}_{s+q}^{R \otimes S}(k, S) \neq 0$，这样由于(1)及上面的等式 $r = s$ 得知 $q \neq 0$ 会导致矛盾.

由于 R 是诺特环，而 S 是有限生成 R—模，$\mathrm{Tor}_q^R(k, S)$ 是有限生成 k—模. 特别地，$\mathrm{Tor}_q^R(k, S)$ 只在有限个内次数上不为 0.

令 q' 为使得 $\mathrm{Tor}_q^R(k,S)_{q'} \neq 0$ 的最大内次数,而 y 是 $\mathrm{Tor}_q^R(k, S)_{q'}$ 的非零元. 由于次数的原因,对 S 的每个多项式生成元 x, $xy = 0$. 由引理 $16(2)$,y 代表了 $E_{s,q}^2$ 的一个非零元.

文献三:《李群论发展中的怪事》①.

1 引言

我和 Joop Kolk 合写了一本李群的书. 这个课题是高度发展的,有很完整的文献可查,其发展历史的主要事实可以根据通常发表于当代的一流刊物(例如 *Mathematische Annalen*)引用的文献来追踪. 当我们发现有几件基本的事实并不像我们认为的那样曾经发生过,这就使我们愈加感到惊讶. 现在使用的术语在一些情形中甚至是完全错误的. 但是,我们并不想改正它,因为它已根深蒂固了. 再说,这些事实大多数数学家都是知道的,所以,我们也不需要装作发现了什么新的东西.

2 李

我还是学生时,就试着读李的一些文章以及他和恩格尔(Engel)合写的三大卷的《变换群》的某些部分,其原因是我想学习相切变换. 然而他那直接的几何推理风格给予我很深的印象.

在他的文章的好几个地方,李对所谓的魏尔斯特拉斯要求,即几何上明显的事实也应该用收敛幂级数的方法加以证明,表示不满. 从这一点来看,令人感到震惊的是在有些文献上向量场经历时间 t 之后的流竟用幂级数

---

① 作者:Hans Duistermaat,原题:*Peculiarities in the Development of the Theory of Lie Groups*. 译自:*Analysis. Et Ceiera*,Research Papers Published in Honor of Jürgen Moser's 60th Birthday,Ed. Paul, H. Rabinowitz and Eduord Zehnder,263-282.

$$e^{t \cdot X} = \sum_{k=0}^{\infty} \frac{t^k}{k!} X^k$$

来给出,这里 X 看成作用在解析函数空间上的一阶线性偏微分算子(一个导子),用这个办法,流在函数空间上的作用被表示成无穷阶的线性偏微分算子. 李把向量场与函数空间上的线性算子等同起来的想法起源于把李括弧与算子的换位子对应起来,后者给出了雅可比恒等式的一个直接证明.

跟他的一贯做法一样没有幂级数的收敛性或算子作用于其上的定义域的讨论. 这是魏尔斯特拉斯(Weierstrass)批评过的异想天开的行径之一,而这种做法在 19 世纪的分析学中是很普遍的,它使得有时很难向近代的读者解释当时的文献. 另外,要说李的群论完全是局部的,也是过于简单化了. 有这样的一个命题:包含单位元的连通分支是子李群. 它只能作为整体性结论来理解.

在 1874 到 1890 年这一时期中,李实际上创造了后来以他的名字命名的理论. 这是已经很好地认识到的. 除了定义了我们现在称为李群与李代数的基本概念外,他还发现了许多基本的定理,例如,李群局部同构,当且仅当它们的李代数是同构的,以及每个抽象的有限维李代数等于一个向量场的李代数(李的第三基本定理). 正如温斯坦(Weinstein)注意到的,关于"伴随群的对偶性"这一章包含了在几何量子化理论中变得十分流行的余伴随轨道的一大部分内容. 必须承认,我没有想起它来. 显然,当我还是学生时,我并没有完全读懂李的书. 最后,我应该提一下李关于可解李代数的特征化工作,它是被他的偏微分方程的伽罗瓦理论推动的. 在这个理论中,伽罗瓦群是作用在解空间上的李群.

李只是在他一生的较晚时才开始了他的数学研究工作. 当时的挪威肯定还不是一个富裕的国家,而作为一个数学家,李当时也还没有什么声誉,然而挪威政府支持他全力从事研究工作,这是非同寻常的! 由于他的大部分论文都是在挪威发表的,他最初的工作只为少数几个数学家知道,其中有克莱因,在 70 年代,李同克莱因合作过.后来,恩格尔进入了李群,作为他的本职工作,他为组织材料提供了帮助.该书使李出了名,并且成了莱比锡的教授,但他却因那里只有很少的学生而十分失望.他的健康恶化了,回到了挪威后不几年他就逝世了,而恩格尔却继续在法国的数学界发挥着影响逾 40 年.

3 弗里德里希·舒尔(Friedrich Schur)

如果 g 是李群 G 的李代数,在单位元与 $X \in g$ 相切的 G 的单参数子群可表示成 $t \to e^{t \cdot X}$. 当 $t = 1$ 时, $e^X \in G$,这就定义了从 g 到 G 的光滑映射 $X \to e^X$,称为指数映射.它的切映射把 g 映到 G 在 e^X 的切空间,以 e^{-X} 右乘,就得到从 g 到 g 的一个线性映射,它由公式

$$Y \mapsto \frac{d}{dt}(e^{X+tY} \cdot e^{-X})\bigg|_{t=0} = \frac{e^{ad\,X}-1}{ad\,X}(Y)$$

给出. 这里 $ad\,X$ 表示从 g 到 g 的线性映射

$$Y \mapsto [X,Y]$$

我们用记号

$$\frac{e^A - 1}{A} := \sum_{k=0}^{\infty} \frac{1}{(k+1)!} A^k$$

表示 $g \to g$ 的线性映射空间中的收敛幂级数. F. 舒尔在 1891 年发现了指数映射的导数公式.随后,他用这个公式在对数坐标中刻画了表示群 G 的乘法的方程

$$e^Z = e^X \cdot e^Y$$

的解 $Z \in g$ 的 X 和 Y 的收敛幂级数.在其他文章中,他也指出了这意味着任意 C^2 群是解析的.

为了证明它,舒尔注意到定义在 g 的原点的某个邻域中的解析向量场

$$X^R : Z \mapsto \frac{ad\,Z}{e^{ad\,Z}-1}(X)$$

表示在原点等于 X 的在对数坐标中右不变的向量场. 由此得到 Z 等于微分方程

$$\frac{dZ}{dt}(t) = X^R(Z(t))$$

在初始条件 $Z(0) = X$ 下, $t = 1$ 时的解.

最后, 舒尔注意到, 对于抽象李代数 g, 映射 $X \to X^R$ 是从 g 到原点的一个邻域中解析向量场空间的李代数同态. 这样, 他得到了一个与李的证明完全不同的, 关于李的第三基本定理的一个十分自然的证明.

Z 可表示成 X 和 Y 的收敛幂级数这一事实是以"坎贝尔 — 豪斯多夫公式"出现在文献中的. 这些作者对系数的递归性质增加了一些评注. 真正明确的公式还是在很晚以后由邓肯 (Dynkin) 给出的. 很难理解舒尔的工作居然被遗忘或被忽视了 (在 1899 年, 庞加莱 (Poincaré) 重新发现了指数映射的导数公式). C^2 李群是解析的这一事实, 已在已有文献中完全描述过, 尽管以略微保留的调子, 而在希尔伯特第五问题中也提到了舒尔的证明.

当他还是魏尔斯特拉斯的学生时, 舒尔学习了用幂级数做研究. 完成了他的学位论文后, 他成了现在叫作 Tartu 的爱沙尼亚的一个小市镇 Dorpat 的大学的数学教授. 在给恩格尔的信中, 他说他十分愉快, 因为在他的函数论的班里有了 20 个学生.

4   基灵

魏尔斯特拉斯的另一个学生基灵在 1884 年系统地提出了按照无穷小自同构的李代数来分类几何学的提纲. 把注意力集中在几何的自同构群上这一想法是受到了克莱因的 Erlangen 纲领的激励. 但是, 正如 Hawkins 指出的, 克莱因用来使几何学更有条理的这个想法来源于空间的直觉和应用问题, 一点也不像不受现实约束的魏尔斯特拉斯学派的纯数学方法.

当时, 基灵正在东普鲁士的勃劳恩斯特的一所训练天主教神父的学校 —— Hosianum 学院任教. 在这个孤立于数学之外的位置上, 他把他的纲领寄给了克莱因. 克莱因建议他注意他的朋友李的工作, 而当时, 在德国, 仅仅只有很少的几个人知道李的工作. 给李写了信后,基灵收到了李发表在挪威的杂志上的关于变换群的论文的重印本. 从这些论文,他断定是李先于他发展了连续群和它们的李代数的概念. 但是,他要用来解决有限维李代数的分类问题的方法却是新的.

正如基灵在后来写给始终鼓励他和他的纲领的恩格尔的信中承认过的,基灵全神贯注的那些问题已经不是李给出来的,例如,他从来也没有对关于可解李代数的李的基本定理融会贯通过. 他很快就把李的重印本退还了,因为他认为他只是把它们借来的. 现在,我猜想,这就可能给李留下这样的印象:基灵终究对他的工作不太感兴趣. 后来,他们的关系变得很坏.

现在我们转向基灵的方法. 他从魏尔斯特拉斯那里学到了从一个有限维向量空间到自身的线性映射 A 的特征分解. 尽管基灵的几何学都被看成是实的,为了简单起见,基灵把自己限制在复数的情况,而这种情况描述起来是最简单的. 对于任意的 $\lambda \in C$, 特征值 λ 的根子空间定义成 $A - \lambda \cdot I$ 的幂零空间,也就是在该映射的适当高的幂作用下得零的向量所成的空间. 这个幂零空间非零,当且仅当 λ 是 A 的特征值,或者说 λ 是特征多项式 $\lambda \rightarrow \det(A - \lambda \cdot I)$ 的零点("根"). 更精确地,这个幂零空间的维数等于特征多项式的根的代数重数,而向量空间等于根子空间的直和. 最后,特征多项式的系数在线性映射 A 的空间上是共轭的不变齐次多项式.

现在基灵把这些事实应用到把李代数 g 映到自身的线性映射 $A = \operatorname{ad} X$ 上,这里 $X \in g$. 他首先注意到,看作 $X \in g$ 的函数的特征多项式的系数是 g 上的齐次多项式 ψ_i, 它在伴随群的作用下是不变的. 函数独立的 ψ_i 的个数称为李代数的秩,基灵建议用秩来给李代数分类. 秩等于零,当且仅当李代数 g 是幂零的,也就是对每个 $X \in g$, $\operatorname{ad} X$ 是幂零线性映射. 自然地,基灵从研究幂零李代数开始,但是,由于那些过分乐观的结论和那些正确的论断的不完全的证明,基灵碰到了许多麻烦.

$\operatorname{ad} X$ 的幂零空间是 g 的李子代数 h. 当 X 属于齐次多项式 ψ_i 之一的零集的补集时,它的维数是最小的,此时,李代数 h 是幂零的. 这意味着对于所有 $H \in h$ 的 $\operatorname{ad} H, g$ 有一个公共的特征分解, 也就是说, g 等于空间

$$g_\alpha := \{X \in g \mid (\operatorname{ad} H - \alpha(H) \cdot 1)^n(X) = 0, \forall H \in h\}$$

的直和. 出现在这里的有限多个 α 是 h 上的复线性函数. 我们有 $g_0 = h$, 使 g_α 非零的 $\alpha \neq 0$ 称为根. 进一步, 相应的 g_α 叫作根空间. g 分解成 h 和根空间称为 g 关于所选的 h 的根空间分解.

如果李代数没有非平凡的理想, 那么称它为单的. 略有游疑, 基灵对李代数引进了术语"半单的", 也就是它等于非交换的单李代数的直和. 在前一阶段, 基灵只研究 h 是交换的, 而且根空间都是一维的李代数 g. 如果 g 是半单的, 它总是这样的. 基灵试图证明这个结论, 但是, 他的论证是不完全的, 而且有时很难读懂.

后来, 基灵的工作中最最奇妙的那部分出现了, 他那著名的单李代数的分类使其达到了顶点. 令 g 单的, 如果 α 是根, 那么 $-\alpha$ 也是根. 进一步, 如果 $X \in g_\alpha, Y \in g_{-\alpha}$, 那么, $H = [X, Y] \in h$. 可以做出安排使 $\alpha(H) = 2$, 此时也可以写 $H = \alpha^\vee$. 对于所有的根 β, 由于 $\operatorname{ad} X(g_\beta) \subset g_{\alpha+\beta}$, 就得到 $\beta(\alpha^\vee)$ 是个整数, 而且对于所有的根 $\alpha, \beta, S_\alpha(\beta) := \beta - \beta(\alpha^\vee)\alpha$ 也是根. 用现代的术语, 也就是根组成一个根系. 从整性和有限性出发, 基灵可以列出所有的根系. 这个分类法令人惊异的特征之一是它的离散性. 它一点也不明显. 魏尔斯特拉斯提出过这样的警告, 关于线性映射的一般性质的过于朴素的认识可能完全被误解. 对于基灵来说, 这正是一个极好的例证. 第二个令人惊讶之处是除了已知的经典李代数系列外, 只有这样少 (仅仅 5 种) 的李代数. 起初, 基灵认为除了特殊线性群和正交群的李代数外, 再没有别的单李代数了.

基灵对整个纲领逐渐感到厌烦, 他越来越多地进入了代数, 而不是几何学. 主要是由于恩格尔的热情鼓励, 这些结果才得以问世, 并使基灵继续前进. 由于他手稿中的不完整证明引起的不愉快, 要不是恩格尔催促他把他的手稿寄给克莱因, 以便在 *Mathematische Annalen* 上发表, 也许他永远也不会发表它们. 克莱因要求恩格尔提出一个审查报告, 恩格尔坦率地回答说, 这

些文章包含了许多重要的结果,但是也有许多部分他看不懂.克莱因接受了这些文章.

以后不久,恩格尔本人就指出了基灵关于幂零李代数的一个结论是错误的,他在莱比锡的一个学生 Umlanf 给出了正确的结果.那个从法国高等师范学院来到莱比锡跟李一起工作的 Tresse 把整个事情告诉了他在巴黎的朋友嘉当,使他开始狂热地研究起基灵的文章.用了两年时间,嘉当完成了他的著名的学位论文 *Sur la structure des groupes de transformation finis et continus*,在该文中他使基灵的一切结果更加明确了,而且添加了几个他自己的原始结果.所以,我想我们必须感谢克莱因做出了发表基灵的文章的勇气及宽宏的决定.

在基灵得到了明斯特的数学教授的位置之后,他没有创造出新的有趣的数学.例如,他没有去寻找其无穷小自同构组成的例外李代数的几何,而这种寻找正是人们自然会期望的.

这里提到的关于基灵的许多事实我是从 Hawkins 那里知道的.

我颇详细地概述了基灵的贡献,为的是弄清楚是否是基灵引进了根空间分解的技术,包括单李代数的分类.长时期来,在谈到这些技术时,几乎完全把它们归于嘉当,人们想要知道到底是怎样变成这样的.一种解释可能是嘉当的学位论文马上就成了学习这一课题的标准工作.这一说法似乎并不能让人完全信服.无论如何,嘉当在引证基灵的工作时是很慷慨的.李也承认基灵的工作是令人激动的,虽然,他是花了好多篇幅指责基灵的很不好的数学态度,尤其是对于李的工作的态度,才描述这个分类的.

在现在用的术语中甚至有些使人哑然失笑,例如,上面谈到的李子代数 h 是由基灵引进的,却被称为嘉当子代数.对于整数 $\beta(\alpha^\vee)$ 可以给出类似的注释.另外,在不变量 ψ_i 中的二次型在基灵的文章中并没有起什么特别的作用(仅仅是人们注意的一条线索).对于嘉当来说,在他的学位论文中,系统地应用了二次型正是他真正感到骄傲的一个主要的贡献,作为一种补偿,现在把

244

这个二次型称为基灵型.

5 外尔

现在我想跳到外尔关于李群的表示的经典文章. 在这不多的几页中, 可以找到紧连通李群的结构理论, 尽管我们只是重新给出了他的论述, 在我们的书中, 仍需要整整一章. (这里我对我们的书是不完全公正的, 因为我们给出的要详细得多, 但是, 对外尔的简短的论述我仍不无美慕.)

设 G 是紧的连通的李群, T 是 G 的极大环面. 外尔开始证明的第一个基本的结果是 G 的每一个元素都共轭于 T 的一个元素.

G 的元素 x 称为正则的, 如果它在 G 中的共轭类的维数是最大的, 等价地, 它在 G 内的中心化子 G^x 的维数是最小的, 这也等价于条件: 把 $G \times G^x$ 映到 G 的映射

$$\gamma : (g, t) \mapsto g \cdot t \cdot g^{-1}$$

的导数在 $g = 1, t = x$ 时是满射. 记 $T^{\mathrm{reg}} := T \bigcap G^{\mathrm{reg}}$, 对每个 $x \in T^{\mathrm{reg}} T$, 在 G^x 内是开的. 由于线性变换环面的作用在 C 上是对角的, 容易得到 T^{reg} 等于 T 内有限多个(平移的)子环面的补集, 尤其是 T^{reg} 在 T 内是紧的.

G 内正则元的集合 G^{reg} 是 G 的开的、紧的共轭不变的子集. $U := \gamma(G \times T^{\mathrm{reg}})$ 是 G^{reg} 的开的、共轭不变的子集, 它在 G 的紧子集 $V = \gamma(G \times T)$ 内是稠的. 这就得到 U 在 G 内的边界包含在 $\gamma(G \times (T \backslash T^{\mathrm{reg}}))$ 内, 从而, 包含在 G 的奇异点集合 $G^{\mathrm{sing}} := G \backslash G^{\mathrm{reg}}$ 内. 所以, 外尔断定, 如果我们能证明 G^{reg} 是连通的, 我们就得到了 $U = G^{\mathrm{reg}}$, 也就是 $V = G$. 为此, 他注意到 G^{sing} 在 G 内的余维数为 3(我们可以说, 它是 G 内余维数大于或等于 3 的有限多个浸入的光滑子流形的并), 这样他就完成了证明.

自然, 外尔是完全正确的, 他的简明性依然是令人吃惊的, 因为, 为了完成这个证明所需要的微分拓扑中的论据在当时不属于数学家必须具备的基本知识范围. 类似地, G 的通用覆盖 $\widetilde{G}$ 被引进了. 甚至连证明的暗示都没有就给出了 $\widetilde{G}$ 是一个李群. 这

个证明是由施赖埃尔 (Schreier) 给出的 (他并没有看过外尔的文章). 一个小小的调查证实了我的猜测: 外尔的文章是难以作为该主题的入门文章来读的. 这就使弗赖登塔尔 (Frendenthal) 为数学进展而写的综述格外受到注意. 他写该综述的时间是在外尔的文章发表的同一年, 当时他只是一个 20 岁的学生. 这篇综述是优秀的.

作为极大环面定理的推论, G 上每个共轭不变的函数都由它在 T 上的限制决定, 对于外尔精确计算过的密度, 它在 G 上的积分等于它在 T 上的积分. 特别地, 这应用于 G 的不可约表示的特征标和 L^2 内积. 应用 Frebenius-Schur 正交性关系, 可导出一个关于特征标 (限制到 T) 的一个漂亮的明确的公式.

6 图

李群论总是带有很强的几何味道, 然而, 在有关这个课题的书和论文中几乎看不到图 (除了根图外), 这使我十分惊讶! 事实上, 当维数增高时, 要想把它们完全显示出来几乎是不可能的, 连最简单的对象也看不到.

为了在这本书里画出李群中我们的几个 "好朋友" 的 "肖像", 今年冬天我学习了怎样在带字模打印机的个人计算机上编排程序.

第一个图 (图 1) 表示行列式等于 1 的实 2×2 矩阵群 $SL(2, R)$. 这个群微分同胚于一个由圆周和单位圆盘内部的笛卡儿积构成的实心环. 圆盘样的面是椭圆元的共轭类, 每一个共轭类都包含一个转动, 这些转动组成了一个单参数子群, 图 1 中以虚线圆表示. 柱面是双曲元的共轭类, 这些双曲元的特征值不等于 ± 1, 它们的乘积等于 1. 对角阵组成一个 "截面" 单参数子群, 在图 1 中用两条虚直线表示. 最后, 幂么元 (所有特征值都等于 1 的元) 组成两个共轭类, 它们都与单位元相连, 就像两个冰淇淋锥那样. 环面的边界在群的无穷远处. 由于负曲率, 群在无穷远是空空洞洞的, 此时, 这个图十分令人迷惑. 我是从阿蒂亚 (Atiyah) 那里学到这个图的.

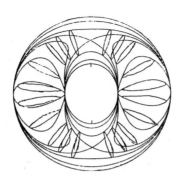

图 1　$SL(2,R)$ 的共轭类

　　所有别的图都限于 $SU(2)$（图 2(a)），相应地，$SU(3)$ 在极大环面的不可约表示的特征标. 第一种情况，T 可以等同于圆圈 R/Z，其上的特征标由 $\sin(k-x)/\sin x$ 给出. 为了指出当 $k \to \infty$ 时它的渐近特性，我把 k 取得充分大. 就像对特征标总是对的那样，它在群的单位元取极大值，其值等于相应的表示的维数. 为了进行比较，我增加了一个图（图 2(b)），它在单位元的值等于 1，而整个图是按比例画出的.

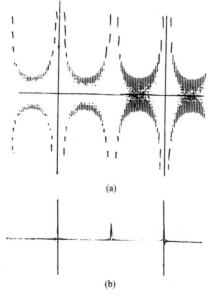

(a)

(b)

图 2　$SU(2)$ 的特征标 64

对于 $SU(3)$，环面 T 是二维的．特征标由它们的所谓最高权决定．有两个基本权，每个最高权等于第一个的 k 倍加上第二个的 l 倍，这里 k, l 是非负整数．所以，常以 k, l 表示这个特征标．

如果 $k \neq l$，那么，特征标不是实的，而是复值的．这就提出了把从平面(T 的通用覆盖 t) 到平面(C) 的映射形象化的问题．开始我曾在两个不同的图里画出特征标的实部和虚部在 t 中的水平曲线，但后来我应用简单得多的想法，画出特征标在 C 中的像点．它们是用 T 内正则光栅的点来赋值的，实际上是 T 的外尔群不变的有限子群．所得到的图令人十分吃惊，即使在最简单的 $1, 0$ 特征标的情况，也使人奇怪它是怎样在一个不太复杂的映射下成为环面的像的．对于 $3, 1$ 特征标，光栅点似乎是沿着曲线的迷人的形状排列的，它在放大和相应地增加光栅点的数目下仍旧保持原来的形状．

如果 $k = l$，那么 $SU(3)$ 的特征标是实值的（图 3），此时，前面的技术在实轴上产生一个区间，看起来它并不那么有趣．所以，在这种情况下，我在平面 t 上画出了水平曲线．容易辨认出双周期结构，实际上它表示有一个环面 T 的函数．同时六边形的外尔群对称也是明显的．

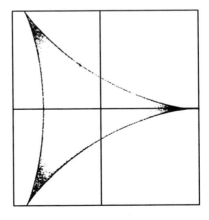

图 3　在复平面上 $SU(3)$ 的基本特征的像

我以 0.5 作为间隔给出水平曲线,但对 $k=4$,特征标在单位元周围上升得这么快,以至于那部分空间被水平曲线完全填满了.由于这些里区域没有提供任何信息,也因为它们将损坏我的打印机,我仅仅画了从 -5 到 $+5$ 的水平曲线之间的曲线(图 4).图中的直线组成零水平,它的解释是,对于 $k=1$(更一般地,当最高权是 p 的一个倍数时),特征标等于作为 t 上某些线性型的函数的秩 1 特征标的乘积.在图 5(e)16,16 特征标中,最大值等于 4 913,好似荷兰那样平坦的风景中一座比勃兰克山更高的摩天大楼.

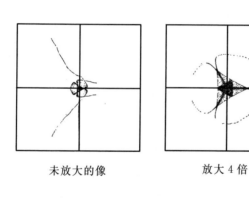

未放大的像   放大 4 倍

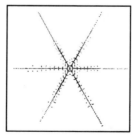

放大 32 倍

图 4 $SU(3)$ 的特征标从 -5 到 $+5$ 之间的水平曲线

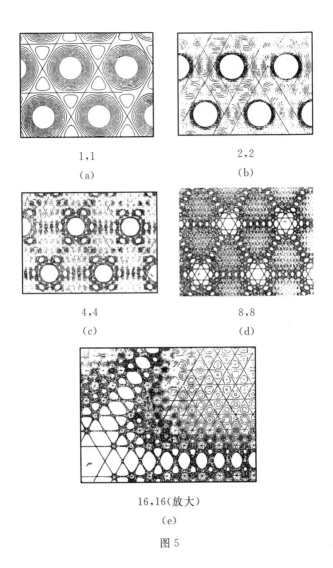

1,1

(a)

2,2

(b)

4,4

(c)

8,8

(d)

16,16(放大)

(e)

图 5

在数学与物理的结合方面,近代做得比较出色的是德国数学家外尔.他的工作之一外尔张量在本书的附录 2 中有体现.

外尔(Weyl,Hermann,1885—1955),德国人.1885 年 11 月 9 日生于什列斯威格—荷尔斯泰因州的埃尔姆斯霍恩.外尔少年时并不显得十分聪明.1904 年由希尔伯特(Hilbert)的一位堂兄介绍进入哥廷根大学学习数学,有一段时间在慕尼黑大学旁听.1908年以积分方程论方面的论文,获得哥廷根大学博士学位,并成为希尔伯特的得意门生.1910年外尔被希尔伯特留在哥廷根大学任兼职讲师.1913年应邀到苏黎世大学任教授.在那里他遇到了爱因斯坦(Einstein),共同研讨过广义相对论.1926年至1927年又回到哥廷根大学任教.1928年至1929年任普林斯顿大学客座教授.1930年外尔接替希尔伯特在哥廷根大学任教授.3年后,外尔辞去了哥廷根大学的职务.1933年被当时初建的普林斯顿高等研究院接纳为教授.1951年退休,辞去该研究所的正教授职务而任名誉教授.晚年,他每年在普林斯顿和苏黎世的时间各一半.1955年12月8日在苏黎世突然逝世.

外尔在数学和理论物理学中有许多开创性、奠基性的业绩.其中最著名的有:

1908年外尔发表了他的博士论文,是关于奇异积分方程解的存在性问题的研究.在此之前,虽然希尔伯特很重视积分方程的研究,而且也有很多人写了论文,但大多数只有短暂的价值,而外尔的论文引起了对分析学一系列问题的研究,有深远的意义.外尔认为,黎曼面不仅只是使解析函数多值性直观化的手段,而且也是解析函数理论的本质部分,是这个理论赖以生长和繁衍的唯一土壤.他在1913年发表的《黎曼面的概念》一书中,不仅给出了黎曼面的精确定义(这是黎曼面最早的几个等价定义之一),而且把以代数函数作为黎曼面上的函数来研究的所谓"解析方法"整理成完美而严密的形式.

在丢番图逼近方面,外尔在1914年提出并研究了所谓"一致分布"问题.他用解析方法(特别是三角级数)作为有效的手段,得到了一致分布的充分必要条件,被称为外尔原理.

在群的表示论方面,外尔做了许多开创性的工作.希尔伯特早年研究的不变量理论,被外尔用来作为李群的线性表示.1925年他研究了紧李群的有限维酉表示.1927年他完成了紧群的理论,并且弄清了殆周期函数理论与群的表示论之间的关系,尤其是拓扑群上的殆周期函数与紧群的表示论之间的关系.1900年至 1930 年间,外尔研究了半单李代数的完全分类和结构,并确定了它们的表示与特征标.1925 年他取得了一个关键性的结果:特征为零的一个代数封闭域上的半单李代数的任何表示都是完全可约的.另外,在李群流形的整体结构研究方面他也做了开创性的工作.

在黎曼几何的推广方面,外尔的研究工作也具有开创性.他在讲授黎曼几何曲面的课程中,引入了流形概念,用抽象方法把研究工作推向新阶段.1918 年,他又引进一类通称为仿射联络空间的几何,使得黎曼几何成为它的一种特例.这一成果向人们展示了,在非黎曼几何中,点与点之间的联系不一定要用依赖于一个度量的方式来规定.这些几何学相互之间也许有极大的差异,但又都像黎曼几何一样有广阔的发展前途.在外尔之后,黎曼几何又有了一些不同的推广.

在泛函分析方面,外尔追随希尔伯特研究了谱论,并将它推广到李群,在弹性力学中也找到了应用.他还研究了希尔伯特空间,特别是希尔伯特空间中的算子.这方面的工作直接启发了冯·诺伊曼公理的提出.

在偏微分方程方面,外尔利用迪利克雷原理证明了,椭圆型线性自伴偏微分方程的边值问题的解存在.

晚年,外尔与其儿子共同研究了由几个亚纯函数所成的亚纯函数组的值分布理论发展而来的亚纯曲线论,并于 1943 年共同发表了专著《亚纯曲线论》.

外尔的数学思想几乎完全被希尔伯特所控制,始终信守希尔伯特的信念:"抽象理论在解决经典问题中,对概念深刻分析的价值大大超过盲目计算."甚至人们称外尔是希尔伯特的"数学儿子".但是,在研究数学基础时,他们却分道扬镳了.外尔属于布劳威尔(Brouwer)倡导的直觉主义派,而且发表了许多论述,产生了不小的影响,对倡导形式主义派的希尔伯特也进行了严厉的批评,乃至攻击.

在物理学方面,外尔的贡献也不少.1916 年,当他一接触到爱因斯坦的广义相对论时,便立即用张量积等数学工具给它装上了数学的框架.进而,1918年外尔又提出了最初的统一场论.

　　外尔写了150多种著作.除了数学、物理学著作外,还包括哲学、历史方面的著作以及各种评论.有许多名言被人引用,诸如"逻辑是指导数学家保持其思想观念强健的卫生学""如果不知道远溯到古希腊各代前辈所建立和发展的概念、方法和成果,我们就不可能理解近50年来数学的目标,也不可能理解近50年来数学的成就."等等.

　　由于本书的阅读门槛较高,所以它注定只会服务于极少数读者.那位大范围"统一"了数学和物理学的数学家伊萨多·辛格(Isadore Singe)于2021年2月11日去世,终年96岁.在晚年的采访中,他曾说:"对于大多数人来说,教授数学就像是给聋子解释音乐.而'真正的数学大门,只对少数人敞开'(available for few people)."

刘培杰

2022.7.27
于哈工大

刘培杰数学工作室
已出版(即将出版)图书目录——原版影印

书　名	出版时间	定　价	编号
数学物理大百科全书.第1卷(英文)	2016—01	418.00	508
数学物理大百科全书.第2卷(英文)	2016—01	408.00	509
数学物理大百科全书.第3卷(英文)	2016—01	396.00	510
数学物理大百科全书.第4卷(英文)	2016—01	408.00	511
数学物理大百科全书.第5卷(英文)	2016—01	368.00	512
zeta函数,q-zeta函数,相伴级数与积分(英文)	2015—08	88.00	513
微分形式:理论与练习(英文)	2015—08	58.00	514
离散与微分包含的逼近和优化(英文)	2015—08	58.00	515
艾伦·图灵:他的工作与影响(英文)	2016—01	98.00	560
测度理论概率导论,第2版(英文)	2016—01	88.00	561
带有潜在故障恢复系统的半马尔柯夫模型控制(英文)	2016—01	98.00	562
数学分析原理(英文)	2016—01	88.00	563
随机偏微分方程的有效动力学(英文)	2016—01	88.00	564
图的谱半径(英文)	2016—01	58.00	565
量子机器学习中数据挖掘的量子计算方法(英文)	2016—01	98.00	566
量子物理的非常规方法(英文)	2016—01	118.00	567
运输过程的统一非局部理论:广义波尔兹曼物理动力学,第2版(英文)	2016—01	198.00	568
量子力学与经典力学之间的联系在原子、分子及电动力学系统建模中的应用(英文)	2016—01	58.00	569
算术域(英文)	2018—01	158.00	821
高等数学竞赛:1962—1991年的米洛克斯·史怀哲竞赛(英文)	2018—01	128.00	822
用数学奥林匹克精神解决数论问题(英文)	2018—01	108.00	823
代数几何(德文)	2018—04	68.00	824
丢番图逼近论(英文)	2018—01	78.00	825
代数几何学基础教程(英文)	2018—01	98.00	826
解析数论入门课程(英文)	2018—01	78.00	827
数论中的丢番图问题(英文)	2018—01	78.00	829
数论(梦幻之旅):第五届中日数论研讨会演讲集(英文)	2018—01	68.00	830
数论新应用(英文)	2018—01	68.00	831
数论(英文)	2018—01	78.00	832

刘培杰数学工作室

已出版（即将出版）图书目录——原版影印

书　名	出 版 时 间	定　价	编号
湍流十讲(英文)	2018－04	108.00	886
无穷维李代数:第3版(英文)	2018－04	98.00	887
等值、不变量和对称性(英文)	2018－04	78.00	888
解析数论(英文)	2018－09	78.00	889
《数学原理》的演化:伯特兰·罗素撰写第二版时的手稿与笔记(英文)	2018－04	108.00	890
哈密尔顿数学论文集(第4卷):几何学、分析学、天文学、概率和有限差分等(英文)	2019－05	108.00	891
偏微分方程全局吸引子的特性(英文)	2018－09	108.00	979
整函数与下调和函数(英文)	2018－09	118.00	980
幂等分析(英文)	2018－09	118.00	981
李群,离散子群与不变量理论(英文)	2018－09	108.00	982
动力系统与统计力学(英文)	2018－09	118.00	983
表示论与动力系统(英文)	2018－09	118.00	984
分析学练习.第1部分(英文)	2021－01	88.00	1247
分析学练习.第2部分,非线性分析(英文)	2021－01	88.00	1248
初级统计学:循序渐进的方法:第10版(英文)	2019－05	68.00	1067
工程师与科学家微分方程用书:第4版(英文)	2019－07	58.00	1068
大学代数与三角学(英文)	2019－06	78.00	1069
培养数学能力的途径(英文)	2019－07	38.00	1070
工程师与科学家统计学:第4版(英文)	2019－06	58.00	1071
贸易与经济中的应用统计学:第6版(英文)	2019－06	58.00	1072
傅立叶级数和边值问题:第8版(英文)	2019－05	48.00	1073
通往天文学的途径:第5版(英文)	2019－05	58.00	1074
拉马努金笔记.第1卷(英文)	2019－06	165.00	1078
拉马努金笔记.第2卷(英文)	2019－06	165.00	1079
拉马努金笔记.第3卷(英文)	2019－06	165.00	1080
拉马努金笔记.第4卷(英文)	2019－06	165.00	1081
拉马努金笔记.第5卷(英文)	2019－06	165.00	1082
拉马努金遗失笔记.第1卷(英文)	2019－06	109.00	1083
拉马努金遗失笔记.第2卷(英文)	2019－06	109.00	1084
拉马努金遗失笔记.第3卷(英文)	2019－06	109.00	1085
拉马努金遗失笔记.第4卷(英文)	2019－06	109.00	1086
数论:1976年纽约洛克菲勒大学数论会议记录(英文)	2020－06	68.00	1145
数论:卡本代尔1979:1979年在南伊利诺伊卡本代尔大学举行的数论会议记录(英文)	2020－06	78.00	1146
数论:诺德韦克豪特1983:1983年在诺德韦克豪特举行的Journees Arithmetiques数论大会会议记录(英文)	2020－06	68.00	1147
数论:1985－1988年在纽约城市大学研究生院和大学中心举办的研讨会(英文)	2020－06	68.00	1148

刘培杰数学工作室
已出版(即将出版)图书目录——原版影印

书　　名	出版时间	定　价	编号
数论:1987 年在乌尔姆举行的 Journees Arithmetiques 数论大会会议记录(英文)	2020—06	68.00	1149
数论:马德拉斯 1987:1987 年在马德拉斯安娜大学举行的国际拉马努金百年纪念大会会议记录(英文)	2020—06	68.00	1150
解析数论:1988 年在东京举行的日法研讨会会议记录(英文)	2020—06	68.00	1151
解析数论:2002 年在意大利切特拉罗举行的 C. I. M. E. 暑期班演讲集(英文)	2020—06	68.00	1152
量子世界中的蝴蝶:最迷人的量子分形故事(英文)	2020—06	118.00	1157
走进量子力学(英文)	2020—06	118.00	1158
计算物理学概论(英文)	2020—06	48.00	1159
物质,空间和时间的理论:量子理论(英文)	2020—10	48.00	1160
物质,空间和时间的理论:经典理论(英文)	2020—10	48.00	1161
量子场理论:解释世界的神秘背景(英文)	2020—07	38.00	1162
计算物理学概论(英文)	2020—06	48.00	1163
行星状星云(英文)	2020—10	38.00	1164
基本宇宙学:从亚里士多德的宇宙到大爆炸(英文)	2020—08	58.00	1165
数学磁流体力学(英文)	2020—07	58.00	1166
计算科学:第 1 卷,计算的科学(日文)	2020—07	88.00	1167
计算科学:第 2 卷,计算与宇宙(日文)	2020—07	88.00	1168
计算科学:第 3 卷,计算与物质(日文)	2020—07	88.00	1169
计算科学:第 4 卷,计算与生命(日文)	2020—07	88.00	1170
计算科学:第 5 卷,计算与地球环境(日文)	2020—07	88.00	1171
计算科学:第 6 卷,计算与社会(日文)	2020—07	88.00	1172
计算科学.别卷,超级计算机(日文)	2020—07	88.00	1173
多复变函数论(日文)	2022—06	78.00	1518
复变函数入门(日文)	2022—06	78.00	1523
代数与数论:综合方法(英文)	2020—10	78.00	1185
复分析:现代函数理论第一课(英文)	2020—07	58.00	1186
斐波那契数列和卡特兰数:导论(英文)	2020—10	68.00	1187
组合推理:计数艺术介绍(英文)	2020—07	88.00	1188
二次互反律的傅里叶分析证明(英文)	2020—07	48.00	1189
旋瓦兹分布的希尔伯特变换与应用(英文)	2020—07	58.00	1190
泛函分析:巴拿赫空间理论入门(英文)	2020—07	48.00	1191
卡塔兰数入门(英文)	2019—05	68.00	1060
测度与积分(英文)	2019—04	68.00	1059
组合学手册.第一卷(英文)	2020—06	128.00	1153
＊一代数、局部紧群和巴拿赫＊一代数丛的表示.第一卷,群和代数的基本表示理论(英文)	2020—05	148.00	1154
电磁理论(英文)	2020—08	48.00	1193
连续介质力学中的非线性问题(英文)	2020—09	78.00	1195
多变量数学入门(英文)	2021—05	68.00	1317
偏微分方程入门(英文)	2021—05	88.00	1318
若尔当典范性:理论与实践(英文)	2021—07	68.00	1366
伽罗瓦理论.第 4 版(英文)	2021—08	88.00	1408

刘培杰数学工作室
已出版(即将出版)图书目录——原版影印

书 名	出版时间	定 价	编号
典型群,错排与素数(英文)	2020—11	58.00	1204
李代数的表示:通过 gln 进行介绍(英文)	2020—10	38.00	1205
实分析演讲集(英文)	2020—10	38.00	1206
现代分析及其应用的课程(英文)	2020—10	58.00	1207
运动中的抛射物数学(英文)	2020—10	38.00	1208
2—纽结与它们的群(英文)	2020—10	38.00	1209
概率,策略和选择:博弈与选举中的数学(英文)	2020—11	58.00	1210
分析学引论(英文)	2020—11	58.00	1211
量子群:通往流代数的路径(英文)	2020—11	38.00	1212
集合论入门(英文)	2020—10	48.00	1213
西反射群(英文)	2020—11	58.00	1214
探索数学:吸引人的证明方式(英文)	2020—11	58.00	1215
微分拓扑短期课程(英文)	2020—10	48.00	1216
抽象凸分析(英文)	2020—11	68.00	1222
费马大定理笔记(英文)	2021—03	48.00	1223
高斯与雅可比和(英文)	2021—03	78.00	1224
π与算术几何平均:关于解析数论和计算复杂性的研究(英文)	2021—01	58.00	1225
复分析入门(英文)	2021—03	48.00	1226
爱德华·卢卡斯与素性测定(英文)	2021—03	78.00	1227
通往凸分析及其应用的简单路径(英文)	2021—01	68.00	1229
微分几何的各个方面.第一卷(英文)	2021—01	58.00	1230
微分几何的各个方面.第二卷(英文)	2020—12	58.00	1231
微分几何的各个方面.第三卷(英文)	2020—12	58.00	1232
沃克流形几何学(英文)	2020—11	58.00	1233
彷射和韦尔几何应用(英文)	2020—12	58.00	1234
双曲几何学的旋转向量空间方法(英文)	2021—02	58.00	1235
积分:分析学的关键(英文)	2020—12	48.00	1236
为有天分的新生准备的分析学基础教材(英文)	2020—11	48.00	1237
数学不等式.第一卷.对称多项式不等式(英文)	2021—03	108.00	1273
数学不等式.第二卷.对称有理不等式与对称无理不等式(英文)	2021—03	108.00	1274
数学不等式.第三卷.循环不等式与非循环不等式(英文)	2021—03	108.00	1275
数学不等式.第四卷.Jensen 不等式的扩展与加细(英文)	2021—03	108.00	1276
数学不等式.第五卷.创建不等式与解不等式的其他方法(英文)	2021—04	108.00	1277

刘培杰数学工作室
已出版(即将出版)图书目录——原版影印

书 名	出版时间	定 价	编号
冯·诺依曼代数中的谱位移函数:半有限冯·诺依曼代数中的谱位移函数与谱流(英文)	2021—06	98.00	1308
链接结构:关于嵌入完全图的直线中链接单形的组合结构(英文)	2021—05	58.00	1309
代数几何方法.第1卷(英文)	2021—06	68.00	1310
代数几何方法.第2卷(英文)	2021—06	68.00	1311
代数几何方法.第3卷(英文)	2021—06	58.00	1312

书 名	出版时间	定 价	编号
代数、生物信息和机器人技术的算法问题.第四卷,独立恒等式系统(俄文)	2020—08	118.00	1199
代数、生物信息和机器人技术的算法问题.第五卷,相对覆盖性和独立可拆分恒等式系统(俄文)	2020—08	118.00	1200
代数、生物信息和机器人技术的算法问题.第六卷,恒等式和准恒等式的相等问题、可推导性和可实现性(俄文)	2020—08	128.00	1201
分数阶微积分的应用:非局部动态过程,分数阶导热系数(俄文)	2021—01	68.00	1241
泛函分析问题与练习:第2版(俄文)	2021—01	98.00	1242
集合论、数学逻辑和算法论问题:第5版(俄文)	2021—01	98.00	1243
微分几何和拓扑短期课程(俄文)	2021—01	98.00	1244
素数规律(俄文)	2021—01	88.00	1245
无穷边值问题解的递减:无界域中的拟线性椭圆和抛物方程(俄文)	2021—01	48.00	1246
微分几何讲义(俄文)	2020—12	98.00	1253
二次型和矩阵(俄文)	2021—01	98.00	1255
积分和级数.第2卷,特殊函数(俄文)	2021—01	168.00	1258
积分和级数.第3卷,特殊函数补充:第2版(俄文)	2021—01	178.00	1264
几何图上的微分方程(俄文)	2021—01	138.00	1259
数论教程:第2版(俄文)	2021—01	98.00	1260
非阿基米德分析及其应用(俄文)	2021—03	98.00	1261
古典群和量子群的压缩(俄文)	2021—03	98.00	1263
数学分析习题集.第3卷,多元函数:第3版(俄文)	2021—03	98.00	1266
数学习题:乌拉尔国立大学数学力学系大学生奥林匹克(俄文)	2021—03	98.00	1267
柯西定理和微分方程的特解(俄文)	2021—03	98.00	1268
组合极值问题及其应用:第3版(俄文)	2021—03	98.00	1269
数学词典(俄文)	2021—01	98.00	1271
确定性混沌分析模型(俄文)	2021—06	168.00	1307
精选初等数学习题和定理.立体几何.第3版(俄文)	2021—03	68.00	1316
微分几何习题:第3版(俄文)	2021—05	98.00	1336
精选初等数学习题和定理.平面几何.第4版(俄文)	2021—05	68.00	1335
曲面理论在欧氏空间 E_n 中的直接表示(俄文)	2022—01	68.00	1444
维纳—霍普夫离散算子和托普利兹算子:某些可数赋范空间中的诺特性和可逆性(俄文)	2022—03	108.00	1496
Maple 中的数论:数论中的计算机计算(俄文)	2022—03	88.00	1497
贝尔曼和克努特问题及其概括:加法运算的复杂性(俄文)	2022—03	138.00	1498

刘培杰数学工作室
已出版(即将出版)图书目录——原版影印

书　名	出版时间	定　价	编号
复分析:共形映射(俄文)	2022—07	48.00	1542
微积分代数样条和多项式及其在数值方法中的应用(俄文)	2022—08	128.00	1543
蒙特卡罗方法中的随机过程和场模型:算法和应用(俄文)	2022—08	88.00	1544
狭义相对论与广义相对论:时空与引力导论(英文)	2021—07	88.00	1319
束流物理学和粒子加速器的实践介绍:第2版(英文)	2021—07	88.00	1320
凝聚态物理中的拓扑和微分几何简介(英文)	2021—05	88.00	1321
混沌映射:动力学、分形学和快速涨落(英文)	2021—05	128.00	1322
广义相对论:黑洞、引力波和宇宙学介绍(英文)	2021—06	68.00	1323
现代分析电磁均质化(英文)	2021—06	68.00	1324
为科学家提供的基本流体动力学(英文)	2021—06	88.00	1325
视觉天文学:理解夜空的指南(英文)	2021—06	68.00	1326
物理学中的计算方法(英文)	2021—06	68.00	1327
单星的结构与演化:导论(英文)	2021—06	108.00	1328
超越居里:1903年至1963年物理界四位女性及其著名发现(英文)	2021—06	68.00	1329
范德瓦尔斯流体热力学的进展(英文)	2021—06	68.00	1330
先进的托卡马克稳定性理论(英文)	2021—06	88.00	1331
经典场论导论:基本相互作用的过程(英文)	2021—07	88.00	1332
光致电离量子动力学方法原理(英文)	2021—07	108.00	1333
经典域论和应力:能量张量(英文)	2021—05	88.00	1334
非线性太赫兹光谱的概念与应用(英文)	2021—06	68.00	1337
电磁学中的无穷空间并矢格林函数(英文)	2021—06	88.00	1338
物理科学基础数学.第1卷,齐次边值问题、傅里叶方法和特殊函数(英文)	2021—07	108.00	1339
离散量子力学(英文)	2021—07	68.00	1340
核磁共振的物理学和数学(英文)	2021—07	108.00	1341
分子水平的静电学(英文)	2021—08	68.00	1342
非线性波:理论、计算机模拟、实验(英文)	2021—06	108.00	1343
石墨烯光学:经典问题的电解解决方案(英文)	2021—06	68.00	1344
超材料多元宇宙(英文)	2021—07	68.00	1345
银河系外的天体物理学(英文)	2021—07	68.00	1346
原子物理学(英文)	2021—07	68.00	1347
将光打结:将拓扑学应用于光学(英文)	2021—07	68.00	1348
电磁学:问题与解法(英文)	2021—07	88.00	1364
海浪的原理:介绍量子力学的技巧与应用(英文)	2021—07	108.00	1365
多孔介质中的流体:输运与相变(英文)	2021—07	68.00	1372
洛伦兹群的物理学(英文)	2021—08	68.00	1373
物理导论的数学方法和解决方法手册(英文)	2021—08	68.00	1374
非线性波数学物理学入门(英文)	2021—08	88.00	1376
波:基本原理和动力学(英文)	2021—07	68.00	1377
光电子量子计量学.第1卷,基础(英文)	2021—07	88.00	1383
光电子量子计量学.第2卷,应用与进展(英文)	2021—07	68.00	1384
复杂流的格子玻尔兹曼建模的工程应用(英文)	2021—08	68.00	1393
电偶极矩挑战(英文)	2021—08	108.00	1394
电动力学:问题与解法(英文)	2021—09	68.00	1395
自由电子激光的经典理论(英文)	2021—08	68.00	1397

刘培杰数学工作室
已出版(即将出版)图书目录——原版影印

书　名	出 版 时 间	定　价	编号
曼哈顿计划——核武器物理学简介(英文)	2021—09	68.00	1401
粒子物理学(英文)	2021—09	68.00	1402
引力场中的量子信息(英文)	2021—09	128.00	1403
器件物理学的基本经典力学(英文)	2021—09	68.00	1404
等离子体物理及其空间应用导论.第1卷,基本原理和初步过程(英文)	2021—09	68.00	1405
拓扑与超弦理论焦点问题(英文)	2021—07	58.00	1349
应用数学:理论、方法与实践(英文)	2021—07	78.00	1350
非线性特征值问题:牛顿型方法与非线性瑞利函数(英文)	2021—07	58.00	1351
广义膨胀和齐性:利用齐性构造齐次系统的李雅普诺夫函数和控制律(英文)	2021—06	48.00	1352
解析数论焦点问题(英文)	2021—07	58.00	1353
随机微分方程:动态系统方法(英文)	2021—07	58.00	1354
经典力学与微分几何(英文)	2021—07	58.00	1355
负定相交形式流形上的瞬子模空间几何(英文)	2021—07	68.00	1356
广义卡塔兰轨道分析:广义卡塔兰轨道计算数字的方法(英文)	2021—07	48.00	1367
洛伦兹方法的变分:二维与三维洛伦兹方法(英文)	2021—08	38.00	1378
几何、分析和数论精编(英文)	2021—08	68.00	1380
从一个新角度看数论:通过遗传方法引入现实的概念(英文)	2021—07	58.00	1387
动力系统:短期课程(英文)	2021—08	68.00	1382
几何路径:理论与实践(英文)	2021—08	48.00	1385
论天体力学中某些问题的不可积性(英文)	2021—07	88.00	1396
广义斐波那契数列及其性质(英文)	2021—08	38.00	1386
对称函数和麦克唐纳多项式:余代数结构与Kawanaka恒等式(英文)	2021—09	38.00	1400
杰弗里·英格拉姆·泰勒科学论文集:第1卷.固体力学(英文)	2021—05	78.00	1360
杰弗里·英格拉姆·泰勒科学论文集:第2卷.气象学、海洋学和湍流(英文)	2021—05	68.00	1361
杰弗里·英格拉姆·泰勒科学论文集:第3卷.空气动力学以及落弹数和爆炸的力学(英文)	2021—05	68.00	1362
杰弗里·英格拉姆·泰勒科学论文集:第4卷.有关流体力学(英文)	2021—05	58.00	1363

刘培杰数学工作室
已出版(即将出版)图书目录——原版影印

书　　名	出版时间	定　价	编号
非局域泛函演化方程:积分与分数阶(英文)	2021—08	48.00	1390
理论工作者的高等微分几何:纤维丛、射流流形和拉格朗日理论(英文)	2021—08	68.00	1391
半线性退化椭圆微分方程:局部定理与整体定理(英文)	2021—07	48.00	1392
非交换几何、规范理论和重整化:一般简介与非交换量子场论的重整化(英文)	2021—09	78.00	1406
数论论文集:拉普拉斯变换和带有数论系数的幂级数(俄文)	2021—09	48.00	1407
挠理论专题:相对极大值,单射与扩充模(英文)	2021—09	88.00	1410
强正则图与欧几里得若尔当代数:非通常关系中的启示(英文)	2021—10	48.00	1411
拉格朗日几何和哈密顿几何:力学的应用(英文)	2021—10	48.00	1412
时滞微分方程与差分方程的振动理论:二阶与三阶(英文)	2021—10	98.00	1417
卷积结构与几何函数理论:用以研究特定几何函数理论方向的分数阶微积分算子与卷积结构(英文)	2021—10	48.00	1418
经典数学物理的历史发展(英文)	2021—10	78.00	1419
扩展线性丢番图问题(英文)	2021—10	38.00	1420
一类混沌动力系统的分歧分析与控制:分歧分析与控制(英文)	2021—11	38.00	1421
伽利略空间和伪伽利略空间中一些特殊曲线的几何性质(英文)	2022—01	68.00	1422
一阶偏微分方程:哈密尔顿—雅可比理论(英文)	2021—11	48.00	1424
各向异性黎曼多面体的反问题:分段光滑的各向异性黎曼多面体反边界谱问题:唯一性(英文)	2021—11	38.00	1425
项目反应理论手册.第一卷,模型(英文)	2021—11	138.00	1431
项目反应理论手册.第二卷,统计工具(英文)	2021—11	118.00	1432
项目反应理论手册.第三卷,应用(英文)	2021—11	138.00	1433
二次无理数:经典数论入门(英文)	2022—05	138.00	1434
数,形与对称性:数论,几何和群论导论(英文)	2022—05	128.00	1435
有限域手册(英文)	2021—11	178.00	1436
计算数论(英文)	2021—11	148.00	1437
拟群与其表示简介(英文)	2021—11	88.00	1438
数论与密码学导论:第二版(英文)	2022—01	148.00	1423

书 名	出版时间	定 价	编号
几何分析中的柯西变换与黎兹变换:解析调和容量和李普希兹调和容量、变化和振荡以及一致可求长性(英文)	2021—12	38.00	1465
近似不动点定理及其应用(英文)	2022—05	28.00	1466
局部域的相关内容解析:对局部域的扩展及其伽罗瓦群的研究(英文)	2022—01	38.00	1467
反问题的二进制恢复方法(英文)	2022—03	28.00	1468
对几何函数中某些类的各个方面的研究:复变量理论(英文)	2022—01	38.00	1469
覆盖、对应和非交换几何(英文)	2022—01	28.00	1470
最优控制理论中的随机线性调节器问题:随机最优线性调节器问题(英文)	2022—01	38.00	1473
正交分解法:涡流流体动力学应用的正交分解法(英文)	2022—01	38.00	1475
芬斯勒几何的某些问题(英文)	2022—03	38.00	1476
受限三体问题(英文)	2022—05	38.00	1477
利用马利亚万微积分进行 Greeks 的计算:连续过程、跳跃过程中的马利亚万微积分和金融领域中的 Greeks(英文)	2022—05	48.00	1478
经典分析和泛函分析的应用:分析学的应用(英文)	2022—03	38.00	1479
特殊芬斯勒空间的探究(英文)	2022—03	48.00	1480
某些图形的施泰纳距离的细谷多项式:细谷多项式与图的维纳指数(英文)	2022—05	38.00	1481
图论问题的遗传算法:在新鲜与模糊的环境中(英文)	2022—05	48.00	1482
多项式映射的渐近簇(英文)	2022—05	38.00	1483
一维系统中的混沌:符号动力学,映射序列,一致收敛和沙可夫斯基定理(英文)	2022—05	38.00	1509
多维边界层流动与传热分析:粘性流体流动的数学建模与分析(英文)	2022—05	38.00	1510
演绎理论物理学的原理:一种基于量子力学波函数的逐次置信估计的一般理论的提议(英文)	2022—05	38.00	1511
R^2 和 R^3 中的仿射弹性曲线:概念和方法(英文)	2022—08	38.00	1512
算术数列中除数函数的分布:基本内容、调查、方法、第二矩、新结果(英文)	2022—05	28.00	1513
抛物型狄拉克算子和薛定谔方程:不定常薛定谔方程的抛物型狄拉克算子及其应用(英文)	2022—07	28.00	1514
黎曼-希尔伯特问题与量子场论:可积重正化、戴森-施温格方程(英文)	2022—08	38.00	1515
代数结构和几何结构的形变理论(英文)	2022—08	48.00	1516
概率结构和模糊结构上的不动点:概率结构和直觉模糊度量空间的不动点定理(英文)	2022—08	38.00	1517

刘培杰数学工作室
已出版(即将出版)图书目录——原版影印

书　名	出版时间	定　价	编号
反若尔当对:简单反若尔当对的自同构	2022—07	28.00	1533
对某些黎曼—芬斯勒空间变换的研究:芬斯勒几何中的某些变换	2022—07	38.00	1534
内诣零流形映射的尼尔森数的阿诺索夫关系	即将出版		1535
与广义积分变换有关的分数次演算:对分数次演算的研究	即将出版		1536
强子的芬斯勒几何和吕拉几何(宇宙学方面):强子结构的芬斯勒几何和吕拉几何(拓扑缺陷)	即将出版		1537
一种基于混沌的非线性最优化问题:作业调度问题	即将出版		1538
广义概率论发展前景:关于趣味数学与置信函数实际应用的一些原创观点	即将出版		1539
纽结与物理学:第二版(英文)	2022—09	118.00	1547
正交多项式和 q—级数的前沿(英文)	即将出版		1548
算子理论问题集(英文)	即将出版		1549
抽象代数:群、环与域的应用导论:第二版(英文)	即将出版		1550
菲尔兹奖得主演讲集:第三版(英文)	即将出版		1551
多元实函数教程(英文)	即将出版		1552

联系地址:哈尔滨市南岗区复华四道街 10 号　哈尔滨工业大学出版社刘培杰数学工作室
网　　址:http://lpj.hit.edu.cn/
邮　　编:150006
联系电话:0451—86281378　　13904613167
E-mail:lpj1378@163.com